Numerical Modelling in Robotics

Editor: Edgar Alonso Martínez García

Numerical Modelling in Robotics

Editor/Author: Edgar Alonso Martínez García[1]

Authors: Joaquín Rivero[1], César García Sariñaga[1], Nilda G. Villanueva Chacón[1], Omar Ramírez[1], Erik Lerín García[1], Manuel Vega Heredia[1], Diana R. Uribe Escalera[1], Jaime Candelaria Solís[1], Jaichandar K. Sheba[1,2], Rajesh E. Mohan[3], Le Tan-Phuc[2], Julio Reyes Muñoz[1], Angel A. Maldonado Ramírez[4], L. Abril Torres Méndez[4], Marco Elizalde Ceballos[1]

[1]Laboratorio de Robótica, Institute of Engineering and Technology, Universidad Autónoma de Ciudad Juárez, Mexico.

[2] Singapore Polytechnic, Singapore.

[3] Singapore University of Technology and Design, Singapore.

[4] Robotics and Advanced Manufacturing Group, CINVESTAV Campus Saltillo, Coahuila, Mexico.

ISBN: 978-84-942118-8-1
DOI: http://dx.doi.org/10.3926/oss.22
© OmniaScience (Omnia Publisher SL) 2015
© Cover design: OmniaScience
© Cover images: Edgar Alonso Martínez García

Preface

In this book the author presents an approach on numerical and analytical mathematical models to control and formally represent robotic engineering systems. Most of the chapters contents were simulated and experimentally tested in the laboratory of the author's institution. Applied computational mathematics has mainly been focused on describing sensing and control algorithms, and the formal models of both aspects combined for robotic research. Robot design and its applications presented scope the three control modalities: wheeled and walking, aerial, and underwater robots. Throughout this book, the author presents academic material on robotic applications divided into 17 chapters organised in five sections: Robot Sensing Models, Robot Navigation, Trajectory Control, Modelling Walking Robots, and 3D Robot Modelling. Its contents makes particular highlight on establishing deterministic mathematical formulations and solutions. Although, a diversity of computational algorithms were developed to obtain the experimental and simulation results that support the book, those algorithms are not explicitly analysed because the approach of the book regards only the deterministic numerical models.

Each chapter material is the result of academic final projects and teachings discussed in the classroom regarding computational mathematics applied to experimental mobile robotics. The research material explained has been produced by deploying home-made robotic platforms built in our Robotics Laboratory. The computer programs that support the theory behind the chapters were developed under C++ language, Open Source libraries, and our set of coded libraries KatanaLibs, which run in our Linux-based robotic OS (SAMURAI) developed during several years by our research group. The purpose of applying deterministic mathematical modelling is to find analytical and numerical solutions of diverse robotics engineering problems. The algebraic and integrate-differential equations are mathematical models that allow the engineer to develop computer algorithms for controlling the robot's sensors and actuators, and execute sophisticated intelligent missions in real-time. The present topics include projective geometry, linear algebra and matrices properties for establishing various sensing models to measure world attributes meaningful to the robot itself. Including the mathematical description of odometers, accelerometers, ultrasonic sonar, light detection and ranging sensors (LiDARs), image processing techniques, and feature extraction. Methods to solve systems of linear and non linear equations are used, such as systems of Jacobian matrices (squared and non squared) analysed by means of matrix properties: determinants, pseudo-inverse and decomposition of singular values. Furthermore, different types of integrate-differential, and partial differential equations are proposed as basis for navigation and trajectory control. The fitting of curves by polynomial interpolations to model schemes for planning, tracking, kinematics and dynamics control of robotic platforms are discussed as well. The models for actuation are obtained adaptively through Taylor series expansion, and polynomial regressions.

The book provides understanding on numerical modelling methods for different types of mobile robots. Its contents is written for graduate, and undergraduate students of advanced courses that are related to engineering sciences with a background on computational Maths and Physics. Finally, I would like to thank all my undergraduate and graduate students that have co-worked and collaborated with me as their academic advisor in the Robotics Laboratory. Students such as Alejandra Marín, Juan Carlos Solís, Karen Cangas, Maribel Bailón, Ramón Esparza, Eder Jimenez, Oscar Payán, Ellian Herrera, Joab Retana, Diana Torres, Nestor Santos, Lídice Castro, Dulce Torres, whom helpfully supported by setting up experiments, coded software tools, collected dataset for the Lab's users, tested algorithms, fixed robotic platforms or helped configuring the technological infrastructure of the Robotics Laboratory. In addition, I thankfully mention the staff of my institution UACJ that provided me the administrative support for the production of this book, Jesús Valencia, Lisbeyli Dominguez, and Luis Gutierrez from the Research and Postgraduate Coordination Office.

Assoc. Prof. Edgar Alonso Martínez García (PhD Eng.)

Head of the Robotic Laboratory

Institute of Engineering and Technology

Universidad Autónoma de Ciudad Juárez

Cd. Juárez, Chihuahua, Mexico

Foreword

Right from its origin, robotics and automation has capitalized on advancements on computing and data management to support the designing, building and testing of advanced robotic systems. The impressive development of technology and its contribution to the information-based society, is undoubtedly reshaping every aspect of our human daily life. Just as processor power is no longer a constraint factor for building computational expensive applications, robotics has been a recipient of new modern ubiquitous systems that include a wide integration of several data communication protocols, high definition graphic displays, and the availability of amazing amounts of computing memory. The overall field has grown to become a mature and growing subject. History began with industrial manipulators five decades ago but spans to our days as robotic explorers that have been sent to Mars to explore the planet, perform experiments and send their results back to Earth. Other advancements include lots of research about autonomous operation in complex environments to develop a driver-less car while humanoid robots have been sent into space and a number of advanced legged mobile robots have been developed by several high Tech companies. Unmanned aerial vehicles have a wide variety of military and commercial uses and new robotic applications now include robots that might hold no bolts or metallic parts on their construction whatsoever; new materials are being used to imitate several nature-inspired locomotion and sensing mechanisms. Despite their clear differences, these robotic applications require the solution of common tasks like controlling the robot's physical behaviour, sensing the environment, building up world models and tools for interacting with the environment. Such issues precisely uphold the foundations of this book.

The education of future engineers and scientist around robotics and automation subjects must include a deep and solid understanding of mathematical modelling fundamentals and a clear vision of control engineering, computational intelligence and advanced processing of several signal types such as voice, images and real-time video. Modern robotics demands a deep integration of such subjects, making space for a common ground between electronics, computer science, mechanics, pervasive computing, among others. However, among all subjects, the mathematical modelling of each component and its clear interpretation of the overall robot mechanism's behaviour must be appropriately addressed. For such a purpose, the author of this book has embraced a carefully developed mathematical framework to support the learning of fundamentals for modern robotics and automation. The book's perspective relays over a step-by-step approach for introducing core concepts of advanced robot modelling. It supports a second or third year undergraduate course in advanced robotics or can be a compulsory text for a first year postgraduate course. It is suitable for a range of different areas including mechanical and electrical engineering, applied mathematics and computer science.

The book has seventeen chapters devoted to the key mathematical topics for basic and advanced robotics, all divided into five sections. The first chapter begins by reviewing basic mathematical concepts of linear algebra and numerical methods, supporting the reader with a smooth transition between introductory and deeper concepts, all required by subsequent chapters. The section I presents a close review of relevant robotic sensing models that include visual methods, odometry-based procedures and multi-sensor registration. The section II focuses on the dynamical behaviour and navigation principles. Four chapters are devoted to directional derivatives, vector fields and task planning respectively. Section III presents three chapters that are completely devoted to the analysis of trajectory control of wheeled and walking legged robots, the required dynamics analysis and the control for trajectory tracking. This section included studies of the complexity of the kinematic analysis for a four-wheeled active suspension. The book's section IV is fully devoted to walking robotic structures, ranging from Klann linkages, Jansen-based quadrupeds, the mixed Hoeken-Jansen bipedal robot and the complex kinematics of self-configuration of heel-leg robots. The last section V is devoted to discuss robots kinematic for robots that navigate in 3D spaces or fluid environments such as underwater vehicles and hover craft mobile robots. This section is a remarkable contribution of the textbook since only few manuscripts in the literature include a study on such an issue.

The organization of each chapter builds on basic concepts to demonstrate the use of classic modelling methods for a wide variety of robotic plants. At the same time, several methodologies are used for the computer-based analysis of modern robot structures. It is expected that students will appreciate the ability to run the simulation exemplars in order to reinforce their understanding of the mathematical derivations. Adopting a simulation-based approach to learning basic and advanced robotic concepts, just after a careful review of theoretical issues, ensures that students approach all concepts in an enjoyable and interactive fashion.

Finally, it is important to share that the problem set included for each robotic architecture has been carefully selected in order to assure a full inclusion of previously presented theoretical concepts. It is expected that this manuscript will become a remarkable tool for undergraduate, postgraduate and researchers alike. Such expectancy grows from its broad, careful and deep study of a considerable number of modern robotic plants, whose concepts are still handy to analyse classic robotic devices. Such coverage assure its usefulness and appreciation among the robotic related community.

Prof. Dr. Marco Antonio Perez-Cisneros, MIEEE MIET MSNI

Robotics, Computer Vision and Automatic Control Research Group

Dean of Science Division, CU TONALÁ

University of Guadalajara, MEXICO

March 16th, 2015

Contents

List of Figures

List of Tables

Chapter 1

INTRODUCTION

Edgar A. Martínez García

Laboratorio de Robótica, Institute of Engineering and Technology
Universidad Autónoma de Ciudad Juárez, Mexico.

This chapter contains a toolbox for the mathematical methods that are used along the book. This is intended to be a general guide to introduce the reader through some fundamental deterministic mathematical concepts generally applied in robotics engineering such as control and sensor models.

1.1   Fundamentals of vector algebra and geometry

A vector is a physical quantity having magnitude, and angular direction. Let $\mathbf{v}$, and $\mathbf{u}$ be vectors in $\mathbb{R}^n$,

$$\mathbf{u} = \begin{pmatrix} u_1 \\ u_2 \\ \vdots \\ u_n \end{pmatrix} ; \quad \mathbf{v} = \begin{pmatrix} v_x \\ v_y \\ \vdots \\ v_n \end{pmatrix}$$

For practical purpose, let us assume $\mathbf{u}, \mathbf{v} \in \mathbb{R}^2$, the scalar product of two vectors, the inner or dot product.

$$\mathbf{u}^\top \cdot \mathbf{v} = u_1 v_1 + u_2 v_2 \tag{1.1}$$

or

$$\mathbf{u}^\top \cdot \mathbf{v} = \|\mathbf{u}\|\|\mathbf{v}\| \cos(\theta) \tag{1.2}$$

$$\theta = \arccos\left(\frac{\mathbf{u} \cdot \mathbf{v}}{\|\mathbf{u}\|\|\mathbf{v}\|}\right); \quad 0 \le \theta \le \pi \tag{1.3}$$

Thus, the following dot product properties are valid. The commutation is denoted by

$$\mathbf{u}^\top \cdot \mathbf{v} = \mathbf{v}^\top \cdot \mathbf{u}$$

the associative property of vectors is given,

$$(\mathbf{v} + \mathbf{u})^\top \cdot \mathbf{w} = \mathbf{u}^\top \cdot \mathbf{w} + \mathbf{v}^\top \cdot \mathbf{w}$$

and let c be a scalar,

$$(c\mathbf{u}^\top) \cdot \mathbf{v} = c(\mathbf{u}^\top \cdot \mathbf{v}) = \mathbf{u}^\top \cdot (c\mathbf{v})$$

likewise,

$$\mathbf{u}^\top \cdot \mathbf{u} \ge 0$$

However, for the case where

$$\mathbf{u}^\top \cdot \mathbf{u} = 0; \quad \Leftrightarrow \quad \mathbf{u} = 0$$

The following definition considers the right-sided depiction of the figure 1.1.

Definition 1.1.1. The length (or norm) of $\mathbf{b}$ is non negative scalar $\|\mathbf{b}\|$ defined by

$$\|\mathbf{b}\| = \sqrt[2]{\mathbf{c}^\top/2 \cdot \mathbf{h}} = \sqrt[2]{c^2/2 + h^2}; \quad \|\mathbf{b}\|^2 = \mathbf{c}^\top/2 \cdot \mathbf{h} \tag{1.4}$$

Which also implies

$$\|c/2\mathbf{b}\| = \left|\frac{c}{2}\right| \|\mathbf{b}\|$$

In general, two vectors $\mathbf{u}$ and $\mathbf{v} \in \mathbb{R}^n$ are orthogonal to each other, if $\mathbf{u}^\top \cdot \mathbf{v} = 0$. Therefore,

Theorem 1.1.2 (The Pitagorean). *Two vectors $\mathbf{u}$ and $\mathbf{v}$ are orthogonal if and only if*

$$\|\mathbf{u} + \mathbf{v}\| = \|\mathbf{u}\| + \|\mathbf{v}\| \tag{1.5}$$

For angles in $\mathbb{R}^2$ and $\mathbb{R}^3$,

$$\mathbf{u}^\top \cdot \mathbf{v} = \|\mathbf{u}\| \|\mathbf{v}\| \cos(\theta) \tag{1.6}$$

Proof,

$$\|\mathbf{u} - \mathbf{v}\|^2 = \|\mathbf{u}\|^2 + \|\mathbf{v}\|^2 - 2\|\mathbf{u}\| \|\mathbf{v}\| \cos(\theta)$$

and

$$\|\mathbf{u}\| \|\mathbf{v}\| \cos(\theta) = \frac{1}{2} \left[\|\mathbf{u}\|^2 + \|\mathbf{v}\|^2 - \|\mathbf{u} - \mathbf{v}\|^2 \right]$$

and

$$= \frac{1}{2} \left[u_1^2 + u_2^2 + v_1^2 + v_2^2 - (u_1 - v_1)^2 + (u_2 - u_2)^2 \right]$$

finally,

$$= u_1 v_1 - u_2 v_2 = \mathbf{u}^\top \cdot \mathbf{v}$$

Given the figure 1.1, we have two triangles$^!$, For figure 1.1 next theorems apply,

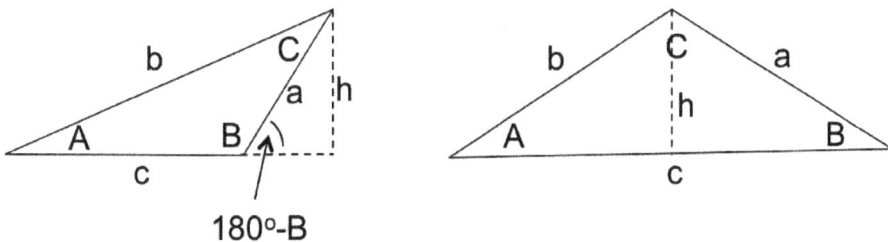

Figure 1.1: For any triangle $\angle A + \angle B + \angle C = 180°$.

Theorem 1.1.3 (Cosine law). *The cosine law :*

$$a^2 = b^2 + c^2 - 2bc \cos(\angle A) \tag{1.7}$$

$$b^2 = a^2 + c^2 - 2ac \cos(\angle B) \tag{1.8}$$

$$c^2 = a^2 + b^2 - 2ab \cos(\angle C) \tag{1.9}$$

In addition,

Theorem 1.1.4 (Sine law).

$$\frac{a}{\sin(\angle A)} = \frac{b}{\sin(\angle B)} = \frac{c}{\sin(\angle C)} \tag{1.10}$$

$$h = b \sin(\angle A) = a \sin(\angle B) \tag{1.11}$$

Accordingly to 1.1 right-sided figure, $b = \sqrt[2]{(c/2)^2 + h^2}$, and

$$\sin(A) = \frac{2h}{c}; \quad \cos(A) = \frac{c}{2b}; \quad \tan(A) = \frac{2h}{c}; \quad \sin(B) = \frac{c}{2b}$$

1.1.1 Outer product

The cross or vectorial product

$$\mathbf{u} \times \mathbf{v} = (u_1 \quad u_2 \quad \dots \quad u_n)^\top \times (v_1 \quad v_2 \quad \dots \quad v_n)^\top \tag{1.12}$$

for three dimensions, where each orthogonal axis is denoted by the unit vectors $\hat{\mathbf{i}}, \hat{\mathbf{j}}, \hat{\mathbf{k}} \in \mathbb{R}^3$

$$\mathbf{u} \times \mathbf{v} = \left(u_1\hat{\mathbf{i}} + u_2\hat{\mathbf{j}} + u_3\hat{\mathbf{k}}\right) \times \left(v_1\hat{\mathbf{i}} + v_2\hat{\mathbf{j}} + v_3\hat{\mathbf{k}}\right) \tag{1.13}$$

defined by

$$\mathbf{u} \times \mathbf{v} = \|\mathbf{u}\|\,\|\mathbf{v}\| \sin(\theta)\hat{\eta} \tag{1.14}$$

where the unit vector states the vector directions w.r.t. the axes,

$$\hat{\eta} = \frac{\mathbf{u} \times \mathbf{v}}{\|\mathbf{u} \times \mathbf{v}\|} \tag{1.15}$$

it is not commutative

$$\mathbf{u} \times \mathbf{v} \neq \mathbf{v} \times \mathbf{u}; \qquad \mathbf{u} \times \mathbf{v} = -(\mathbf{v} \times \mathbf{u})$$

but it is distributive

$$\mathbf{u} \times (\mathbf{v} \times \mathbf{w}) = \mathbf{u} \times \mathbf{v} + \mathbf{u} \times \mathbf{w}$$

it is not associative,

$$(\mathbf{u} \times \mathbf{u}) \times \mathbf{w} \neq \mathbf{u} \times (\mathbf{v} \times \mathbf{w})$$

and

$$c\mathbf{u} \times \mathbf{v} = c(\mathbf{u} \times \mathbf{v})$$

and if the vectors are parallel,

$$\mathbf{u} \times \mathbf{v} = 0; \quad \theta = 0.$$

The vectors are perpendicular $\mathbf{u} \perp \mathbf{v}$, when $\mathbf{u}^\top \cdot \mathbf{v} = 0$. Two vectors are parallel $\mathbf{u} \| \mathbf{v}$, when $\mathbf{u} \times \mathbf{v} = 0$.

1.1.2 Unit vectors

$$\hat{\mathbf{u}} = \frac{\mathbf{u}}{\|\mathbf{u}\|} = \frac{\mathbf{u}}{\sqrt[2]{\mathbf{u}^\top \mathbf{u}}} \tag{1.16}$$

where the Cartesian components of a vector $\mathbf{u} \in \mathbf{R}^2$

$$\mathbf{u} = u_1 \hat{\mathbf{i}} + u_2 \hat{\mathbf{j}}$$

The scalar magnitudes of the Cartesian components u_1 and u_2 that are mutually orthogonal,

$$u_1 = \|\mathbf{u}\| \cos(\phi); \quad u_2 = \|\mathbf{u}\| \sin(\phi)$$

where,

$$u = \|\mathbf{u}\| = \sqrt[2]{u_1^2 + u_2^2}$$

the angle w.r.t. the axis u_1,

$$\phi = \arctan \left(\frac{u_2}{u_1} \right)$$

for three dimensions the vector components are

$$\mathbf{u} = u_1 \hat{\mathbf{i}} + u_2 \hat{\mathbf{j}} + u_3 \hat{\mathbf{k}}$$

where each unit vector representing an independent Cartesian axis are defined by,

$$\hat{\mathbf{i}} = (1,0,0)^\top; \quad \hat{\mathbf{j}} = (0,1,0)^\top; \quad \hat{\mathbf{k}} = (0,0,1)^\top$$

which is equally represented by,

$$\mathbf{u} = \|\mathbf{u}\| \cos(\phi_1)\hat{\mathbf{i}} + \|\mathbf{u}\| \cos(\phi_2)\hat{\mathbf{j}} + \|\mathbf{u}\| \cos(\phi_3)\hat{\mathbf{k}}$$

thus, by factorising the common term $\|\mathbf{u}\|$,

$$\mathbf{u} = \|\mathbf{u}\| \left(\cos(\phi_1)\hat{\mathbf{i}} + \cos(\phi_2)\hat{\mathbf{j}} + \cos(\phi_3)\hat{\mathbf{k}} \right)$$

hence, by defining the unit vector $\hat{\lambda}$ as:

$$\hat{\lambda} = \cos(\phi_1)\hat{\mathbf{i}} + \cos(\phi_2)\hat{\mathbf{j}} + \cos(\phi_3)\hat{\mathbf{k}}$$

it is deduced that,

$$\lambda_1 = \cos(\phi_1); \quad \lambda_2 = \cos(\phi_2); \quad \lambda_3 = \cos(\phi_3)$$

The magnitude of the three angles are not independent from each other,

$$\lambda_1^2 + \lambda_2^2 + \lambda_3^2 = \cos^2(\phi_1) + \cos^2(\phi_2) + \cos^2(\phi_3) = 1$$

Thus, the direction cosines are defined as:

$$\cos(\phi_1) = \frac{u_1}{\|\mathbf{u}\|}; \quad \cos(\phi_2) = \frac{u_2}{\|\mathbf{u}\|}; \quad \cos(\phi_3) = \frac{u_3}{\|\mathbf{u}\|}$$

The unit vectors properties are

$$\hat{\mathbf{i}} \times \hat{\mathbf{i}} = 0; \quad \hat{\mathbf{j}} \times \hat{\mathbf{i}} = -\hat{\mathbf{k}}; \quad \hat{\mathbf{k}} \times \hat{\mathbf{i}} = \hat{\mathbf{j}};$$
$$\hat{\mathbf{i}} \times \hat{\mathbf{j}} = \hat{\mathbf{k}}; \quad \hat{\mathbf{j}} \times \hat{\mathbf{j}} = 0; \quad \hat{\mathbf{k}} \times \hat{\mathbf{j}} = -\hat{\mathbf{i}};$$
$$\hat{\mathbf{i}} \times \hat{\mathbf{k}} = -\hat{\mathbf{j}}; \quad \hat{\mathbf{j}} \times \hat{\mathbf{k}} = \hat{\mathbf{i}}; \quad \hat{\mathbf{k}} \times \hat{\mathbf{k}} = 0$$

1.2   Linear matrix algebra

A system of linear equations (or linear system) is a collection of one or more linear equations involving the same variables. A solution of the system is a list $(s_1, s_2, \ldots, s_n)$ of numbers that makes each equation a true statement when the values $s_1, s_2, \ldots, s_n$ are substituted for $x_1, \ldots, x_n$, respectively. Let us consider the following linear model representing a generalized linear system of algebraic equations ,

$$\mathbf{Ax} = \mathbf{b} \tag{1.17}$$

If $\mathbf{A}$ is not an invertible $n \times n$ matrix, then for each $\mathbf{b} \in \mathbb{R}^{\mathbf{n}}$ the equation (1.17) has the unique solution $\mathbf{x} = \mathbf{A}^{-1} \cdot \mathbf{b}$. Therefore, let us define a single linear model equation in the following general form,

$$a_1 x_1 + a_2 x_2 + \cdots + a_n x_n = b_1 \tag{1.18}$$

for a system of linear equations of the form

$$a_{i1} x_1 + a_{i2} x_2 + \cdots + a_{in} x_n = b_1$$
$$a_{j1} x_1 + a_{j2} x_2 + \cdots + a_{jn} x_n = b_2$$
$$\vdots \tag{1.19}$$
$$a_{m1} x_1 + a_{m2} x_2 + \cdots + a_{mn} x_n = b_n$$

Thus, the matrix form of our system of linear equations for $m = n$, it is denoted by as previously defined by eq.(1.17),

$$\begin{pmatrix} a_{i1} + a_{i2} + \cdots + a_{in} \\ a_{j1} + a_{j2} + \cdots + a_{jn} \\ \vdots \\ a_{m1} + a_{m2} + \cdots + a_{mn} \end{pmatrix} \cdot \begin{pmatrix} x_1 \\ x_2 \\ \vdots \\ x_n \end{pmatrix} = \begin{pmatrix} b_1 \\ b_2 \\ \vdots \\ b_n \end{pmatrix} \tag{1.20}$$

For either cases when $m = n$, and $m \neq n$, we are usually interested in solving the vector $\mathbf{x}$.

Let $\mathbf{A}$, $\mathbf{B}$ and $\mathbf{C}$ be matrices of the same size $m \times n$, and let r and s be scalars.

a) $\mathbf{A} + \mathbf{B} = \mathbf{B} + \mathbf{A}$, and $\mathbf{A} + \mathbf{0} = \mathbf{A}$

b) $\mathbf{A} + (\mathbf{B} + \mathbf{C}) = (\mathbf{A} + \mathbf{B}) + \mathbf{C}$

c) $\mathbf{A}(\mathbf{BC}) = (\mathbf{AB})\mathbf{C}$

d) $\mathbf{A}(\mathbf{B} + \mathbf{C}) = \mathbf{AB} + \mathbf{AC}$, or $(\mathbf{B} + \mathbf{C})\mathbf{A} = \mathbf{BA} + \mathbf{CA}$

e) $r(\mathbf{AB}) = (r\mathbf{A})\mathbf{B} = \mathbf{A}(r\mathbf{B})$

f) $\mathbf{IA} = \mathbf{A} = \mathbf{AI}$

The transpose of a product of matrices equals the product of their transposes in the reverse order. Thus, let $\mathbf{A}$ and $\mathbf{B}$ denote matrices where sizes are appropriate for the following sums and products.

a) $(\mathbf{A}^{\mathsf{T}})^{\mathsf{T}}$

b) $(\mathbf{A} + \mathbf{B})^{\mathsf{T}} = \mathbf{A}^{\mathsf{T}} + \mathbf{B}^{\mathsf{T}}$

c) $(r\mathbf{A})^{\mathsf{T}} = r\mathbf{A}^{\mathsf{T}}$

d) $(\mathbf{AB})^{\mathsf{T}} = \mathbf{B}^{\mathsf{T}}\mathbf{A}^{\mathsf{T}}$

The symmetric matrix is equal to its transpose,

$$\mathbf{A}^{\mathsf{T}} = \mathbf{A}$$

The antisymmetric matrix has opposite sign to its transpose,

$$\mathbf{A}^{\mathsf{T}} = -\mathbf{A}$$

In the orthogonal matrix, the inverse of a matrix is equally obtained by its transpose,

$$\mathbf{A}^{\mathsf{T}} = \mathbf{A}^{-1}$$

1.2.1 Determinants

A determinant is a single number that summarises a square numeric matrix (multivariate phenomenon) in a certain way and represents special characteristics of that matrix.

The determinant of a 1-by-1 matrix is

$$\det[a_{11}] = a_{11}$$

for a 2×2 matrix,

$$\det \begin{bmatrix} a_{11} & a_{12} \\ a_{21} & a_{22} \end{bmatrix} = a_{11}a_{22} - a_{12}a_{21}$$

and for a 3×3 matrix,

$$\det \begin{bmatrix} a_{11} & a_{12} & a_{13} \\ a_{21} & a_{22} & a_{23} \\ a_{31} & a_{32} & a_{33} \end{bmatrix} = a_{11}a_{22}a_{33} + a_{12}a_{23}a_{31} + a_{13}a_{21}a_{32} - a_{11}a_{23}a_{32} - a_{12}a_{21}a_{33} - a_{13}a_{22}a_{31}$$

Let us consider $\mathbf{A} = [a_{ij}]$ with $a_{11} \neq 0$. If we multiply the second and third rows of $\mathbf{A}$ by a_{11} and then subtract appropriate multiples of the first row from the other two rows, we find that $\mathbf{A}$ is row equivalent to the following two matrices:

$$\begin{pmatrix} a_{11} & a_{12} & a_{13} \\ a_{21} & a_{22} & a_{23} \\ a_{31} & a_{32} & a_{33} \end{pmatrix} \sim \begin{pmatrix} a_{11} & a_{12} & a_{13} \\ 0 & a_{11}a_{22} - a_{12}a_{21} & a_{11}a_{23} - a_{13}a_{21} \\ 0 & a_{11}a_{32} - a_{12}a_{31} & a_{11}a_{33} - a_{13}a_{31} \end{pmatrix} \tag{1.21}$$

Subsequently, multiply the row 3 by $(a_{11}a_{22} - a_{12}a_{21})$, and then to the new row 3 add $-(a_{11}a_{32} - a_{12}a_{31})$ times row 2. Thus, this will yield

$$\mathbf{A} \sim \begin{pmatrix} a_{11} & a_{12} & a_{13} \\ 0 & a_{11}a_{22} - a12a_{21} & a_{11}a_{23} - a_{13}a_{21} \\ 0 & 0 & a_{11}\Delta \end{pmatrix} \tag{1.22}$$

where,

$$\Delta = a_{11}a_{22}a_{33} + a_{12}a_{23}a_{31} + a_{13}a_{21}a_{32} - a_{11}a_{23}a_{32} - a_{12}a_{21}a_{33} - a_{13}a_{22}a_{31}$$

The terms in Δ can be grouped as

$$(a_{11}a_{22}a_{33} - a_{11}a_{23}a_{32}) - (a_{12}a_{21}a_{33} - a_{12}a_{23}a_{31}) + (a_{13}a_{21}a_{32} - a_{13}a_{22}a_{31}),$$

such algebraic expression is equivalent, if we factorize and regroup the terms,

$$\Delta = a_{11} \cdot \det \begin{bmatrix} a_{22} & a_{23} \\ a_{32} & a_{33} \end{bmatrix} - a_{12} \cdot \det \begin{bmatrix} a_{21} & a_{23} \\ a_{31} & a_{33} \end{bmatrix} + a_{13} \cdot \det \begin{bmatrix} a_{21} & a_{22} \\ a_{31} & a_{32} \end{bmatrix} \tag{1.23}$$

and for brevity, we write with sub matrix notation in the following manner

$$\Delta = a_{11} \cdot \det \mathbf{A}_{11} - a_{12} \cdot \det \mathbf{A}_{12} + a_{13} \cdot \det \mathbf{A}_{13} \tag{1.24}$$

where $\mathbf{A}_{11}$, $\mathbf{A}_{12}$, and $\mathbf{A}_{13}$ are obtained from $\mathbf{A}$ by deleting the first row and one of the columns. For any square matrix $\mathbf{A}$, let a_{ij} denote the sub matrix formed by deleting the ith row and jth column of $\mathbf{A}$. Therefore, the determinant may be defined inductively by the Laplace expansion:

$$\sum_{j=1}^{n}(-1)^{i+j}a_{ij}\det \mathbf{A}_{ij} = \sum_{i=1}^{n}(-1)^{i+j}a_{ij}\det \mathbf{A}_{ij} \tag{1.25}$$

for all $i \leq n, j \leq n$, and this common value is $\det(\mathbf{A})$ or simply $|\mathbf{A}|$. The left-hand side is the Laplace expansion by minors along row i, and the right-hand side is the Laplace expansion along column j.

1.2.2   Trace of a matrix

The trace of a matrix $\mathbf{A}$ Tr($\mathbf{A}$), $\mathbf{A} \in \mathbb{R}^{n \times n}$. The properties of the trace of a matrix,

$$\text{Tr}(\mathbf{A}) = \sum_{i=1}^{n} a_{ii} = \text{Tr}(\mathbf{A}^\top)$$

and

$$\text{Tr}(c\mathbf{A}) = c\text{Tr}(\mathbf{A}), \quad \forall \quad k \in \mathbb{R}$$

$$\text{Tr}(\mathbf{A} + \mathbf{B}) = \text{Tr}(\mathbf{A}) + \text{Tr}(\mathbf{B})$$

and

$$Tr(\mathbf{AB}) = Tr(\mathbf{BA})$$

1.2.3 Matrix inversion

The matrix inversion based on the determinant calculus is a common approach deployed when we have squared matrices $\mathbf{A}_{n \times n}$. An $n \times n$ matrix $\mathbf{A}$ is said to be invertible in fact, if there is an $n \times n$ inverse matrix denoted by $\mathbf{A}^{-1}$, uniquely defined by $\mathbf{A}$. The following are some inverse matrix conditions:

a) If $\mathbf{A}$ is an invertible matrix, then $\mathbf{A}^{-1}$ is invertible,

$$(\mathbf{A}^{-1})^{-1} = \mathbf{A}$$

b) If $\mathbf{A}$ and $\mathbf{B}$ are $n \times n$ invertible matrices, then the product of the inverses is

$$(\mathbf{AB})^{-1} = \mathbf{BA}$$

c) If $\mathbf{A}$ is an invertible matrix, then the inverse of its transpose $\mathbf{A}^{\top}$

$$(\mathbf{A}^{\top})^{-1} = (\mathbf{A}^{-1})^{\top}$$

d) The identity matrix of a matrix $\mathbf{A}$ and its inverse,

$$\mathbf{A}^{-1} \cdot \mathbf{A} = \mathbf{I}, \qquad \mathbf{A} \cdot \mathbf{A}^{-1} = \mathbf{I}$$

A non invertible matrix is sometimes called singular matrix; an invertible matrix is called a non singular matrix where $|\mathbf{A}| \neq 0$ must exist.

Thus, according to the next theorem,

Theorem 1.2.1 (An Inverse Formula). *Let* A *be an invertible* $n \times n$ *matrix, such that*

$$A = \begin{pmatrix} a_1 & a_2 \\ a_3 & a_4 \end{pmatrix}$$

hence, if $a_1 a_4 - a_2 a_3 \neq 0$, *then* A *is invertible and*

$$A^{-1} = \frac{1}{\det(A)} \operatorname{adj}(A) = \frac{1}{a_1 a_4 - a_2 a_3} \begin{pmatrix} a_4 & -a_2 \\ -a_3 & a_1 \end{pmatrix} \tag{1.26}$$

But, if $a_1 a_4 - a_2 a_3 = 0$, A *is not invertible.*

This can be an efficient way to calculate the inverse of small matrices. For $n \geq 2$, the determinant of an $n \times n$ matrix $A = [a_{ij}]$ is the sum of n terms of the form $\pm a_{1j} \det(A_{1j})$, where the entries $a_{11}, a_{12}, \ldots, a_{1n}$ are from the first row of A.

Definition 1.2.2. The determinant of an $n \times n$ matrix A being $n \geq 2$ is

$$|A| = a_{11} \det(A_{11}) - a_{12} \det(A_{12}) + \cdots + (-1)^{1+n} a_{1n} \det(A_{1n}) \tag{1.27}$$

or

$$|A| = \sum_{j=1}^{n} (-1)^{1+j} \det(A_{ij}) \tag{1.28}$$

In addition, the transpose of the matrix of cofactors is known as the adjugate matrix,

$$\operatorname{adj}(A) = C^{\top}$$

Thus, let $\mathbf{A}$ be a square matrix,

$$
\mathbf{A} = \begin{pmatrix}
a_{11} & a_{11} & \ldots a_{1n} \\
a_{21} & a_{21} & \ldots a_{2n} \\
\vdots & & \\
a_{n1} & a_{n1} & \ldots a_{nn}
\end{pmatrix}
\tag{1.29}
$$

And, the cofactors matrix $\mathbf{C}$ is stated as

$$
\mathbf{C} = \begin{pmatrix}
c_{11} & c_{11} & \ldots c_{1n} \\
c_{21} & c_{21} & \ldots c_{2n} \\
\vdots & & \\
c_{m1} & c_{m1} & \ldots c_{mn}
\end{pmatrix}
\tag{1.30}
$$

Where each cofactor matrix terms is formulated by

$$
c_{ij} = (-1)^{i+j} a_{ij}
\tag{1.31}
$$

1.2.4 Cramer's Rule

This is an explicit formula for the solution of a system of linear equations with as many equations as unknowns, valid whenever the system has a unique solution. It expresses the solution in terms of the determinants of the (square) coefficient matrix and of matrices obtained from it by replacing one column by the vector of right hand sides of the equations.

For a general systems of linear equations,

$$
a_{11}x_{11} + a_{12}x_{12} + \cdots + a_{1n}x_{1n} = b_1,
$$

$$
a_{21}x_{21} + a_{22}x_{22} + \cdots + a_{2n}x_{2n} = b_2,
$$

and

$$
a_{m1}x_{m1} + a_{m2}x_{m2} + \cdots + a_{mn}x_{mn} = b_m,
$$

Arranging in the matrix form,

$$
\begin{pmatrix}
a_{11} & a_{12} & \cdots & a_{1n} \\
a_{21} & a_{21} & \cdots & a_{2n} \\
\vdots & & & \\
a_{m1} & a_{m2} & \cdots & a_{mn}
\end{pmatrix}
\cdot
\begin{pmatrix}
x_1 \\ x_2 \\ \vdots \\ x_n
\end{pmatrix}
=
\begin{pmatrix}
b_1 \\ b_2 \\ \vdots \\ b_n
\end{pmatrix}
\tag{1.32}
$$

Given the linear form of the vector $\mathbf{x} \in \mathbb{R}^n$, $\mathbf{x} = (x_1, x_2, \ldots, x_n)^\top$.

$$
\mathbf{A} \cdot \begin{pmatrix} x_1 & x_2 & \cdots & x_n \end{pmatrix}^\top = \mathbf{b}
\tag{1.33}
$$

Thus, let us define and apply the Cramer's theorem,

Theorem 1.2.3 (Cramer's Rule). *Let* $\mathbf{A}$ *be an invertible* $n \times n$ *matrix. For any* $\mathbf{b}$ *in* $\mathbb{R}^n$, *the unique solution* $\mathbf{x}$ *of* $\mathbf{Ax} = \mathbf{b}$ *has the entries given by*

$$
x_i = \frac{\det \mathbf{A}_i(\mathbf{b})}{\det \mathbf{A}}, \qquad i = 1, 2, \ldots, n
\tag{1.34}
$$

thus,

$$
x_1 = \frac{\begin{vmatrix} b_1 & a_{12} & \cdots & a_{1n} \\ b_2 & a_{22} & \cdots & a_{2n} \\ \vdots & & & \\ b_n & a_{n2} & \cdots & a_{nn} \end{vmatrix}}{\begin{vmatrix} a_{11} & a_{12} & \vdots & a_{1n} \\ a_{21} & a_{22} & \vdots & a_{2n} \\ \vdots & & & \\ a_{n1} & a_{n2} & \vdots & a_{nn} \end{vmatrix}},
\quad
x_2 = \frac{\begin{vmatrix} a_{12} & b_1 & \cdots & a_{1n} \\ a_{22} & b_2 & \cdots & a_{2n} \\ \vdots & & & \\ a_{n2} & b_n & \cdots & a_{nn} \end{vmatrix}}{\begin{vmatrix} a_{11} & a_{12} & \vdots & a_{1n} \\ a_{21} & a_{22} & \vdots & a_{2n} \\ \vdots & & & \\ a_{n1} & a_{n2} & \vdots & a_{nn} \end{vmatrix}},
\quad \cdots \quad
x_n = \frac{\begin{vmatrix} a_{11} & a_{12} & \cdots & b_1 \\ a_{21} & a_{22} & \cdots & b_2 \\ \vdots & & & \\ a_{n1} & a_{n2} & \cdots & b_n \end{vmatrix}}{\begin{vmatrix} a_{11} & a_{12} & \vdots & a_{1n} \\ a_{21} & a_{22} & \vdots & a_{2n} \\ \vdots & & & \\ a_{n1} & a_{n2} & \vdots & a_{nn} \end{vmatrix}},
\tag{1.35}
$$

1.2.5 Pseudoinverse

The following equations are used to define a generalised inverse, a reflexive generalized inverse, and a pseudoinverse of $\mathbf{A}$

Proposition 1.2.4. *For* $\mathbf{J} \in \mathbb{R}^{m \times n}$, *and* $\mathrm{rank}(\mathbf{J}) = \mathbf{m}$, *then* $(\mathbf{JJ})^{-1}$ *exists.*

Consider the following Penrose conditions,

$$\mathbf{A} \cdot \mathbf{J} \cdot \mathbf{A} = \mathbf{A} \tag{1.36}$$

$$\mathbf{J} \cdot \mathbf{A} \cdot \mathbf{J} = \mathbf{J} \tag{1.37}$$

$$(\mathbf{A} \cdot \mathbf{J})^{\top} = \mathbf{A} \cdot \mathbf{J} \tag{1.38}$$

$$(\mathbf{J} \cdot \mathbf{A})^{\top} = \mathbf{J} \cdot \mathbf{A} \tag{1.39}$$

Definition 1.2.5. A generalised inverse of a matrix $\mathbf{A} \in \mathbb{R}^{m \times n}$ is a matrix $\mathbf{J} = \mathbf{A}^{-} \in \mathbb{R}^{n \times m}$ satisfying (1.36).

Definition 1.2.6. A reflexive generalised inverse of a matrix $\mathbf{A} \in \mathbb{R}^{m \times n}$ is a matrix $\mathbf{J} = \mathbf{A}_r^{-} \in \mathbb{R}^{n \times m}$ satisfying (1.36) and (1.37).

A pseudo-inverse is sometimes called the Moore-Penrose inverse [1].

Definition 1.2.7. A pseudo-inverse of a matrix $\mathbf{A} \in \mathbb{R}^{m \times n}$ is a matrix $\mathbf{J} = \mathbf{A}_r^{+} \in \mathbb{R}^{n \times m}$ satisfying (1.36) through (1.39).

---

[1] After the pioneering works by Mores (1920, 1935) and Penrose (1955).

Lemma 1.2.8. *For a linear equation* (1.17), *where* $\mathbf{A} \in \mathbb{R}^{m \times n}$, $\mathbf{x} \in \mathbb{R}^n$, *and* $\mathbf{b} \in \mathbb{R}^m$, *a necessary and sufficient condition for the existence of solution* $\mathbf{x}$ *is*

$$\text{rank}([\mathbf{A} \quad \mathbf{b}]) = \text{rank}(\mathbf{A})$$

If previous equation is satisfied then $\mathbf{x} = \mathbf{A}^- \cdot \mathbf{b}$, which is a solution of eq.(1.17).

Lemma 1.2.9. *For an arbitrary* $\mathbf{A} \in \mathbb{R}^{m \times n}$, *there exist at least one generalised inverse* $\mathbf{A}^-$ *and* $\text{rank}\mathbf{A}^- \geq \text{rank}\mathbf{A}$. *Thus,* $\mathbf{A}^-$ *coincides with a reflexive generalized inverse if and only if* $\text{rank}\mathbf{A}^- = \text{rank}\mathbf{A}$

Lemma 1.2.10. *Generally* $\mathbf{A}^-$ *and* $\mathbf{A}_r^-$ *are not unique. If* $\mathbf{A}$ *is square and non singular, then the generalised inverse* $\mathbf{A}^-$ *and the reflexive generalized inverse* $\mathbf{A}_r^-$ *are unique, and* $\mathbf{A}^- = \mathbf{A}_r^- = \mathbf{A}^{-1}$

Lemma 1.2.11. $\mathbf{A} \cdot \mathbf{A}^-$ *and* $\mathbf{A}^- \cdot \mathbf{A}$ *are idempotent (if a square matrix* $\mathbf{A}^2 = \mathbf{A}$).

Using an arbitrary matrix $\mathbf{U} \in \mathbb{R}^{n \times m}$ and a generalised inverse $\mathbf{A}^-$, all the generalised inverses of $\mathbf{A}$ can be represented by the following $\mathbf{J}$:

$$\mathbf{J} = \mathbf{A}^- + \mathbf{U} - \mathbf{A}^- \cdot \mathbf{A} \cdot \mathbf{U} \cdot \mathbf{A} \cdot \mathbf{A}^- \tag{1.40}$$

This result can be readily shown by multiplying $\mathbf{A}$ at the two sides of each term of the equation according to the Penrose condition (1.36),

$$\mathbf{A} \cdot \mathbf{J} \cdot \mathbf{A} = \mathbf{A}(\mathbf{A}^- + \mathbf{U} - \mathbf{A}^- \cdot \mathbf{A} \cdot \mathbf{U} \cdot \mathbf{A} \cdot \mathbf{A}^-)\mathbf{A} \tag{1.41}$$

and by multiplying $\mathbf{A}$ by all terms,

$$\mathbf{A} \cdot \mathbf{J} \cdot \mathbf{A} = (\mathbf{A} \cdot \mathbf{A}^- \cdot \mathbf{A}) + \mathbf{A} \cdot \mathbf{U} \cdot \mathbf{A} - (\mathbf{A} \cdot \mathbf{A}^- \cdot \mathbf{A}) \cdot \mathbf{U} \cdot (\mathbf{A} \cdot \mathbf{A}^- \mathbf{A}) \tag{1.42}$$

and then by applying the condition (1.36) the expression is reduced,

$$A = A + (A \cdot U \cdot A) - (A \cdot U \cdot A) \tag{1.43}$$

where $A \cdot A^- \cdot A = A$ was used. Thus,

$$A = A \tag{1.44}$$

Thus, the pseudo-inverse $A^+ \in \mathbb{R}^{n \times m}$ is unique for a given $A \in \mathbb{R}^{m \times n}$, whereas A^- and A_r^- are not necessarily unique. Let the sets of A^-, A_r^- and A^+ be S^-, S_r^- and S^+, respectively; then, the following inclusion holds:

$$S^+ \subset S_r^- \subset S^- \tag{1.45}$$

Then,

$$(A^+)^+ = A$$

Similarly for reverse order the transpose applies,

$$(A^\top)^+ = (A^+)^\top$$

and

$$A^+ = (A^\top \cdot A)^{\prime} \cdot A^\top = A^\top \cdot (A \cdot A^\top)^+$$

and the identity matrix is a product of

$$AA^+ = I$$

or well

$$(AA^\top)(AA^\top)^{-1} = I$$

regrouping terms according to the associative property of matrices,

$$A\left[A^\top(AA^\top)^{-1}\right] = I$$

Therefore, it follows the next theorems,

Theorem 1.2.12 (The Right-Pseudo-inverse). *For* $\mathbf{A} \in \mathbb{R}^{m \times n}$, *if* $m < n$ *and* rank(A) = **m**, *then* $\mathbf{A}\mathbf{A}^\top$ *is non singular and*

$$\mathbf{A}^+ = \mathbf{A}^\top \cdot (\mathbf{A} \cdot \mathbf{A}^\top)^{-1} \tag{1.46}$$

Here, $\mathbf{A}^+ = \mathbf{A}^\top(\mathbf{A}\mathbf{A}^\top)^{-1}$ is called a right pseudo-inverse of **A**, since $\mathbf{A}\mathbf{A}^+ = \mathbf{I}$. Thus, contrary,

Theorem 1.2.13 (The Left-Pseudo-inverse). *For* $\mathbf{A} \in \mathbb{R}^{m \times n}$ *If* $m > n$ *and* rank(A) = **n**, *then* $\mathbf{A}^\top \cdot \mathbf{A}$ *is non singuar and*

$$\mathbf{A}^+ = (\mathbf{A}^\top \cdot \mathbf{A})^{-1} \cdot \mathbf{A}^\top \tag{1.47}$$

Furthermore,

Theorem 1.2.14 (The Orthogonal Matrix). *If* $m = n$ *and* rank(A) = **m**, *then*

$$\mathbf{A}^\top = \mathbf{A}^{-1} \tag{1.48}$$

Thus, for the general linear model (1.17), if $m < n$ and applying theorem 1.2.12,

$$\mathbf{x} = \mathbf{A}^\top (\mathbf{A} \cdot \mathbf{A}^\top)^{-1} \cdot \mathbf{b} \tag{1.49}$$

Finally, note that $\mathbf{A}^+ \cdot \mathbf{A} \in \mathbb{R}^{n \times n}$, and that in general, $\mathbf{A}^+\mathbf{A} \neq \mathbf{I}$ because matrix multiplication is not commutative.

1.2.6 Singular value decomposition

In order to deal with the case for non-squared matrices $\mathbf{A}_{m \times n}$ where $m \neq n$, for multiple independent variables; the determinant value is not possible. Thus, not all matrices can be factored as $\mathbf{A} = \mathbf{PDP}^{-1}$ with $\mathbf{D}$ diagonal, then a factorization $\mathbf{A} = \mathbf{QDP}^{-1}$ is possible for any $\mathbf{A}_{m \times n}$. A special factorization of this type, called the *singular value decomposition* (SVD) is a linear algebra matrix factorization.

The SVD is based on the fact that the ordinary diagonalizability can be imitated for rectangular matrices. The magnitudes of the eigenvalues λ_i of a symmetric matrix $\mathbf{A}$ measure the amounts that $\mathbf{A}$ stretches or shrinks its eigenvectors. That is, if λ_i is the greatest eigenvalue's magnitude, then a corresponding unit vector $\mathbf{v}_1$ identifies a direction in which the stretching effect of $\mathbf{A}$ is greatest.

The norm of $\|\mathbf{Ax}\| = \|\lambda\mathbf{x}\|$, $\|\mathbf{x}\| = 1$ is maximised when $\mathbf{x} = \mathbf{v}_1$,

$$\|\mathbf{Ax}\| = \|\lambda\mathbf{x}\| = |\lambda|\|\mathbf{x}\| = |\lambda|$$

The following equivalence holds having a non trivial solution.

$$(\mathbf{A} - \lambda\mathbf{I})\mathbf{x} = \mathbf{0} \tag{1.50}$$

Thus, the next theorem is stated,

Theorem 1.2.15 (Singular Value Decomposition). *Let* Rank$(\mathbf{A})_{m \times n} = r$, *then there exist* $\mathbf{\Sigma}_{m \times n}$ *with diagonal entries in* $\mathbf{D}$ *are the first r singular values of* $\mathbf{A}$, $\sigma_1 \geq \sigma_2 \geq \cdots \geq \sigma_r > 0$, *and there exist an orthogonal matrix* $\mathbf{U}_{m \times m}$, *and an orthogonal matrix* $\mathbf{V}_{n \times n}$ *such that*

$$\mathbf{A} = \mathbf{U\Sigma V}^\top$$

Thus, $\mathbf{A}^\top\mathbf{A}$ is symmetric and can be orthogonally diagonalized. And

$$\mathbf{U} = [\mathbf{u}_1, \ldots, \mathbf{u}_m] \in \mathbb{R}^{m \times m}$$

and let $\mathbf{V} = [\mathbf{v}_1, \ldots, \mathbf{v}_n] \in \mathbb{R}^{n \times n}$ be an orthonormal basis consisting of eigenvectors of $\mathbf{A}^\top\mathbf{A}$, where $\mathbf{A}$ is represented as in theorem 1.2.15.

Likewise, let $\lambda_1, \ldots, \lambda_n$ be the associated eigenvalues of $\mathbf{A}^\top \mathbf{A}$. Then, for $1 \leq i \leq n$,

$$\|\mathbf{A}\mathbf{v_i}\|^2 = (\mathbf{A}\mathbf{v}_i)^\top (\mathbf{A}\mathbf{v}_i) = \mathbf{v}_i^\top \mathbf{A}^\top \mathbf{A}\mathbf{v}_i = \mathbf{v}_i^\top (\lambda_i \mathbf{v}_i) = \lambda_i \qquad (1.51)$$

Arranging the corresponding all non negative eigenvalues and renumbering to satisfy the values $\lambda_1 \geq \lambda_2 \geq \cdots \geq \lambda_n > 0$, then $\{\mathbf{A}\mathbf{v}_1, \ldots \mathbf{A}\mathbf{v}_r\}$ is an orthonormal basis for $\mathrm{Col}\,\mathbf{A}$ and $\mathrm{Rank}(\mathbf{A}) = r$, and by expression (1.51), their lengths are the singular values, then $(\mathbf{A}\mathbf{v}_i) \neq \mathbf{0} \Leftrightarrow 1 \leq i \leq r$. The singular values of $\mathbf{A}^\top \mathbf{A}$ are $\sigma_1, \ldots \sigma_n$, arranged in decreasing order. That is,

$$\sigma_i = \sqrt[2]{\lambda_i}, \quad 1 \leq i \leq n$$

For any $\mathbf{y}$ in $\mathrm{Col}(\mathbf{A})$, $\mathbf{y} = \mathbf{A}\mathbf{x}$,

$$\mathbf{x} = c_1 \mathbf{v}_1 + \ldots c_n \mathbf{v}_n \qquad (1.52)$$

and

$$\mathbf{y} = \mathbf{A}\mathbf{x} = c_1 \mathbf{A}\mathbf{v}_1 + \cdots + c_r \mathbf{A}\mathbf{v}_r + c_{r+1} \mathbf{A}\mathbf{v}_{r+1} + \cdots + c_n \mathbf{A}\mathbf{v}_n \qquad (1.53)$$

$$\mathbf{y} = \mathbf{A}\mathbf{x} = c_1 \mathbf{A}\mathbf{v}_1 + \cdots + c_r \mathbf{A}\mathbf{v}_r + \mathbf{0} + \cdots + \mathbf{0} \qquad (1.54)$$

Following that $\dim[\mathrm{Col}\,\mathbf{A}] = \mathrm{Rank}(\mathbf{A}) = r$.

Thus, the diagonal matrix $\mathbf{D} = \mathrm{diag}(\sigma_0, \sigma_1, \ldots, \sigma_n) \in \mathbb{R}^{r \times r}$ is squared and symmetric.

$$\mathbf{D}_{r \times r} = \begin{pmatrix} \sigma_1 & & & \\ & \sigma_2 & & \\ & & \ddots & \\ & & & \sigma_r \end{pmatrix} \qquad (1.55)$$

In addition the condition number k denotes the ratio of the longest and smallest values namely

$$k = \frac{\sigma_1}{\sigma_r}; \quad (k \geq 1) \qquad (1.56)$$

For a small k the matrix is well conditioned; for a large k the matrix is ill conditioned. Thus, numerical computation of an ill-conditioned coefficient matrix may involve large computational errors.

Furthermore, the matrix $\mathbf{\Sigma}$,

$$\mathbf{\Sigma} = \begin{pmatrix} \mathbf{D} & 0 \\ 0 & 0 \end{pmatrix}$$

For example, the singular values σ_i of $\mathbf{J}$ can be used to find the eigenvectors $\mathbf{u}, \ldots, \mathbf{u_m}$ that satisfy the equality $\mathbf{JJ^T} = \mathbf{u_i} = \sigma_i \mathbf{u_i}$. Such eigenvectors comprise the matrix $\mathbf{U} = [\mathbf{u_1}, \mathbf{u_2}, \cdots, \mathbf{u_m}]$. Thus, the system is then rewritten as,

$$\mathbf{JJ^T U} = \mathbf{U\Sigma_m^2} \tag{1.57}$$

Hence, it is defined $\mathbf{V_m} = \mathbf{J^T U\Sigma_m^{-1}}$, and let $\mathbf{V}$ be any orthogonal matrix that satisfies the expression $\mathbf{V} = [\mathbf{V_m}|\mathbf{V_{n-m}}]$. Notice that $\mathbf{V}$ is an $n \times n$ matrix. Then, constructing the right pseudo-inverse of $\mathbf{J}$ using singular value decomposition, the pseudo-inverse $\mathbf{J^+} = \mathbf{V\Sigma^{-1}U^T}$. Therefore, the SVD are given by (12.36), in which $\mathbf{\Sigma}_m^+$ is the inverse (square) matrix of $\mathbf{\Sigma}_m$.

$$\mathbf{\Sigma}_m^+ = \begin{bmatrix} \sigma_1^{-1} & & & \\ & \sigma_2^{-1} & & \\ & & \ddots & \\ & & & \sigma_1^{-1} \end{bmatrix} \tag{1.58}$$

For instance the solution for $\mathbf{x}$ in example $\mathbf{y} = \mathbf{A} \cdot \mathbf{x}$, $\mathbf{A} \in \mathbb{R}^{m \times n}$, such that $\mathbf{x} = \mathbf{A^{-1}} \cdot \mathbf{y}$ is given by,

$$\mathbf{x} = \left(\mathbf{A} \cdot \mathbf{U} \cdot \mathbf{\Sigma_m^{-1}}\right) \cdot \mathbf{\Sigma_m^{-1}} \cdot \left(\left(\mathbf{A^\top} \cdot \mathbf{A}\right)^{-1} \mathbf{U} \cdot \mathbf{\Sigma_m^2}\right)^\top \cdot \mathbf{y} \tag{1.59}$$

1.3   Expansion by Taylor series

Taylor's theorem provides a way of expressing a function as a power series in the independent variable x, known as a Taylor series, but only applied to those functions that are continuous and differentiable within the x-range of interest[.] Difference formulas can be developed using Taylor series. This approach is especially useful for deriving finite difference approximations of exact derivatives (both total derivatives and partial derivatives) that appear in differential equations[7]. The theorem establishes that any smooth function can be approximated by means of a polynomial[8].

Theorem 1.3.1 (Taylor's theorem). *Let $f(x) \in \mathbb{D}$, and $x = x_0$ is any point in $\mathbb{D}$. Thus, a power series exist with centre in x_0 that represents an $f(x)$ in the following form,*

$$f(x) = \sum_{i=0}^{\infty} a_i (x - x_0)^i \tag{1.60}$$

where,

$$a_i = \frac{1}{i!} f^{(n)} x_0$$

Such coefficients satisfy the inequality,

$$|a_i| \leq \frac{\max |f(x)|}{(x - x_0)^i}$$

where $\max |f(x)|$ is the maximal value over the residual circumference $|x - x_0|$.

Difference formulas for functions of a single variable, for example, $f(x)$, can be developed from the Taylor series for a function of a single variable[6],

$$f(x) \cong f(x) + f^{(1)}(x)\Delta x + \frac{f^{(2)}(x)}{2!}\Delta x^2 + \frac{f^{(4)}(x)}{3!}\Delta x^3 + \cdots + \frac{f^{(n)}(x)}{n!}\Delta x^n + R_n \tag{1.61}$$

where $\Delta x = (x - h)$, and the residual term R_n is

$$R_n = \frac{f(\xi)}{(n+1)!}\Delta x^{n+1} \tag{1.62}$$

If the residual R_n is omitted, the equation becomes the approximation of the Taylor polynomial for $f(x)$. Thus, the general algorithmic formula,

$$f(x) \cong \sum_{i=0}^{\infty} \frac{f(x)^{(i)}}{i!}\Delta x^i \tag{1.63}$$

1.3.1   Multivariate expansion by Taylor series

For multivariate functions the Taylor series formulation is stated in the following manner,

$$f(x_0, x_1, \ldots, x_n) \sim \sum_{k=0}^{\infty} \frac{1}{k!} \left(\sum_{i=1}^{n} (x_i - h_i) \frac{\partial}{\partial x_i} \right)^k f(x_0, x_1, \ldots, x_n) \tag{1.64}$$

for n variables given in a vector $\mathbf{x} \in \mathbb{R}^n$, such that $\mathbf{x} = \{x_1, x_2, \ldots, x_n\}$. And the k degree Taylor expansion is developed,

$$f(\mathbf{x}) \approx f(\mathbf{x}) + \left((x_1 - h_1) \frac{\partial}{\partial x_1} + (x_2 - h_2) \frac{\partial}{\partial x_2} + \cdots + (x_n - h_n) \frac{\partial}{\partial x_n} \right)^k f(\mathbf{x}) \tag{1.65}$$

For instance, by developing the expansion of Taylor series for two variables x_1 and x_2, and for $h = 0$, with $k = 2$ degrees (second order derivative), the following expression is provided:

$$f(x_1, x_2) \approx f(x_1, x_2) + \left(x_1 \frac{\partial}{\partial x_1} + x_2 \frac{\partial}{\partial x_2} \right)^2 f(x_1, x_2) \tag{1.66}$$

by algebraically expanding the two degree binomial,

$$f(x_1, x_2) \approx f(x_1, x_2) + \left(x_1^2 \frac{\partial^2}{\partial x_1^2} + 2x_1 x_2 \frac{\partial}{\partial x_1 \partial x_2} + x_2^2 \frac{\partial^2}{\partial x_2^2} \right) f(x_1, x_2) \tag{1.67}$$

thus, the second order derivative approximation is

$$f(x_1, x_2) \approx f(x_1, x_2) + x_1^2 \frac{\partial^2 f(x_1, x_2)}{\partial x_1^2} + 2x_1 x_2 \frac{\partial f(x_1, x_2)}{\partial x_1 \partial x_2} + x_2^2 \frac{\partial^2 f(x_1, x_2)}{\partial x_2^2} \tag{1.68}$$

1.4   Solution of non linear equations

The term root of an equation refers to the values of the independent variable x calculated that make $f(x) = 0$. There exist numerous complex equations where it is not possible to find a direct analytical solution of the variables of interest. There are numerical methods to solve equations where analytical solutions are not possible. This section presents open methods that deploy only an initial value x_0 to find the solution. Sometimes, such methods diverge from finding the real root value as the process of calculation progresses. Nevertheless, when they converge, unlike close methods they are faster and iteratively find the solution.

1.4.1   The Newton-Raphson method

In order to calculate the root of a non linear function, from the known initial value x_i, a tangent line is yielded from the point $(x_i, f(x_i))$ to the X axis, which represent a better approximation of the real root. Form this geometric meaning, it is deduced that the first order derivative w.r.t. x is equivalent to the tangent line slope. The following deduction is made by the expansion of Taylor series,

$$f(x_{i+1}) = f(x_i) + f'(x_i)(x_{i+1} - x_i) + \frac{f''(\xi)}{2!}(x_{i+1} - x_i)^2 \qquad (1.69)$$

where ξ exists in the interval $([x_i, \ldots, x_{i+1})$, just after the term of the first order derivative. Thus, an approximation is provided by

$$f(x_{i+1}) \cong f(x_i) + f'(x_i)(x_{i+1} - x_i) \qquad (1.70)$$

the intersection of $f(x_{i+1})$ with the axis x must be $f(x_{i+1}) = 0$, therefore

$$f(x_i) + f'(x_i)(x_{i+1} - x_i) = 0 \qquad (1.71)$$

such that,

$$f'(x_i) = \frac{f(x_i) - 0}{x_i - x_{i+1}} \qquad (1.72)$$

and dropping-off x_{i+1}, the Newton-Raphson equation is

$$x_{i+1} = x_i - \frac{f(x_i)}{f'(x_i)} \qquad (1.73)$$

Thus, by estimating the error formula,

$$f(x_i) + f'(x_i)(x_{i+1} - x_i) + \frac{f''(\xi)}{2!}(x_{i+1} - x_i)^2 = 0 \qquad (1.74)$$

evaluating x_r in $f(x_r)$ and by substituting in

$$f'(x_i)(x_r - x_i) + \frac{f''(\xi)}{2!}(x_r - x_i)^2 = 0 \qquad (1.75)$$

where

$$E_{t,i+1} = x_r - x_{i+1} \qquad (1.76)$$

The series of Taylor of the error is ideally zero,

$$f'(x_i)E_{t,i+1} + \frac{f''(\xi)}{2!}E_{t,i}^2 = 0 \tag{1.77}$$

hence,

$$E_{t,i+1} = \frac{-f''(x_r)}{2f'(x_r)}E_{t,i}^2 \tag{1.78}$$

Thus, the relative error,

$$\varepsilon_i = \left| \frac{x_i - x_{i-1}}{x_i} \right| \tag{1.79}$$

1.4.2   The secant method

Sometimes exist cases of functions where their derivatives might be difficult to calculate, where the method of Newton-Rapshon may loose efficiency due to divergence, specially when slopes values are nearly 0, making difficult to find roots. Hence, in such cases a potential problem on calculating the derivative, may be overcome by approximating it by backward finite differences. This type of approximation basically substitute the derivative by the use of a secant, which is an extrapolation of the tangent line crossing the x axis.

$$x_{i+1} = x_i - \frac{f(x_i)(x_{i-1} - x_i)}{f(x_{i-1}) - f(x_i)} \tag{1.80}$$

Nevertheless, the secant method may diverge for some types of functions $f(x)$. Instead of using two arbitrary numeric values to approximate the derivative, an alternative modified method considers a fractional change of the independent variable to estimate $f'(x)$,

$$f'(x_i) = \frac{f(x_i + \delta x_i) - f(x_i)}{\delta x_i} \tag{1.81}$$

where δ represents a small fractional change, and substituting it in the derivative function of the Newton-Raphson formula previously stated, the following modified secant method formula is obtained,

$$x_{i+1} = x_i - \frac{\delta x_i f(x_i)}{f(x_i + \delta x_i) - f(x_i)} \tag{1.82}$$

1.4.3   Solution for polynomials

The Müller method is a manner to solve polynomial functions, similarly to the secant method that directs a line until the x axis using two values of the function. The Müller method is very similar, but considering three consecutive points along the polynomial curve. Such segment of three points is assumed to model a parabolic function .

$$f_2(x) = a(x - x_2)^2 + b(x - x_2) + c \tag{1.83}$$

It is desired the parabola curve passes along $(x_0, f(x_0)), (x_1, f(x_1))$, and $(x_2, f(x_2))$. Those three points are substituted in eq.(1.83) to evaluate its coefficients a, b, c. Therefore, the first point

$$f(x_0) = a(x_0 - x_2)^2 + b(x_0 - x_2) + c \tag{1.84}$$

for the second point,

$$f(x_1) = a(x_1 - x_2)^2 + b(x_1 - x_2) + c \tag{1.85}$$

and for the third point,

$$f(x_2) = a(x_2 - x_2)^2 + b(x_2 - x_2) + c \tag{1.86}$$

From third equation, c is easy found

$$f(x_2) = c$$

and substituting c next, two equations with two variables come up,

$$f(x_0) - f(x_2) = a(x_0 - x_2)^2 + b(x_0 - x_2) + c \tag{1.87}$$

and

$$f(x_1) - f(x_2) = a(x_1 - x_2)^2 + b(x_1 - x_2) + c \tag{1.88}$$

being $h_0 = x_1 - x_0$ and $h_1 = x_2 - x_1$ and by algebraic arrangements, a and b are solved, the divided differences are defined

$$\delta_0 = \frac{f(x_1) - f(x_0)}{x_1 - x_0}; \quad \delta_1 = \frac{f(x_2) - f(x_1)}{x_2 - x_1}$$

and substituting $\delta_{0,1}$ in $f(x_0)$, and $f(x_1)$; and dropping off a and b,

$$a = \frac{\delta_1 - \delta_0}{h_1 - h_0} \tag{1.89}$$

$$b = ah_1 + \delta_1 \tag{1.90}$$

and

$$c = f(x_2) \tag{1.91}$$

Therefore, to find either the real or complex roots, the quadratic expression is solved by the general formula,

$$x_3 = x_2 + \frac{-2c}{b \pm \sqrt{b^2 - 4ac}} \tag{1.92}$$

This solution is a direct manner to find the approximation error. In addition, the discriminant value is defined by $d = \sqrt{b^2 - 4ac}$, and it is evaluated by $|b - d| > |b + d|$. In the same way, the absolute error is calculated by

$$\varepsilon_r = \left| \frac{x_3 - x_2}{x_3} \right|$$

Finally, the sign is chosen by matching the sign of factor b. This choice provides a larger numeric denominator, and therefore the root will be very approximated to x_2. Once, x_3 is determined, the process is repeated.

1.5 System of non linear equations

The problem of this section consist of obtaining the roots of a set of simultaneous non linear equations. Where the solution is determined by a set of x_i that simultaneously make that all equations are zero.

$$f_1(x_1, x_2, \ldots, x_n) = 0$$
$$f_2(x_1, x_2, \ldots, x_n) = 0$$
$$\vdots$$
$$f_n(x_1, x_2, \ldots, x_n) = 0$$

There exist a number of methods to solve systems of non linear equations, and most of them are extensions of open methods. In next section the Newthon-Raphson method for systems of non linear equations is presented.

1.5.1　The Newton-Raphson method

The Newton-Raphson method utilises the first order derivative to evaluate the tangent line slope of a function. Such calculation has its foundations on the series of Taylor first order derivative,

$$f(x_{i+1}) = f(x_i) + (x_{i+1} - x_i)f'(x_i) \tag{1.93}$$

Given that we look for solving the systems of equations when the root is approximately found $f(x_{i+1}) = 0$, when the function is evaluated for x_{i+1}. Thus, by dropping-off the independent variable,

$$x_{i+1} = x_i - \frac{f(x_i)}{f'(x_i)}$$

which is the method for a single equation. For multiple equations the method to calculate a root is very similar. However, the series of Taylor for multiple variables must be used, in order to know that more than a single independent variable contributes to determine the root. Considering the case for two variables,

$$u_{i+1} = u_i + (x_{i+1} - x_i)\frac{\partial u_i}{\partial x} + (y_{i+1} - y_i)\frac{\partial u_i}{\partial y} \tag{1.94}$$

and

$$v_{i+1} = v_i + (x_{i+1} - x_i)\frac{\partial v_i}{\partial x} + (y_{i+1} - y_i)\frac{\partial v_i}{\partial y} \tag{1.95}$$

Considering the same principle for the roots that $u_{i+1} = 0$, y $v_{i+1} = 0$. The iterative solution method[5] starts by stating the initial values,

$$\frac{\partial u_i}{\partial x}x_{i+1} + \frac{\partial u_i}{\partial y}y_{i+1} = -u_i + x_i\frac{\partial u_i}{\partial x} + y_i\frac{\partial u_i}{\partial y}$$

and

$$\frac{\partial v_i}{\partial x}x_{i+1} + \frac{\partial v_i}{\partial y}y_{i+1} = -v_i + x_i\frac{\partial v_i}{\partial x} + y_i\frac{\partial v_i}{\partial y}$$

Now previous equations are stated as linear, where x_i, y_i are known, and x_{i+1}, y_{i+1} are the unknown variables to solve. Therefore, by realising algebraic manipulations to equations, and using an algebraic method to solve the system (i.e. Cramer theorem, section 1.2.4).

$$x_{i+1} = x_i - \frac{u_i \dfrac{\partial v}{\partial y} - v_i \dfrac{\partial u_i}{\partial y}}{\dfrac{\partial u_i}{\partial x} \dfrac{\partial v_i}{\partial y} - \dfrac{\partial u_i}{\partial y} \dfrac{\partial v_i}{\partial x}} \tag{1.96}$$

and

$$y_{i+1} = y_i - \frac{v_i \dfrac{\partial v}{\partial x} - u_i \dfrac{\partial v_i}{\partial x}}{\dfrac{\partial u_i}{\partial x} \dfrac{\partial v_i}{\partial y} - \dfrac{\partial u_i}{\partial y} \dfrac{\partial v_i}{\partial x}} \tag{1.97}$$

In addition, the denominator term in both equationsis known as the determinant of the Jacobian matrix of the system of equations, which may be described separately. The Jacobian matrix is defined as

$$J_i = \begin{pmatrix} \dfrac{\partial u_i}{\partial x} & \dfrac{\partial u_i}{\partial y} \\ \dfrac{\partial v_i}{\partial x} & \dfrac{\partial v_i}{\partial y} \end{pmatrix}$$

Hence, its determinant is formulated by

$$|J_i| = \frac{\partial u_i}{\partial x} \frac{\partial v_i}{\partial y} - \frac{\partial v_i}{\partial x} \frac{\partial u_i}{\partial y}$$

Another way to express the solution for the multivariate system of equations is

$$x_{i+1} = x_i - \frac{u_i \dfrac{\partial v_i}{\partial y} - v_i \dfrac{\partial u_i}{\partial y}}{|J_i|}$$

and

$$y_{i+1} = y_i - \frac{v_i \dfrac{\partial u_i}{\partial x} - u_i \dfrac{\partial v_i}{\partial x}}{|J_i|}$$

Furthermore, the sufficient conditions for convergence are the following criteria,

$$\left| \frac{\partial u}{\partial x} \right| + \left| \frac{\partial v}{\partial x} \right| < 1$$

and

$$\left|\frac{\partial u}{\partial y}\right| + \left|\frac{\partial v}{\partial y}\right| < 1$$

1.6 Numerical fitting models

Either the realistic calculation of a point between discrete values, or the simplified version of a complex mathematical function, is known as Curves Fitting. A fitting model is obtained from the process of constructing a curve having its best fit to a series of data points.

1.6.1 Newton polynomial interpolation

If it is known that the points of a dataset or table are very precise, then the basic procedure is to fix a curve or series of curves crossing through each point directly. The estimation of values between discrete points are known as interpolation.

Let be the polynomial of the general form,

$$f_n(x) = b_0 + b_1(x - x_0) + b_2(x - x_0)(x - x_1) + \cdots + b_n(x - x_0)\ldots(x - x_{n-1}) \tag{1.98}$$

where the coefficients are obtained from the next expression,

$$b_0 = f(x_0) \tag{1.99a}$$

$$b_1 = f(x_1, x_0) \tag{1.99b}$$

$$b_2 = f(x_2, x_1, x_0) \tag{1.99c}$$

$$b_n = f(x_n, x_{n-1}, \ldots, x_0) \tag{1.99d}$$

The divided differences are described by,

$$f(x_i, x_j) = \frac{f(x_i) - f(x_j)}{x_i - x_j} \tag{1.100}$$

likewise, the second divided difference is deduced in the next expression,

$$f(x_i, x_j, x_k) = \frac{f(x_i, x_j) - f(x_j - x_k)}{x_i - x_k} \tag{1.101}$$

In such a way, the nth finite divided difference is

$$f(x_n, x_{n-1}, \ldots, x_1, x_0) = \frac{f(x_n, x_{n-1}, \ldots x_1) - f(x_{n-1}, x_{n-2}, \ldots x_0)}{x_n - x_0} \tag{1.102}$$

Such differences are useful to evaluate the coefficients of equations (1.99), which are substituted in equation (1.98) to obtain the interpolative polynomial, which is defined next,

$$f_n(x) = f(x_0) + (x - x_0)f(x_1, x_0) + (x - x_0)(x - x_1)f(x_2, x_1, x_0) + \cdots + (x - x_0)\cdots(x - x_{n-1})f(x_n, \ldots, x_0) \tag{1.103}$$

which is know as the Newton's interpolation polynomial of divided differences. Points equally spaced, or values along abscissa ascendantly ordered are not required.

1.6.2   Lagrange polynomial interpolation

In order to avoid the calculus of the divided differences, the Newton polynomial interpolation is algebraically reformulated to state the Lagrange interpolation, which is concisely represented by the next equation,

$$f(x) = \sum_{i=0}^{n} [L_i(x)y_i] \tag{1.104}$$

where

$$L_i(x) = \prod_{\substack{j=0 \\ i \neq j}}^{n} \frac{x - x_i}{x_j - x_i} \tag{1.105}$$

or well, directly used as

$$f(x) = \sum_{i=0}^{n} \left[\left(\prod_{j=0}^{n} \frac{x - x_i}{x_j - x_i} \right) y_i \right] \tag{1.106}$$

For instance, the linear version $n = 1$ is

$$f_1(x) = \frac{x - x_1}{x_0 - x_1} f(x_0) + \frac{x - x_0}{x_1 - x_0} f(x_1) \tag{1.107}$$

The general model will produce a polynomial equation that fit a table of data of degree $n - 1$.

$$y_i(x_i) = a_o + a_1 x_i + a_2 x_i^2 + \cdots + a_n x_i^n \tag{1.108}$$

1.6.3 Polynomial regression

When data exhibit a significant degree of error, unlike intersecting all points, but a single curve that represent the data trend as a group is known as Regression. The procedure of square least is enhanced to adjust data that fit non linear functions, such as polynomials.

Given the empirical model $y = y_m + \epsilon$, in order to approximate it to a suitable theoretical model y_m, and fitting the numeric data to a second degree polynomial with the general form,

$$y_m = a_0 + a_1x + a_2x^2 \tag{1.109}$$

Thus, the sum of the squared differences yields the residual theorem,

Theorem 1.6.1 (Least square polynomial regression). *The residual of the squared sum of the empirical and theoretical fitting model is $s_r = (y - y_m)^2$*

$$s_r = \sum_{i=1}^{n} \left(y_i - (a_0 + a_1x_i + a_2x_i^2 + \cdots + a_kx_i^k)\right)^2 \tag{1.110}$$

By partial derivatives the rate of change of the function w.r.t. each coefficient is determined by the next three equations. Let us assume a quadratic problem for simplicity purposes.

$$\frac{\partial s_r}{\partial a_0} = -2\sum_{i=1}^{n}(y_i - a_0 - a_1x_i - a_2x^2) \tag{1.111a}$$

$$\frac{\partial s_r}{\partial a_0} = -2\sum_{i=1}^{n}x_i(y_i - a_0 - a_1x_i - a_2x^2) \tag{1.111b}$$

$$\frac{\partial s_r}{\partial a_0} = -2\sum_{i=1}^{n}x_i^2(y_i - a_0 - a_1x_i - a_2x^2) \tag{1.111c}$$

Equating to zero each function and algebraically factorizing them, a set of linear equations in terms of their coefficients a_i are stated for subsequent solution,

$$(n)a_0 + \left(\sum_i x_i\right)a_1 + \left(\sum_i x_i^2\right)a_2 = \sum_i y_i \tag{1.112a}$$

$$\left(\sum_i x_i\right) a_0 + \left(\sum_i x_i^2\right) a_1 + \left(\sum_i x_i^3\right) a_2 = \sum_i x_i y_i \qquad (1.112b)$$

$$\left(\sum_i x_i^2\right) a_0 + \left(\sum_i x_i^3\right) a_1 + \left(\sum_i x_i^4\right) a_2 = \sum_i x_i^2 y_i \qquad (1.112c)$$

Algebraically ordering in the matrix form as a linear system $\mathbf{v} = \mathbf{A} \cdot \mathbf{a}$; see please section 1.2,

$$\begin{pmatrix} \sum_i y_i \\ \sum_i x_i y_i \\ \sum_i x_i^2 y_i \end{pmatrix} = \begin{pmatrix} n & \sum_i x_i & \sum_i x_i^2 \\ \sum_i x_i & \sum_i x_i^2 & \sum_i x_i^3 \\ \sum_i x_i^2 & \sum_i x_i^3 & \sum_i x_i^4 \end{pmatrix} \cdot \begin{pmatrix} a_0 \\ a_1 \\ a_2 \end{pmatrix} \qquad (1.113)$$

by solving the system of equations by the algebraic inverse matrix obtaining its determinant,

$$\mathbf{a} = \mathbf{A}^{-1} \cdot \mathbf{v} \qquad (1.114)$$

for the actual quadratic problem, the coefficients are a_0, a_1, and a_2. In addition, with standardised error $s_{y/x}$, and coefficient of determination $r^2 = (s_t - s_r)/s_t$,

$$s_{y/x} = \sqrt{\frac{s_r}{n - (k + 1)}}$$

1.7 Numerical differentiation and integration

1.7.1 Numerical derivation

The derivative of a function is the means of differentiation that represents the rate of change of the dependent variable (function) w.r.t. an independent variable. Its mathematical definition starts by an approximation by differences,

$$\frac{\Delta y}{\Delta x} = \frac{f(x_i + \Delta x) - f(x_i)}{\Delta x} \qquad (1.115)$$

where y or $f(x)$ are alternative representations of the dependent variable; while x is the independent variable. By approximating Δx to zero, the quotient of the differences becomes a derivation.

$$\frac{dy}{dx} = \lim_{\Delta x \to 0} \frac{f(x_i + \Delta x) - f(x_i)}{\Delta x} \qquad (1.116)$$

For high accuracy derivation formulae, the divided differences can be generated by taking additional terms from the expansion of Taylor series.

$$f(x_{i+1}) = f(x_i) + f^{(1)}(x_i)h + \frac{f^{(2)}(x_i)}{2}h^2 + \cdots \qquad (1.117)$$

from where, drop-off the first order derivative function,

$$f^{(1)}(x_i) = \frac{f(x_{i+1}) - f(x_i)}{h} - \frac{f^{(2)}(x_i)}{2}h + O(h^2) \qquad (1.118)$$

as we are interested in the first derivative term, we truncate the second derivative term,

$$f^{(1)}(x_i) = \frac{f(x_{i+1}) - f(x_i)}{h} + O(h) \qquad (1.119)$$

However, we now keep the second derivative term substituting its next approximation,

$$f^{(2)}(x_i) = \frac{f(x_{i+2}) - 2f(x_{i+1}) + f(x_i)}{h^2} + O(h) \qquad (1.120)$$

thus,

$$f^{(1)}(x_i) = \frac{f(x_{i+1}) - f(x_i)}{h} - \frac{f(x_{i+2}) - 2f(x_{i+1}) + f(x_i)}{2h^2} + O(h) \qquad (1.121)$$

by algebraically ordering,

$$f^{(1)}(x_i) = \frac{-f(x_{i+2}) + 4f(x_{i+1}) - 3f(x_i)}{2h} + O(h^2) \qquad (1.122)$$

1.7.2   High precision numerical derivation

Furthermore, the forward finite divided differences is presented in two version for each derivative. The last version uses more terms of the expansion of Taylor series, and as a consequence is becomes more exact.

The first derivative

$$f^{(1)}(x_i) = \frac{f(x_{i+1}) - f(x_i)}{h} \qquad (1.123)$$

and

$$f^{(1)}(x_i) = \frac{-f(x_{i+2}) + 4f(x_{i+1}) - 3f(x_i)}{2h} \qquad (1.124)$$

Second derivative,

$$f^{(2)}(x_i) = \frac{f(x_{i+2}) - 2f(x_{i+1}) + f(x_i)}{h^2} \tag{1.125}$$

and

$$f^{(2)}(x_i) = \frac{-f(x_{i+3}) + 4f(x_{i+2}) - 5f(x_{i+2}) + 2f(x_i)}{h^2} \tag{1.126}$$

Third derivative

$$f^{(3)}(x_i) = \frac{f(x_{i+3}) - 3f(x_{i+2}) + 3f(x_{i+1}) + f(x_i)}{h^3} \tag{1.127}$$

and

$$f^{(2)}(x_i) = \frac{-3f(x_{i+4}) + 14f(x_{i+3}) - 24f(x_{i+2}) + 18f(x_{i+1}) - 5f(x_i)}{2h^3} \tag{1.128}$$

Fourth derivative

$$f^{(4)}(x_i) = \frac{f(x_{i+4}) - 4f(x_{i+3}) + 6f(x_{i+2}) - 4f(x_{i+1}) + 5f(x_i)}{h^4} \tag{1.129}$$

and

$$f^{(4)}(x_i) = \frac{-2f(x_{i+5}) + 11f(x_{i+4}) - 24f(x_{i+3}) + 26f(x_{i+2}) - 14f(x_{i+1}) + 3f(x_i)}{h^4} \tag{1.130}$$

In addition, the backward divided differences is presented in two versions for each order derivative. The second one poses more terms than the Series of Taylor, hence consequently is more exact. Thus, the first derivative

$$f^{(1)}(x_i) = \frac{f(x_i) - f(x_{i-1})}{h} \tag{1.131}$$

and

$$f^{(1)}(x_i) = \frac{3f(x_i) - 4f(x_{i-1}) + f(x_{i-2})}{2h} \tag{1.132}$$

Second derivative,

$$f^{(2)}(x_i) = \frac{f(x_{i-2}) - 2f(x_{i-1}) + f(x_i)}{h^2} \tag{1.133}$$

and

$$f^{(2)}(x_i) = \frac{-f(x_{i-3}) + 4f(x_{i-2}) - 5f(x_{i-1}) + 2f(x_i)}{h^2} \tag{1.134}$$

Third derivative

$$f^{(3)}(x_i) = \frac{-f(x_{i-3}) + 3f(x_{i-2}) - 3f(x_{i-1}) + f(x_i)}{h^3} \tag{1.135}$$

and

$$f^{(2)}(x_i) = \frac{3f(x_{i-4}) - 14f(x_{i-3}) + 24f(x_{i-2}) - 18f(x_{i-1}) + 5f(x_i)}{2h^3} \tag{1.136}$$

Fourth derivative

$$f^{(4)}(x_i) = \frac{f(x_{i-4}) - 4f(x_{i-3}) + 6f(x_{i-2}) - 4f(x_{i-1}) + f(x_i)}{h^4} \tag{1.137}$$

and

$$f^{(4)}(x_i) = \frac{-2f(x_{i-5}) + 11f(x_{i-4}) - 24f(x_{i-3}) + 26f(x_{i-2}) - 14f(x_{i-1}) + 3f(x_i)}{h^4} \tag{1.138}$$

The central divided differences are also presented in two versions for each derivative. The second derivative formula uses more terms than the series of Taylor, hence it is more exact. The first derivative

$$f^{(1)}(x_i) = \frac{f(x_{i+1}) - f(x_{i-1})}{2h} \tag{1.139}$$

and

$$f^{(1)}(x_i) = \frac{-f(x_{i+2}) + 8f(x_{i+1}) - 8f(x_{i-1}) + f(x_{i-2})}{12h} \tag{1.140}$$

The second derivative

$$f^{(2)}(x_i) = \frac{f(x_{i+1}) - f(x_i) + f(x_{i-1})}{h^2} \tag{1.141}$$

and

$$f^{(2)}(x_i) = \frac{-f(x_{i+2}) + 16f(x_{i+1}) - 30f(x_i) + 16f(x_{i-1}) - f(x_{i-2})}{12h^2} \tag{1.142}$$

The third derivative

$$f^{(3)}(x_i) = \frac{f(x_{i+2}) - 2f(x_{i+1}) + 2f(x_{i-1}) - f(x_{i-2})}{2h^3} \tag{1.143}$$

and

$$f^{(3)}(x_i) = \frac{-f(x_{i+3}) + 8f(x_{i+2}) - 13f(x_{i+1}) + 13f(x_{i-1}) - 8f(x_{i-2}) + f(x_{i-3})}{8h^3} \tag{1.144}$$

The fourth derivative

$$f^{(4)}(x_i) = \frac{f(x_{i+2}) - 4f(x_{i+1}) + 6f(x_i) - 4f(x_{i-1}) + f(x_{i-2})}{h^4} \tag{1.145}$$

and

$$f^{(4)}(x_i) = \frac{-f(x_{i+3}) + 12f(x_{i+2}) - 39f(x_{i+1}) + 56f(x_i) - 39f(x_{i-1}) + 12f(x_{i-2}) + f(x_{i-3})}{6h^4} \tag{1.146}$$

1.7.3 Numerical integration

The Newton-Cotes formulae are common types of numeric integration methods. They replace a complex function or a table of data by an approximated polynomial that is easier to integrate.

$$I = \int_a^b f(x)dx \cong \int_a^b f_n(x)dx \tag{1.147}$$

where $f_n(x)$ is polynomial of the form,

$$f_n(x) = a_0 + a_1 x + \cdots + a_{n\;1} a^{n-1} + a_n x^n \tag{1.148}$$

Before the trapezoidal rule is applied, the previous polynomial may be expressed as

$$f_1(x) = \frac{f(b) - f(a)}{b - a} x + f(a) - \frac{af(b) - af(a)}{b - a} \tag{1.149}$$

algebraically grouping the last terms,

$$f_1(x) = \frac{f(b) - f(a)}{b - a} x + \frac{bf(a) - af(a) - af(b) + af(a)}{b - a} \tag{1.150}$$

and simplifying,

$$f_1(x) = \frac{f(b) - f(a)}{b - a} x + \frac{bf(a) - af(b)}{b - a} \tag{1.151}$$

which can be integrated between the limits $x = a$ and $x = b$ to obtain,

$$I = \frac{f(b) - f(a)}{2(b - a)} x^2 + \frac{bf(a) - af(b)}{b - a} x \Big|_a^b \tag{1.152}$$

Such result is evaluated,

$$I = \frac{f(b) - f(a)}{2(b - a)}(b^2 - a^2) + \frac{bf(a) - af(b)}{b - a}(b - a) \qquad (1.153)$$

In addition, since

$$b^2 - a^2 = (b - a)(b + a),$$

thus,

$$I = [f(b) - f(a)]\frac{b + a}{2} + bf(a) - af(b) \qquad (1.154)$$

By multiplying and algebraically grouping, the Trapezoidal rule is obtained,

$$I = (b - a)\frac{f(a) + f(b)}{2} \qquad (1.155)$$

In order to improve the trapezoidal rule's precision, the interval a to b is divided in $n + 1$ segments. The integration method is applied to each equally spaced $n + 1$ segments. Consequently,

$$h = \frac{b - a}{n} \qquad (1.156)$$

If a and b are designated as x_0 and x_n, respectively, the complete integration is represented by

$$I = \int_{x_0}^{x_1} f(x)dx + \int_{x_1}^{x_2} f(x)dx + \cdots + \int_{x_{n-1}}^{x_n} \qquad (1.157)$$

Therefore, by substituting the trapezoidal rule in each integral,

$$I = h\frac{f(x_0) + f(x_1)}{2} + h\frac{f(x_1) + f(x_2)}{2} + \cdots + h\frac{f(x_{n-1}) + f(x_n)}{2} \qquad (1.158)$$

by algebraically factorizing,

$$I = \frac{h}{2}\left[f(x_0) + 2\sum_{i=1}^{n-1} f(x_i) + f(x_n)\right] \qquad (1.159)$$

or if substituting h, then

$$I = (b - a)\frac{f(x_0) + 2\sum_{i=1}^{n-1}(f(x_i) + f(x_n))}{2n} \qquad (1.160)$$

Bibliography

[1] Soutas-Little R.W., Inman D.J., Balint D.S., *Mechanical engineering statics*, Cengage Learning, 2009.

[2] Lay D.C., *Linear Algebra and its applications*, Pearson Intl. Ed., 2006.

[3] Horn R.A., Johnson C.R., *Matrix analysis*, Cambridge University Press, 1985, doi: 10.1017/CBO9780511810817.

[4] Riley K.F., Hobson M.P., Bence S.J., *Mathematical methods for physics and engineering*, 3rd Ed., Cambridge University Press, 2006, doi: 10.1017/CBO9780511810763.

[5] Chapra S.C., Canale R.P., *Numerical methods for engineers*, 5th Ed., McGrawHill, 2007.

[6] Cheney W., Kincais D., *Numerical mathematics and computing*, 6th Ed., Cengage Learning, 2011.

[7] Hoffman J.D., *Numerical methods for engineers and scientists*, 2nd Ed.,2001.

[8] Kreyszig E., *Advanced engineering mathematics*, 3rd Ed., Limusa Wiley, 2009.

Part I

Mapping and Robot Sensing Models

Chapter 2

SENSING MODELS IN ROBOTICS

Edgar A. Martínez García

Laboratorio de Robótica, Institute of Engineering and Technology
Universidad Autónoma de Ciudad Juárez, Mexico.

The mathematical formulations to describe how sensors commonly used in mobile robotics obtain data is a matter of discussion provided in this chapter. Sensor measurements that are relative to the robot's fixed inertial frame are transformed into global or Cartesian spaces to allow robots to map, navigate, and autonomously perceive attributes of the world. Any type of sensor data is useful to feedback the robot about the changes of the world, and to take smart decisions autonomously. Furthermore, a sensor is a device that detects attributes of the environment provided in forms that usually are readable or understandable by the human users, such as odometers, a colour camera, gyroscope, etc. There exist sensors that provide proprioceptive data, which arise from robot's inner stimulus. Likewise, exteroceptive data provides robot's attributes, which are stimuli measured w.r.t. to external objects surrounding the robot. A sensor model is a mathematical description on how a sensor obtains data from the physical measurements of the environment. Sensing models could be either, to describe passive sensors, which detect energy naturally from the environmental conditions (i.e. vision cameras, gyroscopes, accelerometers, stereo pairs); or active sensors, which pose a traducer to detect the reflection of energy that was previously emitted by the sensor's transmitter (i.e. ultrasonic sonars, light detection and ranging, light strips). A transducer is a device that detects a type of energy and transforms it into another type, commonly electric energy. Some examples of transducers are thermocouples, pulse encoders, optical arrays, quartz crystals, and so forth. One of the main interest of this chapter is to obtain sensing models to infer the instantaneous robot's posture $(x_t, y_t, \theta_t)^\top$, from different sensing means. The robot posture is a fundamental information to register massive data into different spaces, building environmental maps, and trajectory control algorithms.

2.1 Odometer sensing model

An odometer sensing model allows to infer the robot's posture by means of quantization of the instantaneous displacements. An odometric model infers displacement, velocity, and acceleration of a wheeled mobile robot from direct sensing of encoder pulses. An encoder device or transducer poses a rotatory mechanisms fixed to the wheels shaft to count rotational motion. For instance, analysing the robot's posture with a dual asynchronous speeds robot, also known as differential speed control, in which the independent variables are the right and left wheels velocities v_r, and v_l. Figure 2.1 shows the odometer displacement ΔS, which is an average of the right and left wheels displacement ΔS_r and ΔS_l respectively. The odometers kinematic considers the distance between the wheels' encoder b, and their radius r.

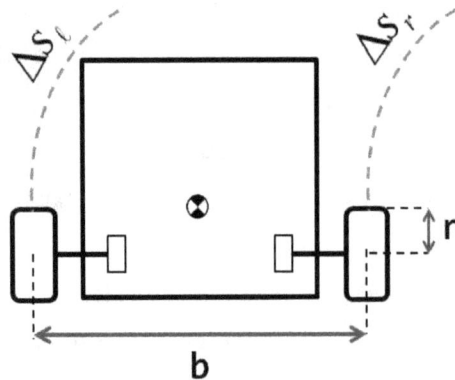

Figure 2.1: Dual wheels speed robot's kinematics inferring odometer's displacement.

The total robot's displacement is an averaged value expressed by S,

$$s = \frac{\Delta s_r + \Delta s_\ell}{2} \tag{2.1}$$

Likewise, the robot's orientation is defined by θ as a function of the wheels displacements,

$$\theta = \frac{\Delta s_r - \Delta s_\ell}{b} \tag{2.2}$$

Where each wheel's displacement is formulated independently, and the right-sided linear displacement is,

$$\Delta s_r = n_r \cdot f_s \tag{2.3}$$

while the left-sided distance,

$$\Delta s_\ell = n_\ell \cdot f_s \tag{2.4}$$

where n_r, and n_ℓ are the measured number of pulses counted by the shafts encoder. The factor f_s represents the encoders resolution defined by the equation (2.5). Where r is the wheel's radius, and k is the total number of pulses within a rotation.

$$f_s = \frac{2\pi r}{k} \tag{2.5}$$

The asynchronous wheels displacement is deduced by substituting the factor of expression (2.5), in equations (2.3) and (2.4).

$$\Delta s_r = \frac{2\pi r}{k} \cdot n_r, \tag{2.6}$$

and

$$\Delta s_\ell = \frac{2\pi r}{k} \cdot n_\ell. \tag{2.7}$$

Thus, the instantaneous robot's position and orientation are provided by substituting the equations (2.6) and (2.7), in expressions (2.1) and (2.2), and algebraically simplifying.

$$s = \frac{\pi r}{k} \cdot (n_r + n_\ell) \tag{2.8}$$

and

$$\theta = \frac{2\pi r}{kb} \cdot (n_r - n_\ell) \tag{2.9}$$

Therefore, the robot's instantaneous posture (position and orientation) is modelled by the following recursive expressions,

$$x_t = x_{t-1} + \Delta s_x, \tag{2.10}$$

$$y_t = y_{t-1} + \Delta s_y, \tag{2.11}$$

and

$$\theta_t = \theta_{t-1} + \Delta\theta. \tag{2.12}$$

Where Δs_x and Δs_y are the Cartesian components of distance, and they are defined in terms of the displacement s, by

$$\Delta s_x = s \cdot \cos(\theta) \tag{2.13}$$

and

$$\Delta s_y = s \cdot \sin(\theta) \tag{2.14}$$

Thus by substituting the expressions (2.8) and (2.9) in equations (2.13) and (2.14), the complete robot's displacements are formulated,

$$\Delta s_x = \frac{\pi r}{k}(n_r + n_\ell) \cdot \cos\left(\frac{2\pi r}{kb}(n_r - n_\ell)\right) \tag{2.15}$$

and

$$\Delta s_y = \frac{\pi r}{k}(n_r + n_\ell) \cdot \sin\left(\frac{2\pi r}{kb}(n_r - n_\ell)\right) \tag{2.16}$$

Thus, the actual robot's posture computed recursively within a common inertial system[5] with origin at robot's initial posture, the following model is provided:

$$\begin{pmatrix} x_t \\ y_t \\ \theta_t \end{pmatrix} = \begin{pmatrix} x_{t-1} \\ y_{t-1} \\ \theta_{t-1} \end{pmatrix} + \begin{pmatrix} \frac{\pi r}{k}(n_r + n_\ell) \cdot \cos\left(\frac{2\pi r}{kb}(n_r - n_\ell)\right) \\ \frac{\pi r}{k}(n_r + n_\ell) \cdot \sin\left(\frac{2\pi r}{kb}(n_r - n_\ell)\right) \\ \frac{2\pi r}{kb} \cdot (n_r - n_\ell) \end{pmatrix} \tag{2.17}$$

Depending on the robot's hardware configuration, we may frequently obtain sequences of sensor data from encoder readings: angular positions, or angular velocities $\dot{\varphi}_0, \ldots, \dot{\varphi}_{t-1}, \dot{\varphi}_t, \ldots, \dot{\varphi}_{t+1}$. For high precision estimation, at least three sensor readings to determine the real actuator's rotational velocity are needed. The next function calculates the angular speed through backward finite divided differences. See section 1.7 for further details.

$$\dot{\varphi}_i = \frac{3\varphi_i - 4\varphi_{i-1} + \varphi_{i-2}}{2\Delta t} \tag{2.18}$$

Similarly, online angular accelerations are possible to measure by a second order numerical differentiation w.r.t. angular positions based on the backward finite divided differences,

$$\ddot{\varphi}_i = \frac{-\varphi_{i-3} + 4\varphi_{i-2} - 5\varphi_{i-1} + 2\varphi_i}{\Delta t^2} \tag{2.19}$$

Likewise, obtaining displacement and speed from angular acceleration sensor data $\ddot{\varphi}$, the inverse calculus is obtained throughout numerical integration. Thus, the trapezoid theorem is one of the close form integration equations of Newton-Cotes (chapter 1.7), where

$$\dot{\varphi}_t = \int_{t_1}^{t_2} \ddot{\varphi}_t \mathrm{dt} \tag{2.20}$$

and

$$\varphi_t = \int_{t_1}^{t_2} \dot{\varphi}_t \mathrm{dt} = \iint_{t_1}^{t_2} \ddot{\varphi}_t \mathrm{dt}^2 \tag{2.21}$$

By obtaining angular speed from acceleration,

$$\dot{\varphi}_i = (t_2 - t_1) \frac{\ddot{\varphi}_0 + 2\sum_{i=1}^{n-1}\left(\ddot{\varphi}_{i+1} + \ddot{\varphi}_n\right)}{2\Delta t} \tag{2.22}$$

A second integration w.r.t. time using the same model (2.22) is applied to obtain the angular position when required.

2.2 Ultrasonic range finding

The ultrasonic sonar is an exteroceptive active type of sensor, which produce an acoustic vibration. The sensor contains a receiver that detects the ultrasonic return, or echo of the objects surrounding the robot that reflected the emitting sound pulses. A sonar is capable to detect a diversity of materials, which do not reflect electromagnetic waves, such as glass.

A robot is frequently instrumented with a ring of i sonar arranged in cylindrical positions $\mathbf{z}_i = (\ell_i, \phi_i)^\top$ w.r.t. to the fixed robot's inertial system (centroid), see figure 2.7-right. Thus, each sonar range finder is located at

$$\begin{pmatrix} \mathbf{z}_{xi} \\ \mathbf{z}_{y_i} \end{pmatrix} = \ell_i \begin{pmatrix} \cos(\phi_i) \\ \sin(\phi_i) \end{pmatrix} \tag{2.23}$$

where ϕ_i is the angle orientation, and ℓ_i is the length on the chassis of the sonar device ith w.r.t. to the robot's centroid. The distance of the nearest object w.r.t. perpendicular sonic emission is the time of a round flight, being the speed of sound $c = 341$m/s at 25°C.

$$c = \frac{2d}{t} \tag{2.24}$$

The total time of flight t in sg includes the emission and returning of echo. Thus, the sensed distance $d(t)$ as a function of time is,

$$d = \frac{ct}{2} \tag{2.25}$$

Let $\mathbf{s}_i$ be the Cartesian coordinates vector of a sensed object at distance d_i by the sonar device i. From the robots body until the sensed object, each sonar device located at position (2.23) poses a bearing direction β_i. Hence, the expression (2.26) completes the real distance from the robot.

$$\mathbf{s}_i^R = d_i(t) \cdot \begin{pmatrix} \cos(\phi_i + \beta_i) \\ \sin(\phi_i + \beta_i) \end{pmatrix} + \begin{pmatrix} z_{xi} \\ z_{yi} \end{pmatrix} \tag{2.26}$$

Therefore, the object position w.r.t. to a global system that is different from the coordinate frame where the robot started motion is given by a homogeneous transformation,

$$\mathbf{s}_i^I = \mathbf{R}(\gamma) \cdot \mathbf{s}_i^R + \boldsymbol{\xi}_i \tag{2.27}$$

For this specific example the coordinates of the robot's position are taken from the posture vector $\boldsymbol{\xi}$, namely px and py as depicted in figure 2.7-centre. By substituting (2.26) in (2.27), the complete expression maps the sensed objects within a common coordinate frame rotated γ degrees by (2.28).

$$\mathbf{s}_i^I = \begin{pmatrix} \cos(\gamma) & -\sin(\gamma) \\ \sin(\gamma) & \cos(\gamma) \end{pmatrix} \cdot \left[d_i(t) \begin{pmatrix} \cos(\phi_i + \theta) \\ \sin(\phi_i + \theta) \end{pmatrix} + \begin{pmatrix} z_{xi} \\ z_{yi} \end{pmatrix} \right] + \begin{pmatrix} px \\ py \end{pmatrix} \tag{2.28}$$

Figure 2.2 depicts some experimental data produced during an experiment deploying a ring of ultrasonic sonar. The figure shows the real scene, the sensors ultrasonic cones by a GUI, the polar form plot, and the Cartesian plot of the objects.

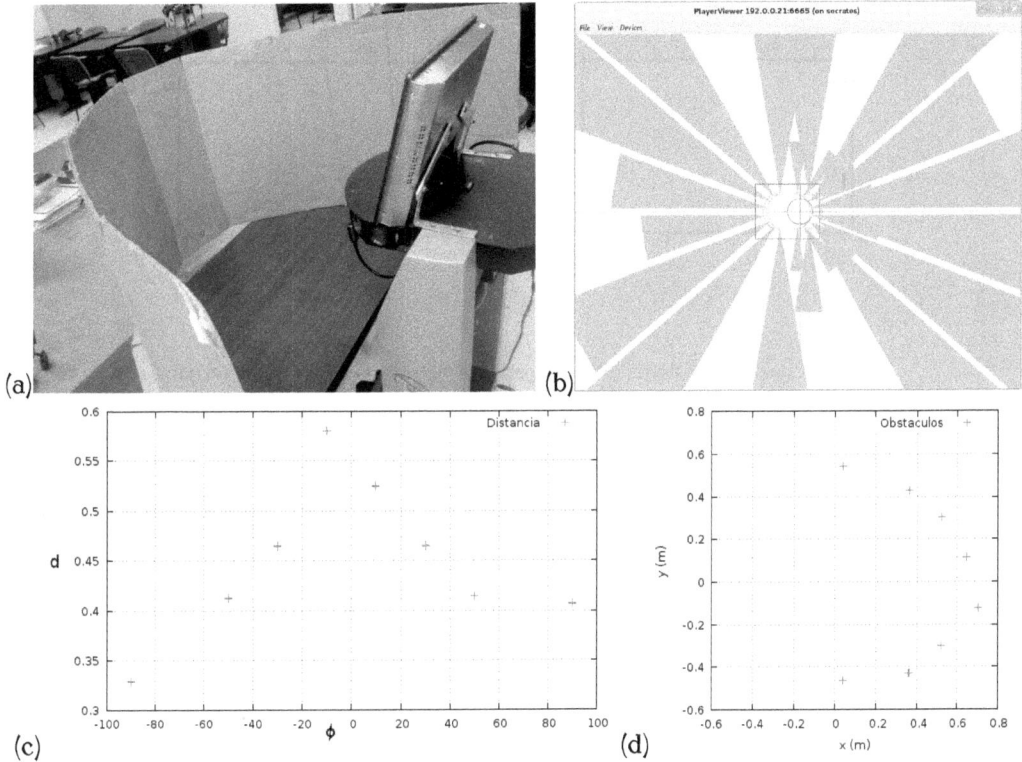

Figure 2.2: (a) Photo of the experimental scene; (b) GUI with sensors conic scope; (c) polar plot of ring of sonars: $d_i(t)$ vs $\phi_i + \beta_i$; (d) Cartesian plot of sensed objects.

2.3 Stereo vision sensing model

Stereo vision[10,11] refers to a visual method to fuse two or more planar intensity images $\mathbf{I}_{n \times m}$ taken from different perspective locations to infer 3D information about the present environmental scene. The importance of deploying stereo vision to mobile robotics regards the metric measure of near obstacles. Stereo vision provides massive 3D data of the environment in the sensors angle of view. This section analyses a stereo sensor comprised of two cameras (binocular), where such visual sensors are physically aligned along a baseline distance b over a vertical epipolar plane (figure 2.3). In this figure p_r and p_ℓ are the pixels in right and left cameras respectively, representing the projection of the same object in the scene. By taking a same measuring reference (i.e. images right-side), the variables x_r and x_ℓ are the number of columns of the pixels $p_{r,\ell}$ with respect to the same images' side.

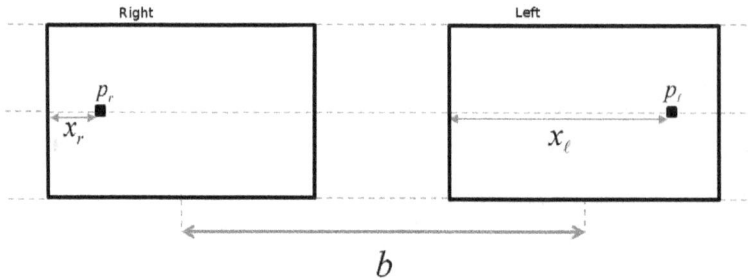

Figure 2.3: Epipolar alignment with baseline b of a binocular system.

The multi view camera system is assumed to have same focal distance f for each camera, and it is depicted in figure 2.4 (top view). From the stereo pair (the two images aligned) we look for correlating the pixels in both images that represent same objects in the scene. As a matter of fact, the magnitude of the arithmetic subtraction of $|x_r - x_\ell|$ is known as the value of disparity d, and is given by equation (2.29).

$$d = |x_\ell - x_r| \tag{2.29}$$

The disparity value is proportional to the distance of the object from the stereo pair centre. The disparity map is a grey-level intensity valued matrix with all pixels disparity values that were calculated from correlated pixels.

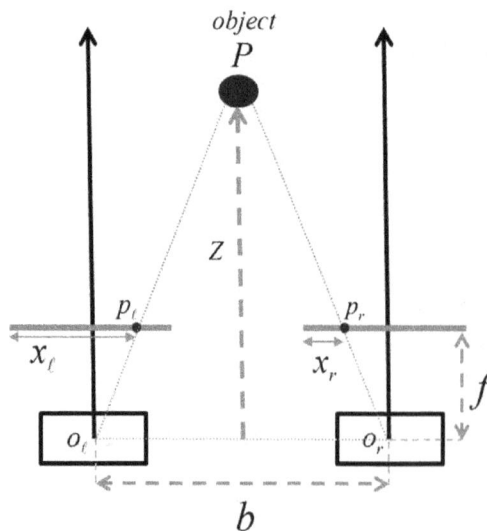

Figure 2.4: Stereo pair top view.

Hence, the further the object, the smaller the numeric value of d; and when the object locates very near from the sensor, the disparity value d approaches the rank($\mathbf{I}$) that is the image wide.

$$\lim_{z \to \infty} |x_r - x_\ell| = 0; \qquad \lim_{d \to rank(\mathbf{I})} z = 0;$$

The disparity values are critical because 3D information is calculated in terms of disparities. From figure 2.4, by trigonometric relationships $\overline{PO_\ell O_r}$, and the triangle $\overline{PP_\ell P_r}$, the object's depth z is involved by the following expression,

$$\frac{b}{z} = \frac{(b + x_r) - x_\ell}{z - f} \tag{2.30}$$

crossing terms,

$$b(z - f) = z[(b + x_r) - x_\ell] \tag{2.31}$$

algebraically expanding in equation both sides,

$$zb - bf = z(b + x_r) - zx_\ell \tag{2.32}$$

and

$$zb - bf = zb + zx_r \quad zx_\ell \tag{2.33}$$

subsequently reducing terms,

$$zx_\ell - zx_r = bf \tag{2.34}$$

the formal expression to calculate the object depth z is

$$z = \frac{bf}{x_\ell - x_r} \tag{2.35}$$

The aligned cameras are separated by a baseline b, and the focal distance (separation between focal plane and the light convergence point). Thus, whith such parameters, the object's 3D postured is geometrically triangulated as decribed by equations (2.36), (2.37), and (2.38). Thus, the depth $z(d)$ is calculated in terms of the object disparity d,

$$z(d) = \frac{bf}{d} \tag{2.36}$$

then, the x component is calculated in terms of d and x_ℓ,

$$x(d, x_\ell) = \frac{z(d)x_\ell}{f} \tag{2.37}$$

and, the component y is obtained by

$$y(d, y_\ell) = \frac{z(d)y_\ell}{f} \tag{2.38}$$

It follows from previous expressions:

- $\sqrt[2]{x^2 + y^2 + z^2}$ is inversely proportional to d, and the coordinates $(x, y, z)^\top$ are measured more accurately for nearer objects than for the farther ones. Making this approach tractable to be used for obstacle avoidance in navigation.

- As b increases, occlusions might occur and such objects would neither have correlation nor disparity values.

- A real 3D point yields a pair of points in focal planes, known as conjugated. For a member of the conjugate pair, the other one exists somewhere along the horizontal epipolar line.

Figure 2.5 depicts an RGB image, and its associated disparity map experimentally taken within the Robotics laboratory.

Figure 2.5:   RGB image and its disparity map.

Furthermore, figure 2.6 depicts three different 3D views, or clouds of points generated from input image 2.5.

Figure 2.6:   Cloud of 3D points of the Robotics Lab.

2.4   Light detection and ranging model

Unlike ultrasonic range finders, the use of light detection and ranging radar (LIDAR), which is a common electro-optic sensor used in robotics. It is used to sense objects by emission of electromagnetic radiation. LIDARs provide more accuracy in measuring distance and much higher resolution than the ultrasonic sonar. LIDARS also use measuring techniques of time of flight of a beam, which reaches (in vacuum) around $v = 299,792,458$m/s. Besides, phase-shift measurement techniques are also used, where phase of reflected beam is compared with the phase of original signal emitted.

$$d = \frac{\Delta\varphi\lambda}{4\pi} = \frac{\Delta\varphi v}{4\pi f} \qquad (2.39)$$

where λ is the wave length of the modulated signal. f is the electromagnetic beam frequency. Thus, for sonars and LIDAR the sensing models and data registration formulation are basically the same. Figure 2.7-centre depicts a diagram of the measuring model . For the x component,

$$l_x = d_j \cdot \cos(\theta + \phi_j) \qquad (2.40)$$

and y component,

$$l_y = d_j \cdot \sin(\theta + \phi_j) \qquad (2.41)$$

where d_l is the measured distance between an object and the LIDAR ad bearing ϕ_j. Likewise, the angle θ is the actual robot's orientation. And, $(p_x, p_y)^\top$ is the actual robot position within a common coordinate system. Thus, the measurement w.r.t. the robot's local frame,

$$\mathbf{s}_j^R = d_j \cdot \begin{pmatrix} \cos(\theta + \phi_j) \\ \sin(\theta + \phi_j) \end{pmatrix} \qquad (2.42)$$

Nevertheless, for the global mapping $(s_j^I)^\top$, we have the following expression:

$$s_j^I = \mathbf{R}(\gamma) \cdot s_j + \xi_t \tag{2.43}$$

where $\mathbf{R}(\gamma)$ is the Euler rotation matrix of equation (2.44), and ξ is the posture vector in the global plane.

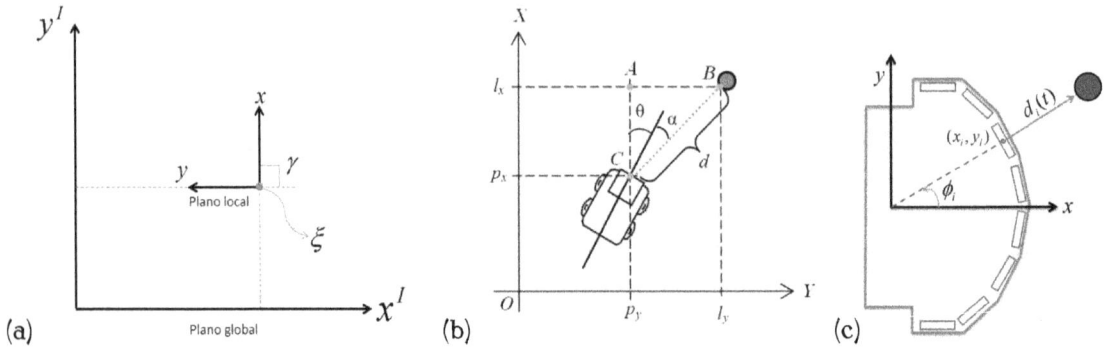

Figure 2.7:   (a) Inertial frames; (b) range measuring w.r.t. the robot's frame; (c) ring of measuring devices w.r.t. the robot's centroid.

Thus, the Euler matrix rotation is defined by

$$\mathbf{R}_z(\gamma) = \begin{pmatrix} \cos\gamma & -\sin\gamma & 0 \\ \sin\gamma & \cos\gamma & 0 \\ 0 & 0 & 1 \end{pmatrix} \tag{2.44}$$

The angle γ represents the difference between the local and global frames. It is assumed a height h that represent the vertical position where the sensor is located $\xi = (x, y, \theta)\top$. By substituting expressions (2.42), (2.44) in (2.43), the final vector expression describes the coordinates[5] of sensed obstacles as formally described by the following expression

$$s_j^I = \begin{pmatrix} \cos\gamma & -\sin\gamma & 0 \\ \sin\gamma & \cos\gamma & 0 \\ 0 & 0 & 1 \end{pmatrix} \left[d_j \cdot \begin{pmatrix} \cos(\theta + \phi_j) \\ \sin(\theta + \phi_j) \\ h/d_j \end{pmatrix} \right] + \begin{pmatrix} p_x \\ p_y \\ 0 \end{pmatrix} \tag{2.45}$$

where the positions of the sensed data points are mapped onto a global inertial system s^I, which origin is as located as far as $(p_x, p_y)^\top$ from the robot, and rotated by a constant angle γ

with respect to the robot's actual orientation. Figure 2.8-(a) depicts a photo of a wheeled mobile robot (PeopleBot) sensing an environment with two short carton-made walls in the Robotics Laboratory. (b) shows the control GUI. (c) is a polar plot of LIDAR's measurements. And (d) shows an environment local Cartesian map.

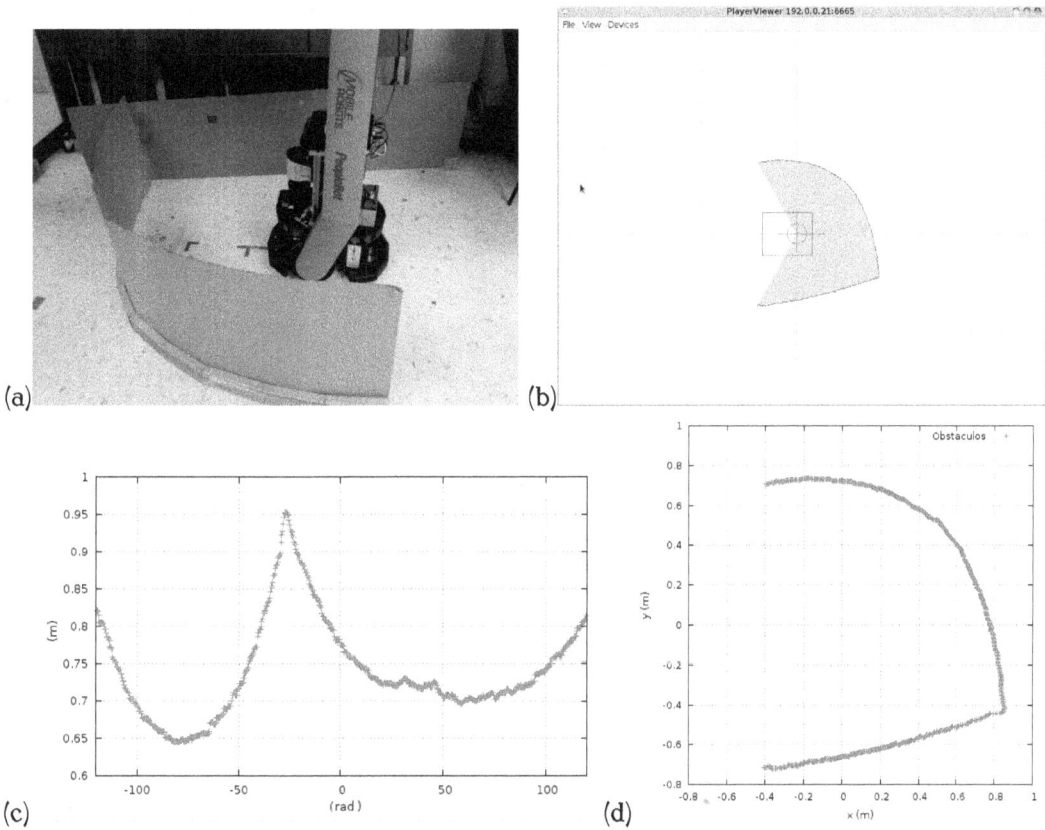

Figure 2.8:   (a) Real experiment in laboratory; (b) the GUI software for robot control; (c) polar form of the LIDAR measurements; (d) local Cartesian map of the environment.

2.5   Robot's orientation model in eigenspace

In scan matching techniques, scans of point s_{ref} and s_k are aligned as to maximize the overlap, so that rotation and translation can be estimated. The angle φ_k is useful to incrementally estimate the robot's heading and to know how much it deviates from the previous robot's angle φ_{k-1}. A range point in polar form is defined by $s_{i,\cdots,N} = (\rho, \theta)$, with distance ρ and angle

θ. The reference and current scans are $\mathbf{S}_{ref}, \mathbf{S}_k$ or $\mathbf{X}_{ref}, \mathbf{X}_k$. In polar or coordinate form $\mathbf{X}$ is the set of sample points. A vector point in Cartesian space $\mathbf{x} = (x, y)$, where the sample points $\mathbf{x}_1, \cdots, \mathbf{x}_N$, for $\mathbb{R}^d$ of d-dimensional space. As showed in figure 2.9 which depicts four consecutive scans distributions (separated components). The points hold similar distribution trends (specially for the $x \in \mathbf{x}_{i,\dots,N}$ component) along the axis despite their roto-translational increments yielded by the robot's motion. The covariance matrix $\mathbf{C}_k$ for the kth range scan

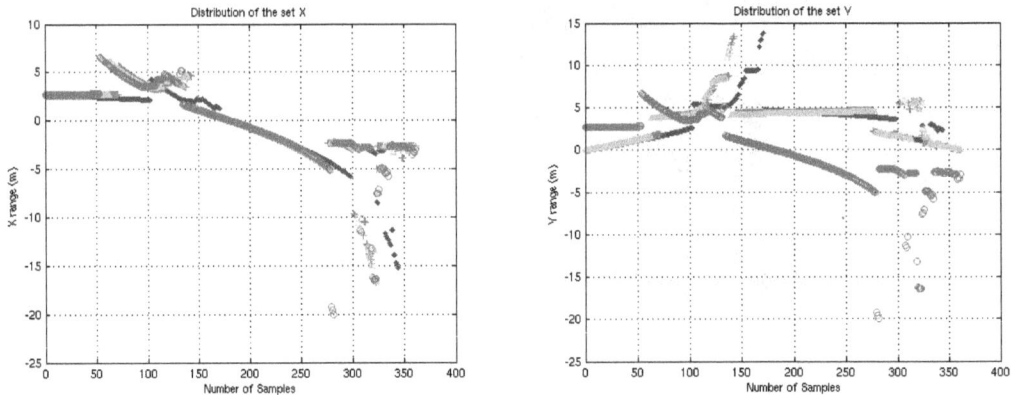

Figure 2.9: The x and y components of 4 consecutive range scan distributions.

is a measure of the degree of statistical dependence between $x, y \in \mathbf{x}_i$. Thus, such statistical information stored in $\mathbf{C}$ geometrically represents the variability of the environment in terms of the sensor data $\mathbf{X} \in \mathbb{R}^d$, resulting that with principal component analysis (PCA) it is possible to compute the environment directions respect to the robot's perspective φ_k projected into a relative eigenspace. In particular, PCA provides linearity and dimensionality reduction, which means that it spans a linear subspace within a minimum number of dimensions. The set of eigenvectors $\mathbf{W}$ computed from the matrix $\mathbf{C}$ characterizes the range data scatter by their local orthonormal orientations, while their magnitudes (eigenvalues) say which direction is stronger. All the column-vectors of $\mathbf{W}$ are orthonormal, no matter how large the $\mathbb{R}^d$ space is. This is important because it means that the sensor information $\mathbf{S}_k$ can be expressed in terms of these orthogonal eigenvectors $\{\mathbf{w}_1, \cdots, \mathbf{w}_d\}$, instead of expressing them in terms of N Cartesian points $\mathbf{x} \in \mathbf{X}$. The principal components $\mathbf{W}$ of the sensor data $\mathbf{C}$ (observation matrix) create an orthogonal basis set $\{\mathbf{w}_1, \cdots, \mathbf{w}_d\}$, which are the eigen-components (eigen-scans) of the Cartesian space. The first principal component is selected by the eigenvector $\mathbf{w}_1$ corresponding to the largest eigenvalue λ_1. The second principal component $\mathbf{w}_2$ is the eigenvector corresponding to the second largest eigenvalue λ_2, and so forth. The data variability (variance) is the spread of deviated N total points in a set of data scans $S_i = \{s_1, \cdots, s_N\}$. Where $x \in X_k$ respect to $\bar{x}_k$, a measure of the spread of the data namely $\hat{x}$ is a crucial parameter that says the average distance of the scan profile to the robot, and how much the points $\{x_1, \cdots, x_N\}$ might vary respect to $\bar{x}$. The sample mean vector $\bar{x}_i$ in $\mathbb{R}^2$ Cartesian space is computed with the Cartesian range scan set X_k of vector points $x \in X_k, (k = 1, \cdots, K)$ of (2.46) as

$$\bar{x}_k = \frac{1}{N} \sum_{x \in X_i} x \tag{2.46}$$

Thus, the mean values $\bar{x}_k = (\bar{x}, \bar{y})$ are adjusted by the expression of equation (2.47) which is also known as the mean-deviation form. It is a subtraction of $\bar{x}$ from the data, and represented by the N-elements in $\hat{X}_k$ with components $\hat{x}_i = (\hat{x}, \hat{y})$, compounded by $\{\hat{x}_1, \cdots, \hat{x}_N\}$

$$\hat{X} = (x - \bar{x})_{x \in X} \tag{2.47}$$

This expression basically accounts for the variability of the objects observed from the robot's location. The sample covariance matrix C_k of the sensor data provides a measure of the correlation between the two (or more) sets of variables. Such that, $C_{1,\dots,K}$. The resulting C_k from the deviated form $\hat{X}$ in equation (2.47) is compounded by two N-elements vectors called $\hat{X}^x$ and $\hat{X}^y$ that are the matrix entry elements to compute C by

$$C = \frac{1}{N-1} \begin{pmatrix} \sum_i \hat{X}_i^x \hat{X}_i^x & \sum_i \hat{X}_i^x \hat{X}_i^y \\ \sum_i \hat{X}_i^y \hat{X}_i^x & \sum_i \hat{X}_i^y \hat{X}_i^y \end{pmatrix} \tag{2.48}$$

It follows that the PCA are obtained by diagonalizing the nonsingular matrix C by $(C - \lambda I)W = 0$ in order to calculate the set of eigenvalues $\{\lambda_1, \cdots, \lambda_d\}$ and its respective set of eigenvectors $\{w_1, \cdots, w_d\}$ of the kth observation matrix C_k, and are called the principal components of the data. It has been important to analyse the PCA formulation and how it works in the context of vector points of 2D laser range scans. A set of different synthetic environments were created that would let us to better understand the laser scan PCA behaviour, and to detect what environment conditions would affect the results. At left-up side of figure 2.10 it shows a single point in an empty environment. A set of 100 points is plotted in local robot's coordinates, forming a unique spatial point. Below are both components xy plotted separately, and up-right side the resulting eigen-space for such data set. The eigenvectors did not yield any dominant direction, as their corresponding set $\{\lambda_1, \lambda_2\}$ had same magnitudes, and then, such values solely express no dominant direction. In the same manner, $\mathbf{w}_{1,2}$ determine always same directions even if the robot moves around such coordinate point (holding $\lambda_1 = \lambda_2$), the angles of $\mathbf{w}_{1,2}$ would simply keep same angle.   Similarly, in figure 2.10-right a set of Cartesian points with

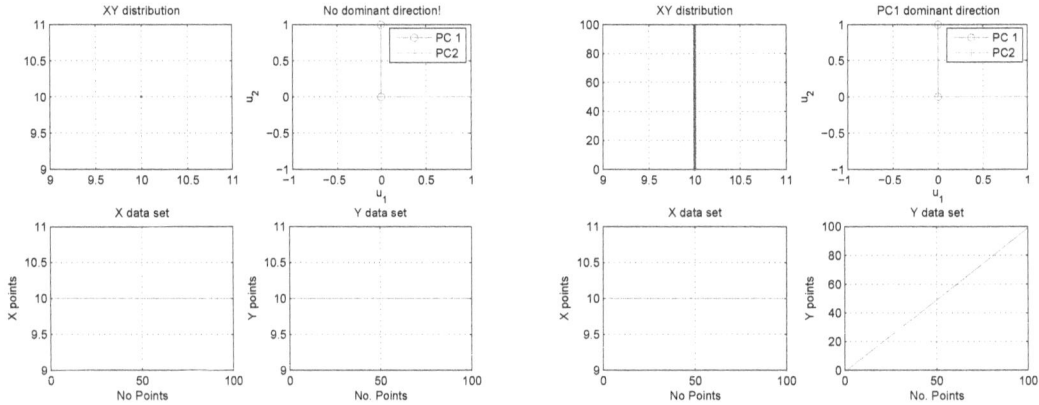

Figure 2.10:   A single point in Cartesian space and eigen-space (left). A vertical line in 2D space and its eigenspace directions (right).

no noise added depict a vertical line which is about 10 metric units from the sensor location at $(0, 0)$ with $360°$ of field of view. It would typically represent a wall in a corridor, in such location the set of points of the component $x \in \mathbf{x}_t$ basically does not variate at all, while the set $y \in \mathbf{x}_t$ has a diagonal trend. For the x-component the mean value is equal to its expected value zero. In this geometric x-component there is no variability and because of that, it is the only information provided that the robot knows about its position respect to the environment. However, the points of the y-component are linearly separated and its mean value is calculated in 50. As the covariance matrix also yields a geometric representation of the variance between variables that geometrically can be depicted as an ellipse, the variability for the x-component would be the minor axis with zero value, but extending its major axis (y-component) as long as the value of its variance. As for the eigenvalues, they only indicate which direction has a stronger influence in the variability of the data. In the legend of the figure 2.10-right up-left hand, the term $PC1$ is the first principal component (the largest one). Thus, $PC1$ has the same direction as the set of points yields the most pronounced variations (vertical component).

Another kind of synthetic environment is a perfect square, generated with 361 vector points and a bearing resolution of 0.5°, assuming a 360° of field of view. The robot's position is at coordinate (0, 0) which is the center of the square. For this, case there is no noise added to the data but for this and previous cases different noise rates were also assumed. In figure

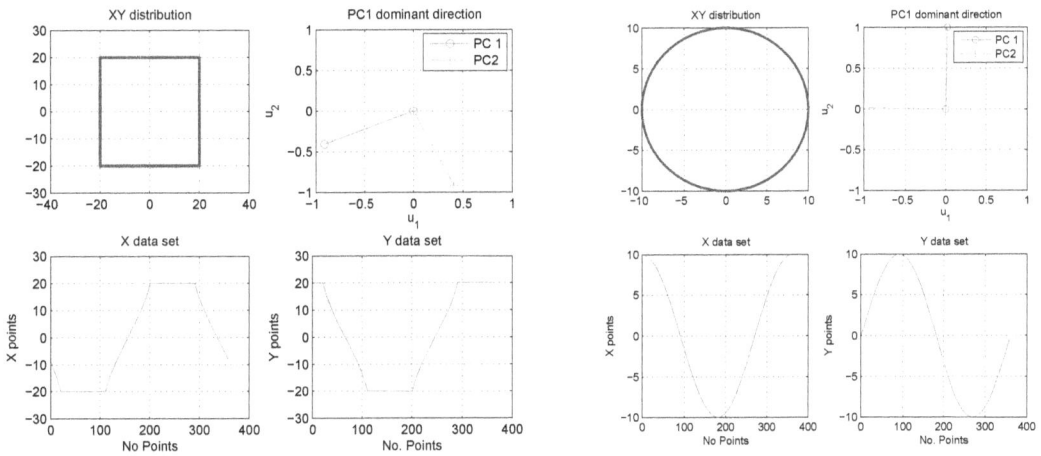

Figure 2.11: A squared area and its eigen-space directions for a robot rotation of 115° (left). A circular area and its eigenspace directions (right).

2.11-left the rotation of 115° is depicted by both eige-spaces and in the Cartesian space. The xy-components practically hold the same magnitudes respect to the mean value, but only rotated in their radial order. Perhaps the most challenging case that could affect the reliability of the PCA approach is for the case of circular rooms, which are symmetric and from any perspective the robot always would see the same geometric shape. However, the method is capable to estimate rotation if the robot turns with no translation because there exist small variability in the data sets. Such a small variability is basically detected at regions were there exist almost no linearity along the circular shape, as shown in the xy-component plots. At a range direction of $0° \pm 5°$ the ranged points denote almost no variation for Δx and increments are smooth, while opposite to this, in Δy their increments have almost no linearity and these regions are located as the peaks or valleys of the xy distribution plots. For this case the robot rotated 120°, such angle rotation has resulted reliably even for a circular area. Although, no noisy data set are being depicted, same analysis was carried out for different rates of noise cases. In values for λ_1 and λ_2 there exist a very slight difference in magnitude, only for the case when the robot solely rotates. However, when the robot performs any coordinate translation, the eigenvalues magnitudes start to differ form each other as the robot approaches any side of the circular wall.

2.6   Segmentation model for 2D range scans

This section introduces a method for segmentation of $2D$ pairwise laser scans, which the is the preamble for scan matching in robot localization. Matching 2D range scans has been a basic component of a diversity of localization and environment mapping techniques[12-15], which has been proposed specially during the last years. Let us define partitions $\mathbf{S}_k$ into groups $\mathbf{D}_l$ of $2D$ points. Only few clusters belonging to $\mathbf{E}_{ref}$ are associated with the clusters found in $\mathbf{E}_k$. Most known procedures either explicitly or implicitly attempt to optimize a global criterion function with a known or assumed number of clusters. Probably the most obvious measure of the similarity (or dissimilarity) between two samples is the distance between them. Many different measures of similarity (and dissimilarity as well) have been proposed[10-13]. Thus, we have made use of the similarity function[17] $\varsigma(\mathbf{x}_i, \mathbf{x}_{i+1})$ between two contiguous vector points namely $\mathbf{x}_i$ and $\mathbf{x}_{i+1}$ to label such points if they may naturally belong to the same groups.

$$\zeta(\mathbf{x}_i, \mathbf{x}_{i+1}) = \nabla_f \left(log \left[\frac{\mathbf{x}_i^T \cdot \mathbf{x}_{i+1}}{\|\mathbf{x}_i - \mathbf{x}_{i+1}\|} \right] \right) \tag{2.49}$$

The logarithmic function that affects the result of the central member in equation (2.49) only reduces the scale of the magnitudes but still preserves the rate of differences between large and small magnitudes. The objects represented by groups of points sharing similar properties (angles and distances) but affected by noise are attenuated. In fact, only the gaps between the objects become the most salient metric values respect to the values representing objects. In addition, a gradient function is applied in order to exaggerate such gaps between objects and attenuate near zero the points representing an object. When c is unknown we can proceed by solving the problem stating a threshold for the criterion of a new cluster. If there is a large gap in the criterion values, it suggests a natural number of clusters. The equation (2.50) is the similarity criterion that automatically calculates a threshold value d_ζ. This allows the partition function equation (2.51) to split the data set in adequate groups essentially separated by a gap. Equation (2.50) relates the statistical mean value of the histogram of $\zeta(\cdot)$ which always

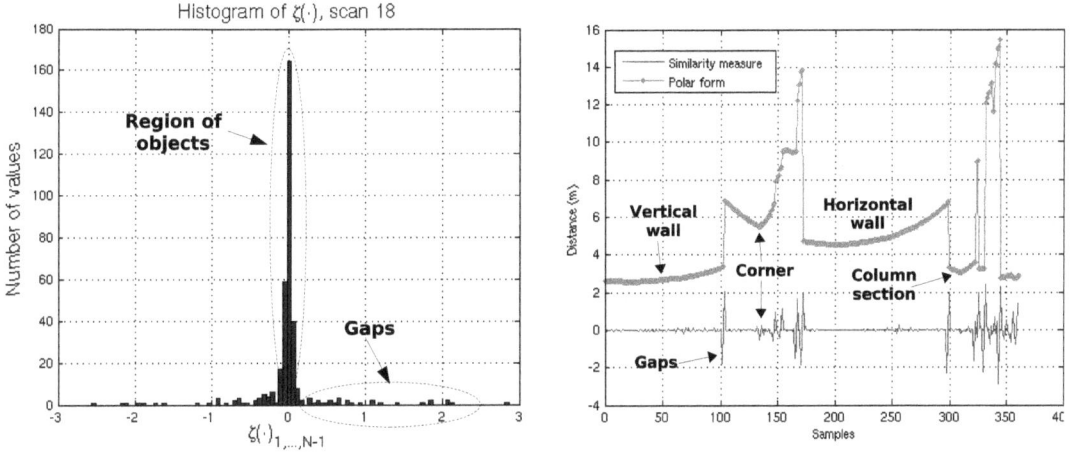

Figure 2.12: Histogram of $\zeta(\cdot)$ (left), and the polar and the similarity linkage (right).

is warranted to be unimodal due to the convergent effect produced by the gradient function $\nabla_f(\cdot)$. As a matter of fact, the array values computed by $\zeta(\cdot)$ resulted unimodal for all the range scan observations in the experiments, with mean value close to zero. Let $\zeta(\mathbf{x}, \mathbf{x}')$ be the similarity function of $\mathbf{x}, \mathbf{x}'$.

$$d_\zeta = \frac{1}{N-1} \sum_{i=1}^{N-1} \zeta_i(\mathbf{x}_i, \mathbf{x}_{i+1}) - \frac{\|\sigma_\zeta\|}{2} \tag{2.50}$$

Given this fact, we established the d_ζ value equivalent to less than half a standard deviation $(\sigma_\zeta/2)$ in equation (2.50) as part of the partition criterion function. From this follows that the threshold to cluster points is determined by equation 2.50

In this context, a raw laser scan $\mathbf{S}_k$ is initially considered a single cluster $D_{c=1} \in \mathbf{D}$ compounded of N-elements, subsequently the scan is split into $c > 1$ different classes (clusters). As can be seen in figure 2.12-right, only the most salient values represent the gaps as depicted by the similarity function plot. The similarity function curve can be compared with its polar form plot, where each object is divided by pairs of sudden large magnitude impulses (negative-positive signs) which highlights the gaps. The criterion for partitioning $\mathbf{X}_k$ into c groups called D_j ($j = 1, \cdots, c$) with n_j as number of points which are labelled by l for each jth group. The groups of points $D_1 \cdots, D_c$ are within c_k clusters of the kth scan. Thus, it is given by the

following expression,

$$D_j(i) = \begin{cases} l, & \zeta(\mathbf{x}_i, \mathbf{x}_{i+1}) \le d_\zeta \\ l+1, & otherwise. \end{cases} \qquad (2.51)$$

After applying the partition criterion to successive range scans, the resulting number of clusters in each scan are depicted in figure 2.13. There is a reduction of the number of clusters based

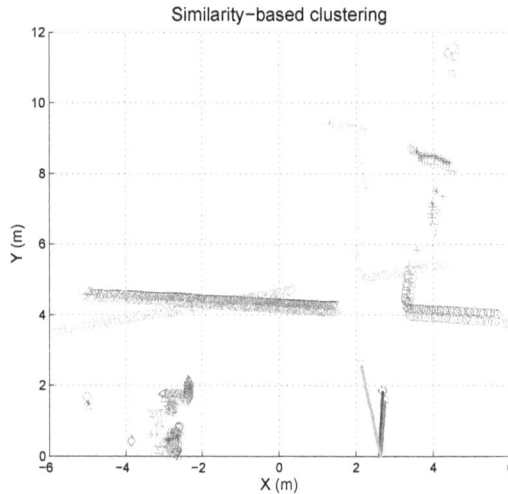

Figure 2.13:   Labelled groups of points in three consecutive range scans.

on a simple statistical summary. Groups containing a covariance value less than 1. Usually such a value numerically represents noisy measurement with too few points (less than 5) or isolated small objects within areas smaller than $0.04m^2$, as depicted in figure 2.14.

2.7   Separability model for 2D range clusters

Linear discriminant analysis (LDA) yields separability and eigen-projection respect to the original data by using their eigenvector directions, and consequently clusters correspondence is improved. The main idea behind using linear discriminant analysis is to find the clusters that have correspondence with their representative ones allocated in a reference scan. With LDA we can project the set of thin c clusters allocated in $\mathbf{D}_k$ by maximizing separability from other clusters through the $\mathbf{S_B}$ matrix and minimizes the distances of all points in a same group by its $\mathbf{S_W}$ matrix. In figure 2.15-(a) two groups of points labelled in the reference scan namely

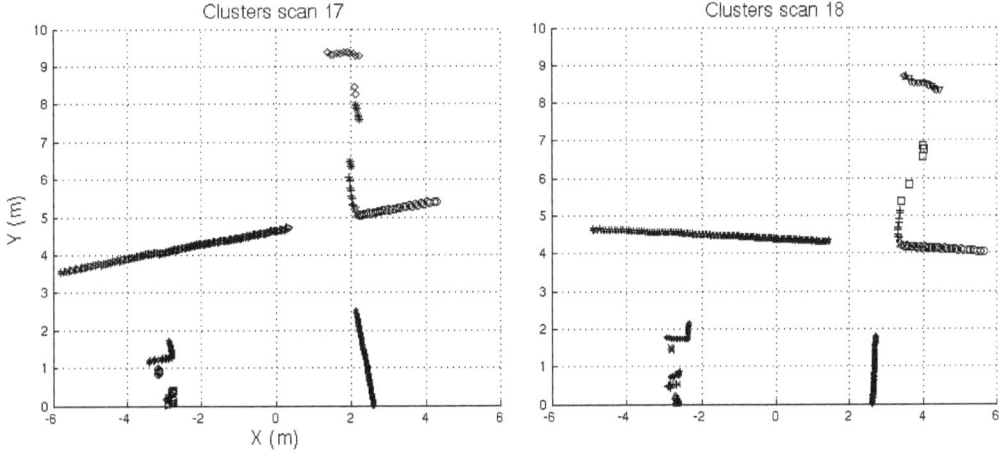

Figure 2.14: Cleaned set of clusters in 2 scans.

$D_i(ref)$ and $D_j(ref)$ are originally projected in X-space. Similarly, both clusters representing same objects were labelled again in the next scan of points, but now called $D_i(k)$ and $d_j(k)$, also projected onto the same X-space but rotated and translated as the robot moved and sensed the environment from different location. Such rigid roto-translation projection makes difficult for the algorithm to correctly associate the correspondent group. Nevertheless, in figure 2.15-(b) same objects are now projected into Y using their eigen-directions, although still misaligned but maximal separability is warranted between clusters, and contrary to it their within-cluster separation is minimized. It makes easier to find correspondences only in some clusters instead of the whole set of vector points. In other words, spurious correspondences are reduced in number and this mechanism assures finding more correspondences, which are essential to ultimately find the best robot's translation that correctly aligns both scans. The figure 2.15-right depicts clusters which belong to objects segmented in three consecutive range scans. Thus, in the correspondence problem of figure 2.15-right the cluster A is approximately close to clusters A' and A'', and their size (number and density of points) are more or less same as cluster A. The problem is to determine which association to A is the correct one. Although PCA finds components that are useful for representing data, there is no reason to assume that these components must be useful for discriminating between data in different classes. However, a close mechanism to PCA could be used. Let $\mathbf{W}$ be the set of eigenvectors of $\mathbf{C}$.

Thus, geometrically, if $|\mathbf{W}| = 1$, each $\mathbf{y}_1$ is the projection of the corresponding $\mathbf{x}_i$ onto a line in the direction of $\mathbf{W}$ by the general equation (2.52). If we form a linear combination of the

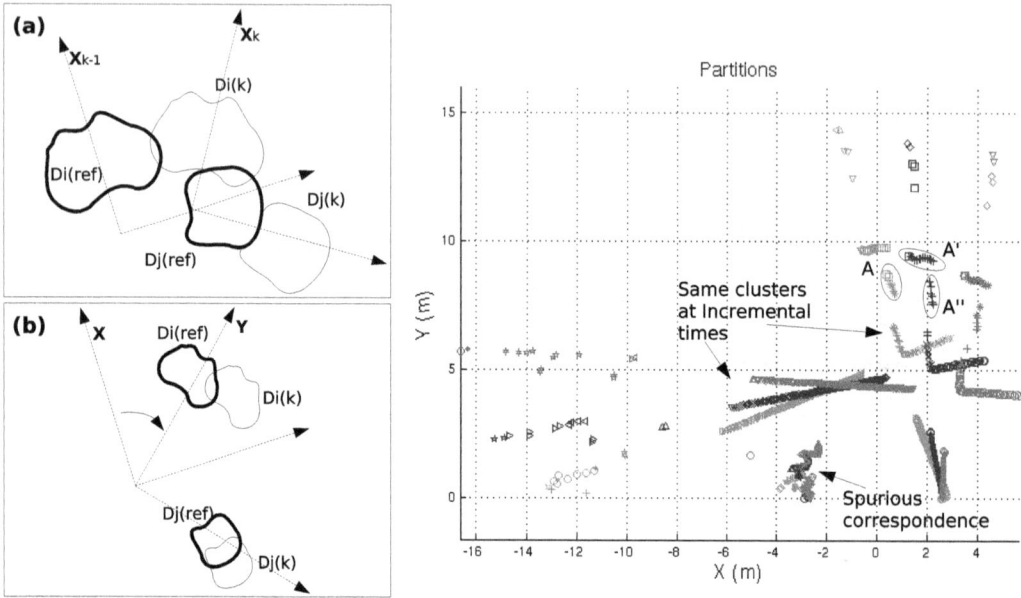

Figure 2.15:   Orthogonal projection of $\mathbf{X_k}$ into $\mathbf{X}_{ref} \mapsto \mathbf{Y}$ (left). Spurious association among clusters (right).

components of $\mathbf{x}_i$, we obtain a corresponding set of N samples $\{\mathbf{y}_1, \cdots, \mathbf{y}_N\}$ divided into the subsets $\mathbb{Y}_1, \cdots, \mathbb{Y}_c$, given a set of N d-dimensional samples $\{\mathbf{x}_1, \cdots, \mathbf{x}_n\}$ and n_i in the subsets $D_1, \cdots, D_C$. Hence, let $\mathbf{Y}$ be the transformation space of vectors, with points $\mathbf{y}_1, \cdots, \mathbf{y}_n$. If the samples labelled l_1 fall more or less into one cluster while those labeled l_2 fall in another, we want the projections falling onto the eigen-direction to be well separated.

$$\mathbf{Y}_k = \mathbf{W}^T \mathbf{X}_k \qquad\qquad (2.52)$$

Thus we now turn to the matter of finding the best such direction $\mathbf{W}$, one that should enable accurate clusters association. A measure of the separation between the projected points is the difference of the sample means $\mathbf{m}_i$ of equation (2.53).

Scatter criteria

Before going farther with this explanation let us firstly define some variables that formulates the scatter criteria. The scatter matrix does not depend on how the set of samples is partitioned into clusters; it depends only on the total set of samples $\mathbf{X}_k$. The within-cluster and between-cluster scatter matrices taken separately do depend on the partitioning. The between-cluster goes up as the within-cluster scatter goes down. Thus, the mean vector for the ith cluster

$$\mathbf{m}_i = \frac{1}{n_i} \sum_{x \in Di} \mathbf{x} \tag{2.53}$$

The total mean vector of all groups, is seen as the general mean value of $\mathbf{X}_k$, and is given by

$$\mathbf{m} = \frac{1}{n} \sum_{Di} \mathbf{x} = \frac{1}{n} \sum_{i=1}^{c} n_i \mathbf{m}_i \tag{2.54}$$

The scatter matrix for the ith cluster

$$\mathbf{S}_i = \sum_{x \in Di} (\mathbf{x} - \mathbf{m}_i)(\mathbf{x} - \mathbf{m}_i)^T \tag{2.55}$$

The within-cluster scatter matrix

$$\mathbf{S}_W = \sum_{i=1}^{c} \mathbf{S}_i \tag{2.56}$$

The between-cluster scatter matrix

$$\mathbf{S}_R = \sum_{i=1}^{c} n_i (\mathbf{m}_i - \mathbf{m})(\mathbf{m}_i - \mathbf{m})^T \tag{2.57}$$

The total scatter matrix is given by the expression

$$\mathbf{S}_T = \sum_{x \in Di} (\mathbf{x} - \mathbf{m})(\mathbf{x} - \mathbf{m})^T \tag{2.58}$$

Thus, it follows that the total scatter matrix is the sum of the within-cluster scatter matrix and the between-cluster scatter matrix $\mathbf{S}_T = \mathbf{S}_W + \mathbf{S}_B$.

Scatter separability

Perhaps the simplest scalar measure of a scatter matrix is its trace. The trace measures the square of the scattering radius, because it is proportional to the sum of the variances in the coordinate directions. Thus, an obvious criterion function to minimize is the trace of $\mathbf{S}_W$. In fact, this criterion is nothing more or less than the sum-of-squared-error criterion.

$$tr[\mathbf{S}_W] = \sum_{i=1}^{c} tr[\mathbf{S}_i] = \sum_{i=1}^{c} \sum_{\mathbf{x} \in D_i} \| \mathbf{x} - \mathbf{m}_i \|^2 = J_e \tag{2.59}$$

Because $tr[\mathbf{S}_T] = tr[\mathbf{S}_W] + tr[\mathbf{S}_B]$ and $tr[\mathbf{S}_T]$ is independent of how the samples are partitioned, and it is important to know that in seeking to minimize the within-cluster criterion $J_e = tr[\mathbf{S}_W]$ we are also maximizing the between-cluster criterion

$$tr[\mathbf{S}_B] = \sum_{i=1}^{c} n_i \| \mathbf{m}_i - \mathbf{m} \|^2 \tag{2.60}$$

The eigenvalues $\lambda_1, \cdots, \lambda_d$ of $\mathbf{S}_W^{-1}\mathbf{S}_B$ are invariant under nonsingular linear transformation of the data. Indeed, these eigenvalues are the basic linear invariants of the scatter matrices. Their numerical values measure the ratio of between-cluster to within-cluster scatter in the direction of the eigenvectors, and partitions that yield large values are usually desirable.

Ortho-projection in Y-space

In order to project the set of vector points onto the Y-space by a slightly adapted version of the general equation (2.52), then we define the sample mean for the $\mathbf{y}_i$ projected points by

$$\tilde{m}_i = \frac{1}{n_i} \sum_{y \in Y_i} y = \frac{1}{n_i} \sum_{x \in D_i} \mathbf{w}^T \mathbf{x} = \mathbf{w}^T \mathbf{m}_i \tag{2.61}$$

It is a simple transformation of $\mathbf{m}_i$, and it follows that the distance between the projected means of two clusters (figure 2.16) is

$$|\tilde{m}_1 - \tilde{m}_2| = |\mathbf{w}^T(\mathbf{m}_i - \mathbf{m}_2)| \tag{2.62}$$

To obtain good separation of the projected data the difference between the mean values must be large relative to some measure of the standard deviations for each class. Rather than forming sample variances, we define the projected scatter matrix $\tilde{s}_i^2$ for the projected samples labelled l_i by

$$\widetilde{s}_i^2 = \sum_{y \in \mathcal{Y}_i} (y - \widetilde{m}_i)^2 \tag{2.63}$$

Thus, $(\widetilde{s}_1^2 + \widetilde{s}_2^2)$ is called the total within-class scatter of the projected samples. The Fisher linear discriminant employs that linear function $\mathbf{w}^T\mathbf{x}$ for which the criterion function $J(\cdot)$.

$$J(\mathbf{w}) = \frac{|\widetilde{m} - \widetilde{m}_1|^2}{\widetilde{s}_1^2 + \widetilde{s}_2^2} \tag{2.64}$$

While the $\mathbf{w}$ maximizing $J(\cdot)$ leads to the best separation between the two projected sets, a threshold criterion would give a good separation between groups. Before we have a classifier. To obtain $J(\cdot)$ as an explicit function of $\mathbf{w}$, we defined the scatter matrices $\mathbf{S}_i$ and $\mathbf{S}_i$. Then we can write

$$\widetilde{s}_i^2 = \mathbf{w}^T\mathbf{S}_i\mathbf{w} \tag{2.65}$$

therefore the sum of these scatters is written

$$\widetilde{s}_1^2 + \widetilde{s}_2^2 = \mathbf{w}^T\mathbf{S}_W\mathbf{w} \tag{2.66}$$

Similarly, the separations of the projected means obeys

$$(\widetilde{m}_1 - \widetilde{m}_2)^2 = \mathbf{w}^T\mathbf{S}_B\mathbf{w} \tag{2.67}$$

In particular, for any $\mathbf{w}$, $\mathbf{S}_B\mathbf{w}$ is in the direction of $\mathbf{m}_1 - \mathbf{m}_2$. The columns of the optimal $\mathbf{W}$ are generalized eigenvectors that correspond to the largest eigenvalues in

$$\mathbf{S}_B\mathbf{w}_i = \lambda_i\mathbf{S}_W\mathbf{w}_i \tag{2.68}$$

The equation (2.69) is the Fisher's discriminant rule on a projection of the the set of observations X onto a Y-space such that a good separation between clusters is achieved. The algorithm studied in [13] to solve the problem of mapping a two-class problem was adapted to solve for e_k clusters contained in the kth scan. Discrimination in the present scan matching algorithm is being used to discriminate false correspondences with the groups labelled in the reference cluster-scan with e_{ref} groups. Firstly, let us project X with respect to the eigen-directions of

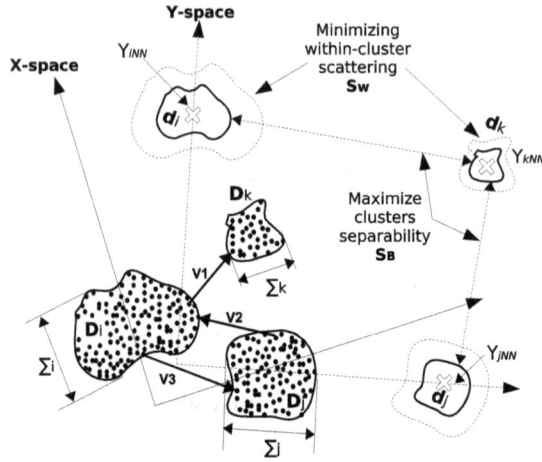

Figure 2.16: S_W minimization and S_B maximization.

S_W, which is equivalent to whiten the data (normalization) by

$$\mathbf{Y} = \Lambda^{-1/2}\mathbf{\Phi X} \tag{2.69}$$

Where Λ and $\mathbf{\Phi}$ are the eigenvalue and the eigenvector matrices of $\mathbf{S_W}$. Although, the vectors spanned by $\mathbf{\Phi}$ are orthonormal, it is merely a scale and they provide the directions, while Λ gives the scale for such a projection. And in the $\mathbf{Y}$-space compute the between-clusters S_b by

$$S_b = \sum_{i=1}^{C} \frac{1}{N} \sum_{j=1}^{n_i} \omega_i (\mathbf{y}_j^i - \mathbf{m}_i)(\mathbf{y}_j^i - \mathbf{m}_i)^T \tag{2.70}$$

The weighting function ω_i of the ith cluster has the property that vector points near the classification boundary such as $\mathbf{v}1, \mathbf{v}2, \mathbf{v}3$ showed in figure 2.16, it takes on values close to 0.5 and drops off to zero as we move away from the classification boundary.

$$\omega_i = \frac{\min\left\{\delta(y_k, y_{jNN}^i), \cdots, \delta(y_k, y_{jNN}^i)\right\}}{\delta(y_k, y_{jNN}^i) + \cdots + \delta(y_k, y_{jNN}^i)} \tag{2.71}$$

The jNN distance function $\delta(\cdot)$ is a procedure for voting the j-nearest neighbour and jNN is defined as

$$y_{jNN} = \frac{1}{n_i} \sum_{\mathbf{y} \in Y_i} \mathbf{y}$$

According to this definition, a sample vector is assigned to the cluster represented by a majority of its j-nearest neighbours in the set (see figure 2.16). In the voting jNN criterion it implicitly assumes the j nearest neighbours of a data point $\mathbf{y}_i$ to be contained in a region of relatively small volume, so that sufficiently good resolution in the estimates of the different conditional densities can be obtained. Furthermore, the criterion used to compute the distance between vector points is defined by the expression

$$\delta(\mathbf{y}_i, \mathbf{y}_j) = (\mathbf{y}_i - \mathbf{y}_j)^T \Sigma_j^{-1} (\mathbf{y}_i - \mathbf{y}_j) \tag{2.72}$$

It is the Mahalanobis distance $\delta(\cdot)$ which was used in this weighting criterion because it differ from Euclidean distance in that it takes into account the correlations of the data set (Σ_j) and is scale-invariant (does not dependent on the scale of measurements). Nevertheless, another criterion can be used accordingly. Moreover, projecting the clusters onto a different space the principal components of only the m-eigenvectors of interest of S_W namely $\psi_1, \cdots, \psi_m$ are then selected, which correspond to the m largest eigenvalues. Thus, the optimum linear labelling (mapping) called $\mathbf{Z}$ that involves only the most representative eigenvectors $\mathbf{\Psi}_m = [\psi_1, \cdots, \psi_m]$

$$\mathbf{Z} = \mathbf{\Psi}_m^T \mathbf{Y} \tag{2.73}$$

In figure 2.17, shows separated and scaled segmented clusters which are mapped in $\mathbf{Z}$-space, while projected along eigen-directions (below) by the relation $\mathbf{y} = \Phi \mathbf{x}_i$. This expression was only deployed to project the groupings along the eigenvectors.

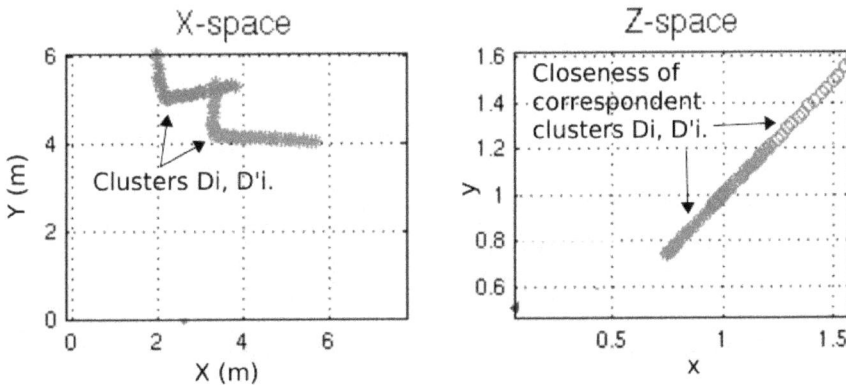

Figure 2.17:   Clusters projected onto $\mathbf{Y}$-space.

2.8 Eigen-space data association model

In order to establish correspondence between two closest groups namely E_j^{ref} and E_l^k, there is a pair relation $\Re\{E_j^{ref} \leftrightarrow E_l^k\}$ that associates them by solely considering their minimal distance $\delta(E_j^{ref}, E_l^k)$ between both centroids.

$$\Re(E_j^{ref}, E_l^k) = \min_l \{\delta(E_j^{ref}, E_1^k), \cdots, \delta(E_j^{ref}, E_{e_i}^k)\} \tag{2.74}$$

Thus, in X-space the Cartesian distance between two closest clusters is denoted by the norm of their central values that are vector points denoted by $\|\mathbf{e}_j - \mathbf{E}_{lNN}\|$ (nearest neighbour centroid), and we can calculate the displacement in x by $\Delta x_{j,lNN}$ which is the horizontal displacement obtained by the magnitude $(x, y) \in (\mathbf{e}_j^{ref} - \mathbf{e}_{lNN}^k)$, and the total displacement based on the associated clusters $\Re(\mathbf{E}_j^{ref}, \mathbf{E}_l^k)$ is average of the displacements in $\Delta x_{1,\cdots,e_k}$ and $\Delta y_{1,\cdots,e_k}$ by,

$$\begin{pmatrix} \Delta X \\ \Delta Y \end{pmatrix} = \frac{1}{e_k} \sum_{l=1}^{e_k} \begin{pmatrix} \Delta x_l^{(i,jNN)} \\ \Delta y_l^{(i,jNN)} \end{pmatrix} \tag{2.75}$$

In the world coordinate system the kth robot state is defined by $\mathbf{R}_k = [x, y, \phi]^T$, and $\mathbf{R}_k = \mathbf{R}_{k-1} + \Delta\mathbf{R}_k$ is the general equation that localizes the robot in world coordinates, being $\Delta\mathbf{R}_k = [\Delta X, \Delta Y, \varphi]^T$ the relative displacement estimated between two consecutive observations, then equation (26) and equation (27) are involved in the general equation for $\mathbf{R}_k$ which is also expressed as

$$\mathbf{R}_k = \begin{pmatrix} x_k \\ y_k \\ \phi_k \end{pmatrix} = \begin{pmatrix} x_{k-1} + \dfrac{1}{e_k} \sum_l^{e_k} \Delta x_l^{(j,iNN)} \\ y_{k-1} + \dfrac{1}{e_k} \sum_l^{e_k} \Delta y_l^{(j,iNN)} \\ \phi_{k-1} + \varphi_k \end{pmatrix} \tag{2.76}$$

The angle φ_k which is the relative rotation of the robot respect to its previous state, and it is calculated directly from the set of eigenvectors $\mathbf{W}_{k-1}$ and $\mathbf{W}_k$. In fact only the rotative difference found in either of the principal components is enough because all the components are orthogonal in the eigen-space, so this simple relation is given by

$$\varphi_k = \arccos\left(\frac{\mathbf{w}_{k-1}^1 \cdot \mathbf{w}_k^1}{\|\mathbf{w}_{k-1}^1\| \|\mathbf{w}_k^1\|}\right) \tag{2.77}$$

Although, these two eigen-spaces are given in separated observations, the eigenvectors are projected in local spaces solely expressing how much the environment is statistically rotated respect to the robot's sensor field of view. Thus, φ_k is the angle either of first or second principal components, we chose the first principal components $\mathbf{w}_{k-1}^1$ and $\mathbf{w}_k^1$. Thus, mapping the environment on to a global coordinate frame we define the new sensor data world coordinates as sx_i and sy_i $(i = 1, \cdots, N)$ computed for the kth observation by the expression

$$\begin{pmatrix} sx_i \\ sy_i \end{pmatrix} = \begin{pmatrix} \mathbf{x}_k + x_i \cos(\theta_i + \phi_k) - y_i \sin(\theta_i + \phi_k) \\ \mathbf{y}_k + y_i \sin(\theta_i + \phi_k) + \cos(\theta_i + \phi_k) \end{pmatrix} \tag{2.78}$$

Where x_i, y_i are the components $(x, y) \in \mathbf{x}_i$ of the local sensor observations with their respective bearing angle θ_i, and x_k, y_k are the world coordinates robot location with angle ϕ_k of equation (2.76) The scan matching method finds the directions of the environment respect to the robot by means of vectorial directions which are a compressed version of the Cartesian sensor data and its covariance matrix. As opposite from geometric models where we are limited to extract only the set of most descriptive feature models (e.g. points, lines, curves, etc.) Instead, the present algorithm finds natural clusters that represent objects of the environment which are used to estimate the robot's translation. Association between clusters of two successive scans can be found by using LDA whereby projects the objects orthogonally along the natural most important directions of the environment and spurious correspondences are reduced as it assures maximal scatter separability.

2.9 LIDAR-based localization

This section describes a scan matching approach based on the ICP (iterative closest point) algorithm as a fundamental to perform SLAM (simultaneous localization and mapping). SLAM consist of building a map of the environment, and deploy it to simultaneously to estimate its posture . The localization process is a procedure to estimate the robot's Cartesian position (x, y), and its instantaneous orientation θ. Map building is the process of sensor data registration of the objects on a same Cartesian space, which exist in the environment . A main advantage on using SLAM is that it estimates robots position from sensor readings. For instance, for LIDAR based data points, robot's posture incrementally become uncertain as the robot navigates for long terms cumulating pose errors. An illustrative experimental approach is by using the C/C++ libraries with the driver MRICP (Map Reference ICP) of player/stage, to build environmental maps and localize a robot online. These libraries used the ICP[⁶] with LIDAR data and odometry correction [⁸⁰]. For the LIDAR based scan matching process, two consecutive sensor measurements with k observed points in cylinder form are taken, $\mathbf{z}_t = (\delta, \phi)^\top$, and $\mathbf{z}_{t+1} = (\delta, \phi)^\top$.

$$\mathbf{s}(\delta, \phi) = \delta \begin{pmatrix} \cos(\phi) \\ \sin(\phi) \end{pmatrix}$$

The measurement $\mathbf{s}_{t-1}$ is always represented onto the global plane,

$$\mathbf{s}^I_{t-1} = \mathbf{R}(\gamma) \cdot \mathbf{s}^R_{t-1} + \mathbf{t}$$

and it is used as reference to match the last data scan measured by obtaining a correlation factor by equation (2.79)

$$f_c = \sum_{i=1}^{k} \sum_{j=1}^{k} |\mathbf{s}(i)_{t-1} - (\mathbf{R}(\psi) \cdot \mathbf{s}(j)_t + \mathbf{t})|^2 \qquad (2.79)$$

If the matrix $\mathbf{R}(\psi)$, and the vector $\mathbf{t}$, satisfy for $\mathbf{s}_t = \mathbf{s}_{t-1}$, then the correlation factor is zero. Hence, all points converge. The ICP algorithm iteratively compute all measured points correspondence. Each iteration computes $\mathbf{R}(\psi)$ and $\mathbf{t}$ that minimises the equation (2.79). It is assumed that in the last iteration the correspondence between points is correct, as shown in figure 2.18.

In practice this localization and mapping algorithm is very sensitive to problems of inter-obstacles occlusion, and hence is prone to fail after short navigation in too cluttered environments. When this occurs the scanmatching algorithm no longer compute accurately the robot's posture. Nevertheless, this problem is solved by an algorithmic proposal. The ICP-SLAM algorithm is restarted in-situ just in the very last posture $\boldsymbol{\xi}_s$ before the fail occurred. Thus, the new initial posture is reset by $\boldsymbol{\xi}_0 = \boldsymbol{\xi}_s$. The subsequent new LIDAR scanlines $\boldsymbol{\xi}_s$ are processed with ICP-SLAM as usually, w.r.t. the new global coordinates origin by the following expression,

$$\boldsymbol{\xi}_g = \mathbf{K}(\psi_s) \cdot \boldsymbol{\xi}_s + \mathbf{t}_g \qquad (2.80)$$

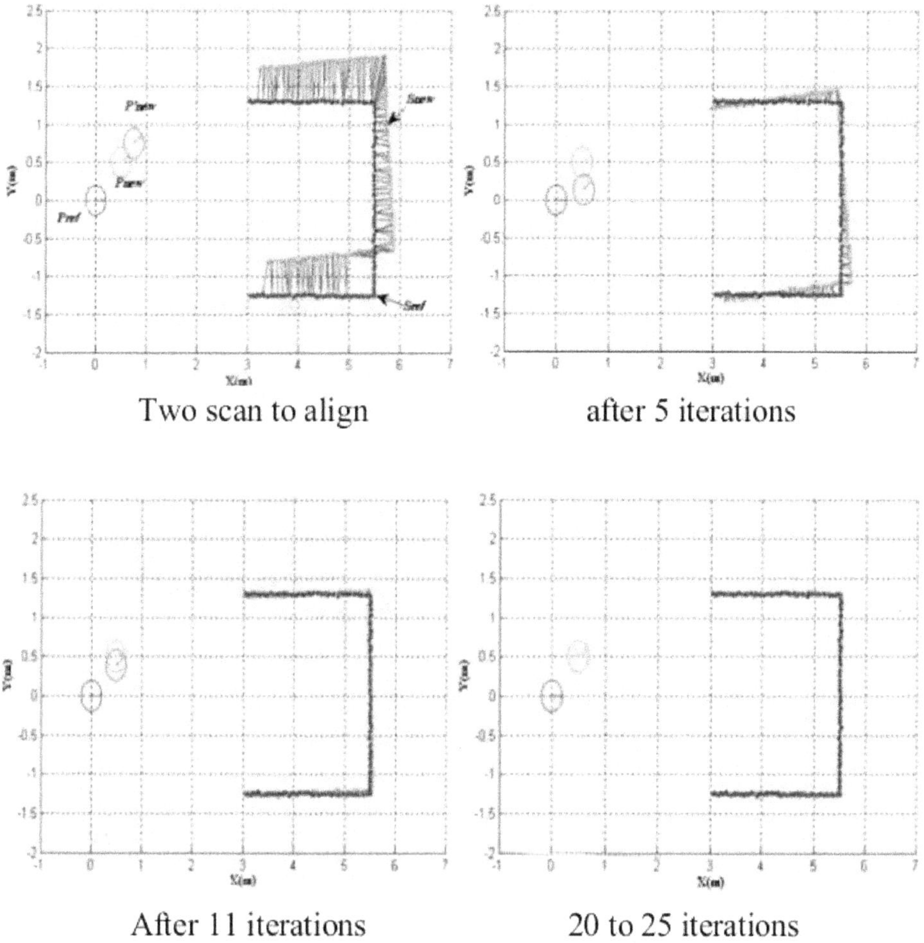

Figure 2.18: Alignment of two scanlines in a common coordinates frame.

Where ξ_g is the robot's posture in global the frame. The new angle $\psi_s = \psi_0 + \theta_s$ is reset using the previous inertial frame angular reference and the last robot's orientation last correct orientation of the robot before the fail. And ξ_0 is once again the new robot's initial posture in global coordinate frame.

$$\mathbf{t}_g = \xi_0 + \xi_r \tag{2.81}$$

Meanwhile the scanmatching does not fail, the current robot's posture ξ_s is preserved. Otherwise, ξ_s is obtained as given by the equation (2.82).

$$\xi_s = \begin{cases} \xi_g & f_c < f_{c_{min}} \\ \xi_s & \text{otro} \end{cases} \qquad (2.82)$$

Figure 2.19 illustrates how previous correction works formulated by expressions (2.80)-(2.82). ξ_{r_1} and ξ_{r_2} are failing points, which if not corrected, the robot's posture accuracy diverges.

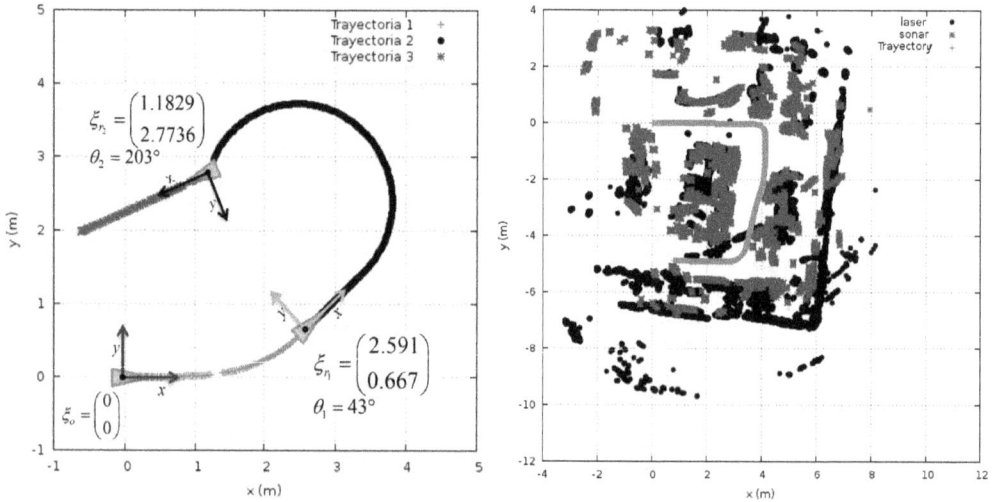

Figure 2.19:   Robot's posture with reset in-situ to fix divergences (left). Robot's localization and mapping mixing ultrasonic sonar and LIDAR data (right).

By using data from LIDAR and a ring of ultrasonic sonar, multi-sensor registration of the Robotics Lab was carried out. The experiments built environmental maps by deploying the ICP-SLAM algorithm, as shown in figure 2.20. It was found out that the map built with ICP-SLAM was quite accurate, being very near to the real environment metrics. Finally, an illustration of sensor data registration is depicted in figure 2.19-right that shows a map of the Robotics Lab mixing ultrasonic sonar and LIDAR data.

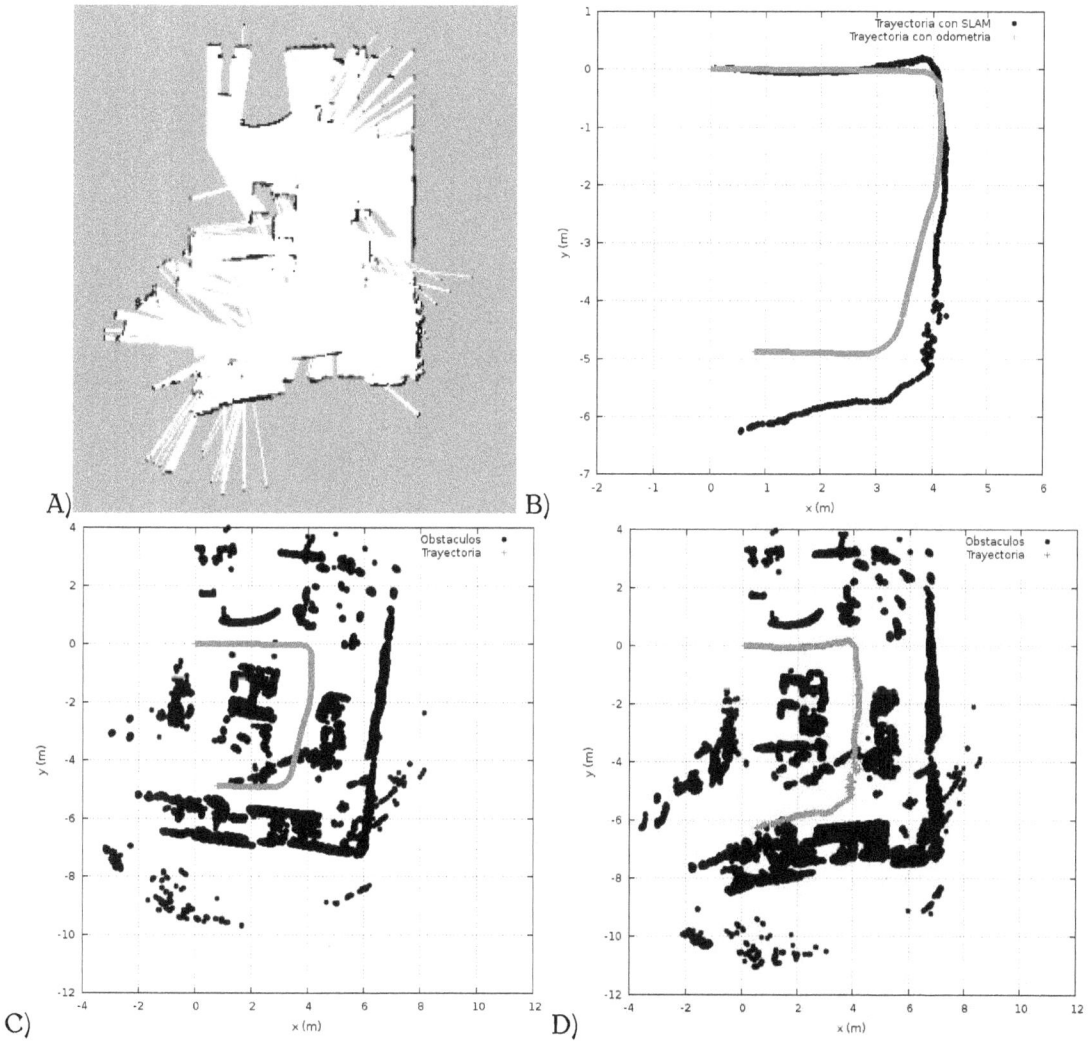

Figure 2.20: Online robot's localization and mapping. A) probabilistic grid-based map. B) robot's trajectories with odometry and ICP-SLAM. C) odometry only mapping. D) mapping with ICP-SLAM.

Bibliography

[1] S. Kumar, *Introduction to robotics*, 1st Ed. McGraw Hill, 2010.

[2] B. Siciliano, and O. Khatib (Eds.), *Handbook of robotics*, Springer, 2008.

[3] H.B. Mitchell, *Multi-sensor data fusion*, Springer 2007.

[4] A. Nüchter, *3D robotic mapping*, Springer, 2009.

[5] P. Corke, *Robotics, vision and control: fundamental algorithms in matlab*, Springer, 2011.

[6] P. Besl, and N. McKay, *A method for registration of 3-D shapes*, IEEE Trans. on Pattern analysis and machine intelligence, vol. 14, no. 2, pp. 239–256, Feb 1992.

[7] H. Choset, K. Lynch, S. Hutchinson, G. Kantor, W. Burgard, L. Kavraki, and S. Thrun, *Principles of robot motion: theory, algorithms and implementations*, MIT Press 2005.

[8] R. Siegwart and I. R. Nourbakhsh, *Introduction to autonomous mobile robots*, MIT Press, 2004.

[9] S. Thurn, W. Burgard, and D. Fox, *Probabilistic robotics*, MIT Press, 2005.

[10] B. Cyganek, and J.P. Siebert, *An introduction to 3D computer vision: techniques and algorithms*, Wiley & Sons, 2009.

[11] O. Faugeras, *Three-dimensional computer vision: a geometric viewpoint*, MIT Press, 1993.

[12] Lu F., Milios E., *Robot pose estimation in unknown environments by matching 2D rangescans*, Journal of Intelligent and Robotic systems, vol. 18, pp. 249–275, Kluwer Academic Pubs., 1997.

[13] Diosi A., Kleeman L., *Laser scan matching in polar coordinates with application to slam*, IEEE/RSJ IROS 2005, pp. 3317–3322, 2005, doi: 10.1109/IROS.2005.1545181.

[14] Gutmann J.S., Schlegel C., *Comparison of scan matching approaches for self-localization in indoor environments*, In Proc.1st Euromicro Work. on Advanced Mobile Robots, pp. 61–67, 1996, doi: 10.1109/EURBOT.1996.551882.

[15] Nunez P., Vazquez-Martin R., Del Toro J.C., Bandera A., Sandoval F., *Feature extraction from laser scan data based on curvature estimation for mobile robotics*, In Proceedings IEEE Intl. Conf. on Robotics and Automation, 2006, doi: 10.1109/ROBOT.2006.1641867.

[16] Kulich M., Mazl R., Preucil L., *Statistical and feature-based methods for mobile robot position localization*, In LNCS, no. 2113, pp. 517–526, Springer-Verlag, Berlin Heidelberg, 2001, doi: 10.1007/3-540-44759-8_51.

[17] Hart P., Duda R., Stork D., *Pattern Classification Wiley-Interscience*, second edition edition, 2001..

[18] Fukunaga K., *Introduction to Statistical Pattern Recognition*, Academic Press, 2nd edition, 1990.

Chapter 3

Multisensor registration

Joaquín Rivero and Edgar A. Martínez-García

Laboratorio de Robótica, Institute of Engineering and Technology
Universidad Autónoma de Ciudad Juárez, Mexico.

This chapter presents a model to register heterogeneous 3D data obtained from three types of sensors: a ring of eight ultrasonic sonar; a high density data LiDAR (light detection and ranging); and from three visual sensors radially placed. One of the contributions is the fusion model to provide a radial multi-stereo geometric system to yield 3D data. All deployed sensors are geometrically placed on-board a wheeled mobile robot platform, and data registration is carried out navigating indoors. The sensor devices in discussion are coordinated and synchronized by a homen-made distributed sensor suite system. Mathematical deterministic formulation for data registration is used to obtain experimental and numerical results on global. Data registration relies on a geometric model to compute depth information from a divergent trinocular stereo sensor w.r.t. a common origin point. Sensor fusion is an engineering research field of study about the process to combine measurements from different sensors, or single sensor with spatio-temporal frames to provide a robust and a complete description about environmental objects. Sensor fusion is used to yield sensor redundancy in order to reduce uncertainty of measurements, to improve the perception of the world in order to take smart decisions. Data registration is a field that search for models to accurately store data obtained from sensors at different spatio-temporal sensor measurements. In the present context, 3D heterogeneous data refers to depth information from different types of sensors with different sensing modalities.

3.1   Sensors suite

A sensors suite (SS) is a device comprised of multiple interconnected sensors that are controlled, coordinated, and synchronized to accomplish detection of relevant environmental percepts through information synthesis [1-4]. Since it manages sensors with different kinds of transducers, the types of energies are also diverse. Therefore, a SS provides distinct sensing modalities, and it is purposed to obtain reliable information through physical and logical redundancy. In the present research we are deploying a home-made apparatus with a distributed computer system for data registration (see figure 3.14 left and center). The sensor devices instrumenting the SS are concretely summarised in table 1 classified by their data types. Each sensor device was labelled for identification with their symbolic variables that represent the types of data.

Table 3.1: Sensor suite devices and their types of data.

Sensor	ID	Modality	Type	variable
Stereo Vision	S1	Vision/range	Pasive	$P = (x,y,z)^T$, I_{nxmx3}
Spherical Vision	S2	Vision/multiple	Pasive	I_{nxmx3}, J_{nxmx3}, K_{nxmx3}
IMU	S3	Linear acceleration & angular velocity	Pasive	$\ddot{x}$, $\ddot{y}$, ω
GPS	S4	Position	Pasive	x, y, z
LIDAR	S5	Range	Active	δ_j, ϕ_j
Encoder	S6	Position	Pasive	v, s
Compass	S7	Angle	Pasive	θ
Ultrasonic Sonar	S8	Range	Active	δ
Binocular Multi-function	S9	Vision/Range	Pasive	$P = (x,y,z)^T$, I_{nxmx3}, J_{nxmx3}

The sensor S1 represents a binocular stereo sensor with maximal resolution of 1600 x 1200pixels, at 15fps (frames per second), with a baseline configuration of 63mm. The sensor device S2 represents a ring of visual sensor, which are geometrically arranged as a cylindrical array set up as a multi-stereo system. It is compounded of three colour cameras connected through an IEEE-1394 port centralised to the SS computer host [5,6]. Device S3 is a 2-DOF gyroscope, with a 2-axis accelerometer integrated. The S4 is a GPS receiver with USB interface, with an accuracy of 5m 2D RMS, when WAAS is enabled. It uses a GPS protocol NMEA 0183 and SiRF binary as secondary protocol. The S5 is a LiDAR sensor device with a scanning area of 240°, angular resolution of 0.36°, and an accuracy range from 60 - 4,095mm. Multiple S6 can be present in the SS, which are quadrature encoders with 90 pulses per revolution. Sensor S7 is a magnetic compass with accuracy of 0.5° and works with an I^2C interface. The S8 are ultrasonic sonar sensors ranging 100 - 5,000 mm. S9 are two visual sensor calibrated as a stereo pair, but configured with an embedded vision processor. Both are set up to either work individually, or in combination as a binocular stereo sensor. Both visual sensors process colour images with resolution of 352x288 pixels. This chapter focuses on providing a mathematical formulation for data registration by deploying several sensors: three S2, one S5 and eight S8. Although, S2 are in principle 2D images, a radial multi-stereo model is formulated in the present context. Thus, 3D information inferred from S2 is then homogenised with S5 and S8.

3.2 Active sensing models

3.2.1 Sonar model

A sonar sensor is an electro-acoustic device that measures range of the nearest orthogonal point by using a time of-flight ranging technique. Sensitivity range of traditional used ultrasonic sonar ranges from $0.10 \leq s < 5m$, is typically deployed in mobile robotics to measure obstacles range. In this work, a ring of eight sonar sensors radially arranged were deployed in our robotic platform . Depth information w.r.t. environmental objects is measured through sound (see 2.2), fundamentally the speed of sound in general is modelled by,

$$s = \frac{ct}{2} \tag{3.1}$$

where c is the sound speed; s is the distance an acoustic vibration travelled over an elapsed period of time t. Measurement data are treated by homogeneous transformations to represent the environment from the robot's fixed coordinate system, according to fig 3.1-a,

$$\mathbf{s}_{sonar_j}^R = l_j \begin{pmatrix} \cos(\phi_j) \\ \sin(\phi_j) \\ 0 \end{pmatrix} + d_j(t) \cdot \begin{pmatrix} \cos(\theta_j) \\ \sin(\theta_j) \\ 0 \end{pmatrix} \tag{3.2}$$

where $d_j(t)$ is the measurement value, and l_j is the Cartesian distance between the robot's geometric centre to the jth sonar. ϕ_j is the angle yielded by the robot's X-axes and line l_j where the sonar is located. Angle θ_j is the orientation of the sonar (see figure 3.1.a).

$$
\mathbf{s}^R_{sonar_j} = \begin{pmatrix} l_j \cos\left(\phi_j\right) + d_j(t) \cos\left(\theta_j\right) \\ l_j \sin\left(\phi_j\right) + d_j(t) \sin\left(\theta_j\right) \\ 0 \end{pmatrix}
\tag{3.3}
$$

Furthermore, by transforming onto a global Cartesian coordinate system for $\xi_t = (x, y, \theta)^\top$, the following postulate is stated,

Postulate 3.2.1 (Global sonar-based data representation).

$$
\mathbf{p}^I_{sonar_j} = \mathbf{R}\left(\gamma\right) \cdot \mathbf{s}^R_{sonar_j} + \xi_t
\tag{3.4}
$$

and by substituting each expression terms the equation is stated in global inertial frame by

$$
\mathbf{p}^I_{sonar_j} = \begin{pmatrix} \cos\gamma & -\sin\gamma & 0 \\ \sin\gamma & \cos\gamma & 0 \\ 0 & 0 & 1 \end{pmatrix} \cdot \begin{pmatrix} l_j \cos\left(\phi_j\right) + d_j(t) \cos\left(\theta_j\right) \\ l_j \sin\left(\phi_j\right) + d_j(t) \sin\left(\theta_j\right) \\ 0 \end{pmatrix} + \begin{pmatrix} x \\ y \\ \theta \end{pmatrix}
\tag{3.5}
$$

3.2.2   Light detection and range model

A Light detection and ranging (LIDAR) sensor is an electro-optic device deployed to measure range of points by electromagnetic signal time-of-flight technique (see 2.4). For a LiDAR sensor, range data are collected with cylindrical order, where points are referenced by distance and known bearing. We deployed a Hokuyo UBG-04LX-F01, with a scanning area of 240°, and was configured with beam angular resolution of 0.36°, which includes a range accuracy of $60 - 4,095$mm. See figure 3.1-b), where $\delta_i(t)$ is the ith measurement range value. In addition, l is the Cartesian distance w.r.t. robot's geometric centre to any LiDAR radial measurement (see figure 3.1.b).

a)
b)
c)

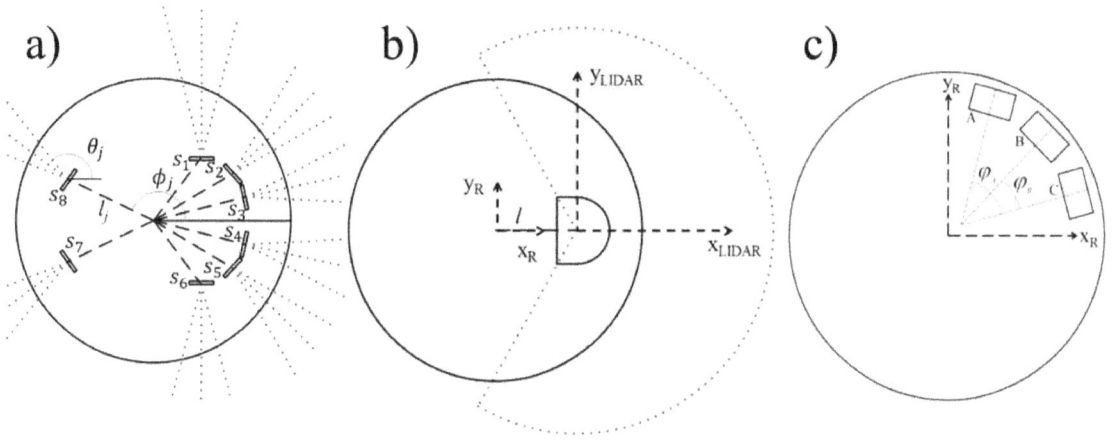

Figure 3.1:   a) Ultrasonic sonar configuration; b) LIDAR location and orientation; c) Radial trinocular stereo vision system.

Likewise, $\Delta\phi$ is the angular resolution, and ϕ_0 is the minimum angle in the scan[8].

$$\mathbf{s}_{\text{LIDAR}}^{R} = \begin{pmatrix} \delta_i(t)\cos\left(\Delta\phi\,(i-1)+\phi_0^{\ell}\right)+l \\ \delta_i(t)\sin\left(\Delta\phi\,(i-1)+\phi_0^{\ell}\right) \\ 0 \end{pmatrix}_{i=1}^{k} \tag{3.6}$$

Thus, by transforming into a global inertial coordinate system, the homogeneous rigid roto-translation model is postulated by

Postulate 3.2.2 (Global LIDAR-based data representation).

$$\mathbf{p}'_{\text{LIDAR}} = \mathbf{R}(\gamma)\cdot\mathbf{s}_{\text{LIDAR}\,i=1}^{k} + \boldsymbol{\xi}_t \tag{3.7}$$

and by substituting rotation and translation terms accordingly the next expression is produced:

$$\mathbf{p}'_{LIDAR}(t) = \begin{pmatrix} \cos\gamma & -\sin\gamma & 0 \\ \sin\gamma & \cos\gamma & 0 \\ 0 & 0 & 1 \end{pmatrix}\cdot\begin{pmatrix} \delta_i(t)\cos\left(\Delta\phi\,(i-1)+\phi_0^{\ell}\right)+l \\ \delta_i(t)\sin\left(\Delta\phi\,(i-1)+\phi_0^{\ell}\right) \\ 0 \end{pmatrix}_{i=1}^{k} + \begin{pmatrix} x \\ y \\ \theta \end{pmatrix} \tag{3.8}$$

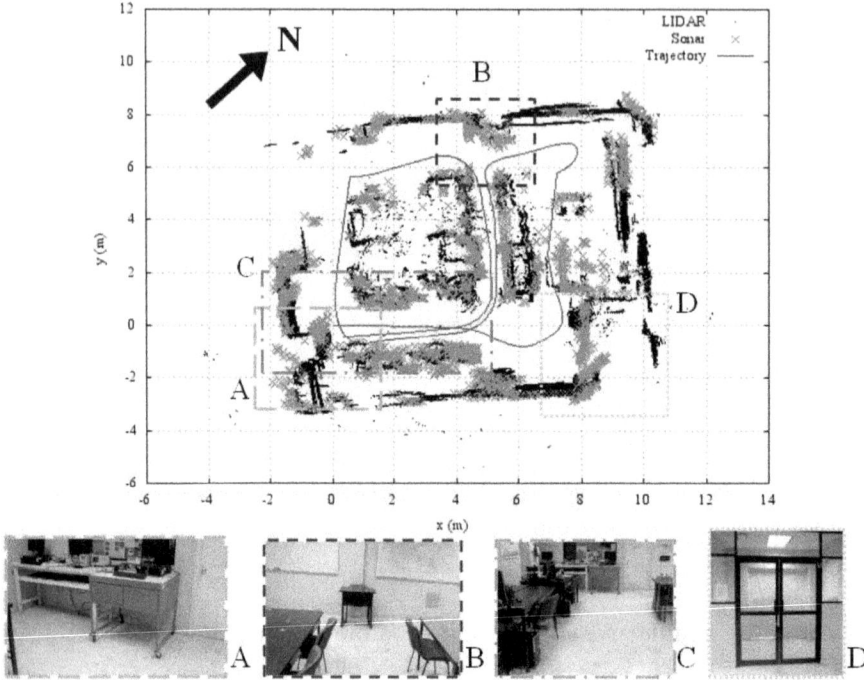

Figure 3.2: Data registration on a global map, using a ring of eight sonar, and a LiDAR.

Indoor experimental results were carried out in the Robotics Lab that consisted of tele-operated explorations within dynamic situations. A wheeled mobile robotic platform was deployed and its speed model was used to estimate and predict positions, in order to match the real observations. Besides, the robot was instrumented with the sensor suite, using only a ring of ultrasonic sonar, and the LiDAR sensor ($240°$ of sensing angle). Along the navigation path, data registration on-line was carried out over a global map[9,10], experimental results are depicted in figure 3.2. Thus, data registration formulation is provided eqs. (3.5)-(3.8). The general 3D data registration model consists of union of datasets into a global Cartesian space from heterogeneous sensory sources: sonars $S = \{\mathbf{p}_i^S\}$ where $\mathbf{p}_i^S = (x, y, h_S)^T$, laser $\mathcal{L} = \{\mathbf{p}_i^L\}$ where $\mathbf{p}_i^L = (x, y, h_L)^T$, and trinocular radial stereo $\mathcal{C}_{AB} = \{\mathbf{p}_i^{AB}\}$ and $\mathcal{C}_{BC} = \{\mathbf{p}_i^{BC}\}$ where $\mathbf{p}_i^{AB} = (x, y, z)^T$ and $\mathbf{p}_i^{BC} = (x, y, z)^T$ respectively. Likewise, h_S and h_L are sonar and LiDAR metric heights, respectively. At this stage, we only considered a deterministic model to unify eight sonar sensors, one laser range finder scan of 681 measurements, and depth information of a trinocular stereo sensor.

$$\mathcal{M} = S \cup \mathcal{L} \cup \mathcal{C}_{AB} \cup \mathcal{C}_{BC} \tag{3.9}$$

3.3 Stereo vision

3.3.1 Image transformations for correction

Affine transformations preserve collinearity, relative distance, parallelism, and proportion rate. Scale (3.10), rotation (3.11), translation (3.12) and skew (3.13) of an image are considered affine transformations[11,12]. s_x and s_y of the diagonal matrix $\mathbf{E}$ are scaling parameters (see 1.2). If s_x and s_y are same rate, then scale is uniform.

$$\mathbf{E} = \begin{pmatrix} s_x & 0 & 0 \\ 0 & s_y & 0 \\ 0 & 0 & 1 \end{pmatrix} \tag{3.10}$$

The next orthogonal matrix $\mathbf{R}$ is the rotation homogeneous transformation matrix, where the angle θ is the rotation angle between two inertial systems. For orthogonal matrices, its transpose is equivalent to its inverse, and conversely.

$$\mathbf{R} = \begin{pmatrix} \cos\theta & \sin\theta & 0 \\ -\sin\theta & \cos\theta & 0 \\ 0 & 0 & 1 \end{pmatrix} \tag{3.11}$$

An image may be translated on the plane with matrix $\mathbf{T}$ where t_x and t_y is the displacement over x and y respectively.

$$\mathbf{T} = \begin{pmatrix} 1 & 0 & t_x \\ 0 & 1 & t_y \\ 0 & 0 & 1 \end{pmatrix} \tag{3.12}$$

$$\mathbf{S} = \begin{pmatrix} 1 & \lambda_x & 0 \\ \lambda_y & 1 & 0 \\ 0 & 0 & 1 \end{pmatrix} \tag{3.13}$$

The set of above operations (3.14) is considered an affine transformation.

$$\mathbf{T}_{\text{affine}} = \mathbf{T} \cdot \mathbf{R} \cdot \mathbf{E} \tag{3.14}$$

thus,

$$
\mathbf{T}_{\text{affine}} = \begin{pmatrix} 1 & 0 & t_x \\ 0 & 1 & t_y \\ 0 & 0 & 1 \end{pmatrix} \begin{pmatrix} \cos\theta & \sin\theta & 0 \\ -\sin\theta & \cos\theta & 0 \\ 0 & 0 & 1 \end{pmatrix} \begin{pmatrix} s_x & 0 & 0 \\ 0 & s_y & 0 \\ 0 & 0 & 1 \end{pmatrix}
\tag{3.15}
$$

and

$$
\mathbf{T}_{\text{affine}} = \begin{pmatrix} s_x \cos\theta & s_y \sin\theta & t_x \\ -s_x \sin\theta & s_y \cos\theta & t_y \\ 0 & 0 & 1 \end{pmatrix}
\tag{3.16}
$$

Figure 3.3 centre is the raw image acquired, the affine transformation is depicted in the middle, and at right-sided the perspective transformation is shown. The perspective transformation is

Figure 3.3: Raw image (left), affine transformation (centre), and perspective transformation (right).

compounded of a 4×4 matrix, a rotation matrix, a translation vector, and a projection vector, thus its collinearity properties are preserved. Using four points from the plane of original image and four points from the plane of resulting image, a matrix of perspective is calculated[11-13].

$$
\mathbf{x'} = \mathbf{H}_p \cdot \mathbf{x} = \begin{pmatrix} h_{11} & h_{12} & h_{13} \\ h_{21} & h_{22} & h_{23} \\ h_{31} & h_{32} & h_{33} \end{pmatrix} \cdot \begin{pmatrix} x \\ y \\ 1 \end{pmatrix}
\tag{3.17}
$$

Four pairs of points generates eight linear equations. Solving this linear equations system, the elements of matrix of perspective transformation are obtained.

$$
\begin{aligned}
x' (h_{31}x + h_{32}y + h_{33}) &= h_{11}x + h_{12}y + h_{13} \\
y' (h_{31}x + h_{32}y + h_{33}) &= h_{21}x + h_{22}y + h_{23}
\end{aligned}
$$

3.3.2 Geometric stereo model

A stereo sensor is a set of two cameras aligned along the x-axis collinear, and paraller w.r.t. yz axis (see 2.3). The baseline (**b**) is the distance between the origin of right camera and left camera[13,14]. A point **P** is projected on the plane of left image $\mathbf{P_l}$, and over the plane of right image $\mathbf{P_r}$. Geometrically, the point **P** is located at the intersection of the ray $\mathbf{LP_l}$ and the ray $\mathbf{RP_r}$. From similar triangles the following is obtained,

$$
\begin{aligned}
z/f &= x/x_l \\
z/f &= (x - b)/x_r \\
z/f &= y/y_l = y/y_r
\end{aligned}
\tag{3.18}
$$

The coordinates y_l and y_r are assumed with identical distance from the centre. A solution for unknown x and y coordinates of point **P** is obtained by algebraic substitution. Thus,

$$
\begin{aligned}
z &= fb/(x_l - x_r) = fb/d \\
x &= x_l z/f = b + x_r z/f \\
y &= y_l z/f = y_r z/f
\end{aligned}
\tag{3.19}
$$

Disparity is the difference between x_l and x_r coordinates from left and right images. It is used to calculate depth along Z.

3.3.3 Calibration and rectification

Stereo calibration is a process to calculate the geometrical relationship between two cameras. After calibration, the intrinsic and extrinsic parameters are obtained. The intrinsic parameters are focal length, principal point, and distortion coefficients. The extrinsic parameters involve rotation and translation within a matrix that relates the camera's coordinates system, and the global coordinates system[12,15].

$$
M_{Rect\_l,r} = \begin{pmatrix} f_{x\_l,r} & \alpha_{l,r} & c_{x\_l,r} \\ 0 & f_{y\_l,r} & c_{y\_l,r} \\ 0 & 0 & 1 \end{pmatrix}
\tag{3.20}
$$

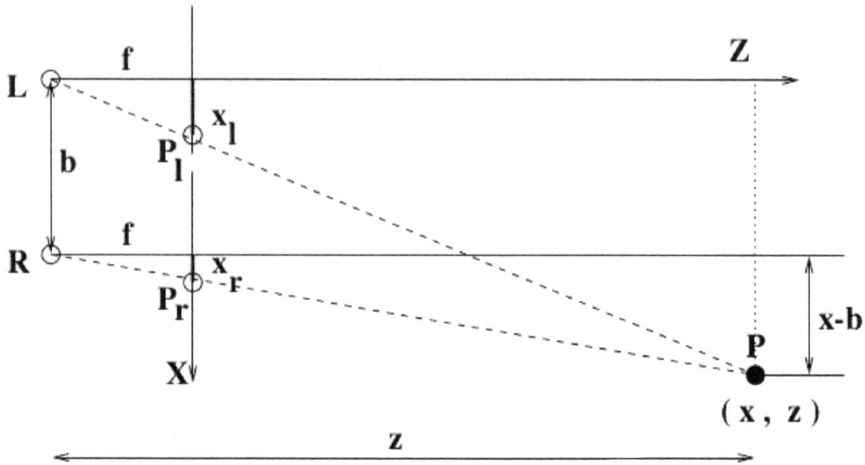

Figure 3.4: Depth geometric model of a binocular stereo sensor.

and

$$\text{Distorsion} = \begin{bmatrix} k_1 & k_2 & p_1 & p_2 & k_3 \end{bmatrix}^\top \tag{3.21}$$

The rotation matrix **R**, the translation vector **T**, the essential matrix **E**, and the fundamental matrix **F** are required to be obtained. **R** and **T** denote rotation and translation of both: the left camera's coordinates system, and the right camera's coordinates system. The essential matrix **E** relates the location of a point, which is located between the left camera, and right camera w.r.t. the global coordinates. the fundamental matrix **F** is similar to the essential matrix **E**, however the former is provided in pixel coordinates. At this process, various images of a chessboard with different perspectives are obtained (figure 3.5).

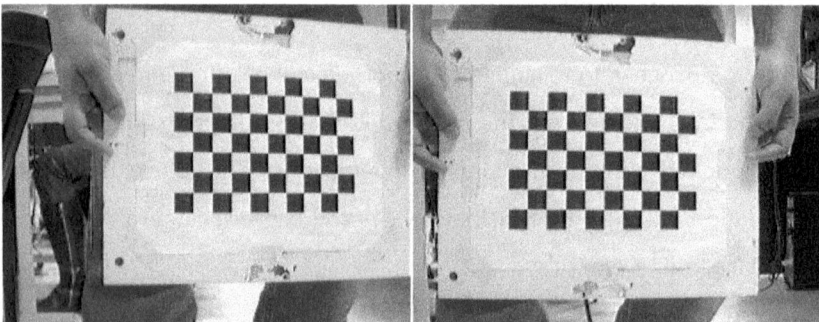

Figure 3.5: Cameras calibration using a chessboard.

The distortion vector consists of three radial distortion coefficients (k_1, k_2, k_3), as well as two tangential distortion coefficients (p_1, p_2). The radial distortion is produced by the form of the lens.

$$x_{\text{corrected}} = x\left(1 + k_1 r^2 + k_2 r^4 + k_3 r^6\right) \tag{3.22}$$

$$y_{\text{corrected}} = y\left(1 + k_1 r^2 + k_2 r^4 + k_3 r^6\right)$$

In addition, tangential distortion is an effect produced by lenses manufactured with defects, and such defects affect the projection of light not being parallel with the plane of image.

$$x_{\text{corrected}} = x + \left[2p_1 xy + p_2\left(r^2 + 2x^2\right)\right] \tag{3.23}$$

$$y_{\text{corrected}} = y + \left[p_1\left(r^2 + 2y^2\right) + 2p_2 xy\right]$$

After distortion parameters are obtained, the next step is the rectification process, which correct each individual image by reducing the effects of radial and tangential lens distortions. It produces misalignment of rows in stereo pairs. The matrix $\mathbf{R}_l$, $\mathbf{R}_r$, $\mathbf{P}_l$, $\mathbf{P}_r$, and $\mathbf{Q}$ are obtained from the rectification process. $\mathbf{R}_l$ and $\mathbf{R}_r$ are the rotation matrices for each camera. $\mathbf{P}_l$ and $\mathbf{P}_r$ are the projection matrices of the rectified system of the left and right cameras.

$$\mathbf{P}_l = \begin{pmatrix} f_{x\_l} & \alpha_l & c_{x\_l} \\ 0 & f_{y\_l} & c_{y\_l} \\ 0 & 0 & 1 \end{pmatrix} \begin{pmatrix} 1 & 0 & 0 & 0 \\ 0 & 1 & 0 & 0 \\ 0 & 0 & 1 & 0 \end{pmatrix} \tag{3.24}$$

$$\mathbf{P}_r = \begin{pmatrix} f_{x\_r} & \alpha_r & c_{x\_r} \\ 0 & f_{y\_r} & c_{y\_r} \\ 0 & 0 & 1 \end{pmatrix} \begin{pmatrix} 1 & 0 & 0 & T_x \\ 0 & 1 & 0 & 0 \\ 0 & 0 & 1 & 0 \end{pmatrix} \tag{3.25}$$

A pair of images after rectification process is showed at figure 3.6.

3.3.4 Disparity calculation

Stereo matching establishes coincidences between left and right rectified images in order to produce a disparity map (see 2.3). To find the correspondence of a section between left and right images, similarity or dissimilarity measures are used[13-15]. An experiment for measuring a

Figure 3.6: A rectified stereo pair.

numerical degree of correlation was developed with an image window of 5×5 pixels in size. And it was slide along the horizontal axis, from the left-sided image until the right-sided image edge. Further, it was compared with a second window simultaneously sliding horizontally too, but in the right-sided image (figure 3.7). With such process, the three different similarity/dissimilarity measures were applied.

Figure 3.7: Epipolar alignment of a stereo pair.

Similarity measurements are generally used to measure the similarity between two datasets. A high value indicates high correspondence[15]. In this application, compared datasets are gray scale intensities in both: left and right images. One of the applied similarity measures is the Pearson correlation coefficient value, it numerically varies between -1 and +1. When $\bar{X}$ and $\bar{Y}$ are equal, it means that the Pearson correlation coefficient resulted $r = +1$. This is known as a perfect positive correlation.

Contrary if $\bar{X}$ is equal to negative value of $\bar{Y}$, the Pearson correlation coefficient results $r = -1$. This is known as a perfect negative correlation.

$$r = \frac{\sum_{i=1}^{n} (x_i - \bar{x})(y_i - \bar{y})}{\left[\sum_{i=1}^{n} (x_i - \bar{x})^2\right]^{\frac{1}{2}} \left[\sum_{i=1}^{n} (y_i - \bar{y})^2\right]^{\frac{1}{2}}} \tag{3.26}$$

Figure 3.8 shows the experimental results when using the Pearson correlation coefficient. Numerous dissimilarity measurements indicate numerical differences between two datasets. For this case, a dissimilarity with a high numeric value basically indicates a low matching between two image regions.

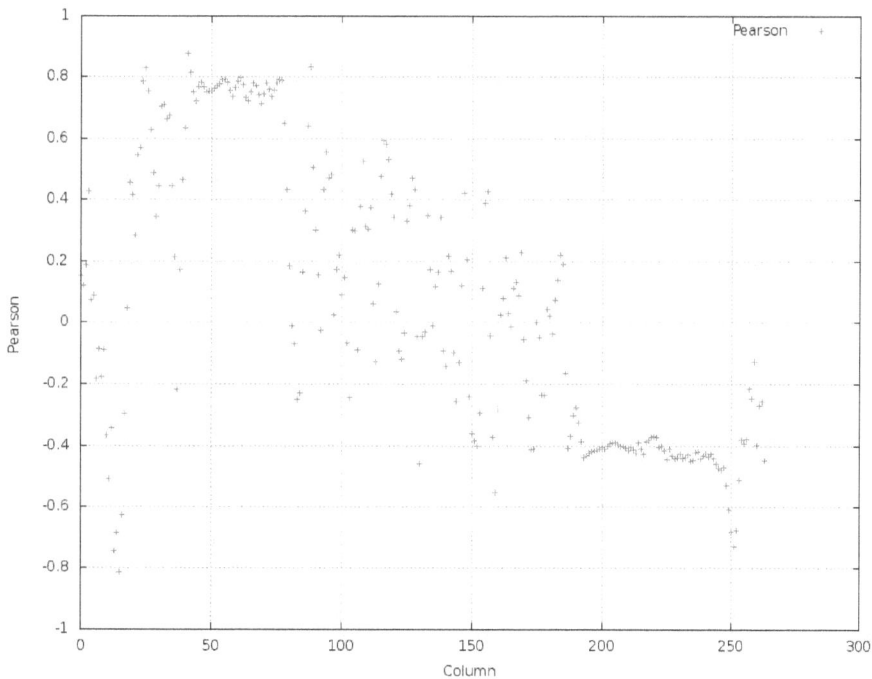

Figure 3.8: Experimental results of dissimilarity for the Pearson correlation coefficient.

Some dissimilarity measures are L_1-Norm, which some well known approaches are the Manhattan norm, or the sum of absolute differences applied to image intensities. Those, are typical dissimilarity measures that are used to traditionally compare similarity between images. When the images are obtained from the same sensor device, and under the same environmental conditions, and if the sensor has a high signal-noise relationship, then this measurement may produce matching results more precisely than other methods further complex usually provide.

The Next equation is the functional form of the sum of the absolute differences:

$$L_1 = \sum_{i=1}^{n} |x_i - y_i| \tag{3.27}$$

Some matching results using L_1-Norm are shown in figure 3.9. The L_2^2-Norm, or Euclidean

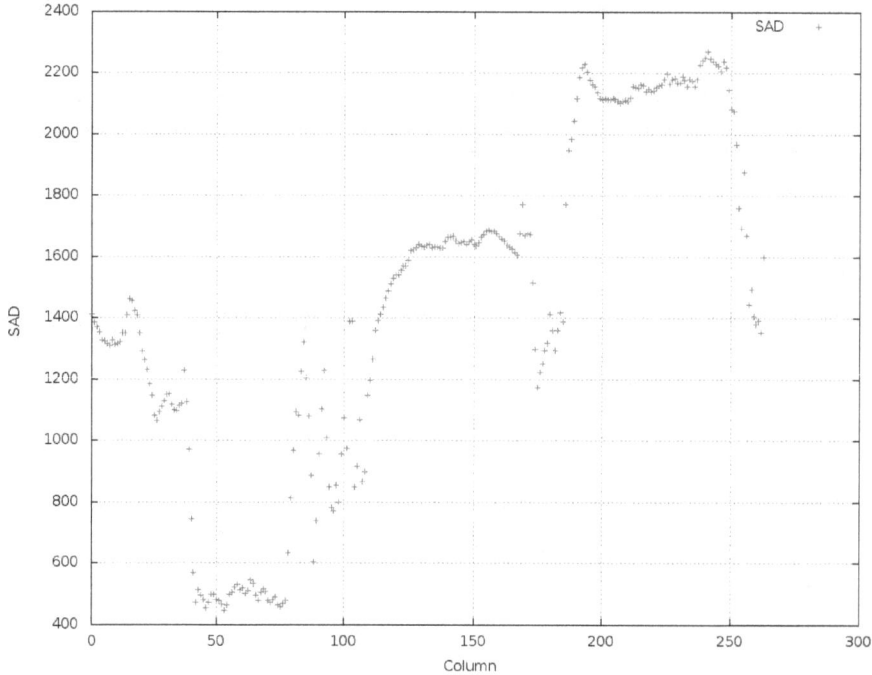

Figure 3.9: Experimental dissimilarity results of the L_1-Norm measure.

distance or sum of squared difference of intensities is a measurement that is more sensible than the Pearson correlation coefficient. However, the results are poorer than the Pearson correlation coefficient when the images are obtained under different lighting conditions.

$$L_2^2 = \sum_{i=1}^{n} (x_i - y_i)^2 \tag{3.28}$$

The matching results by using the L_2^2-Norm is shown at figure 3.10. A disparity map is a 2D image which use values of gray scale to indicate disparity or difference between the features at left and right images. An example of a disparity map is shown in figure 3.11. The light gray areas indicate that the objects in the scene are closer than the dark gray regions.

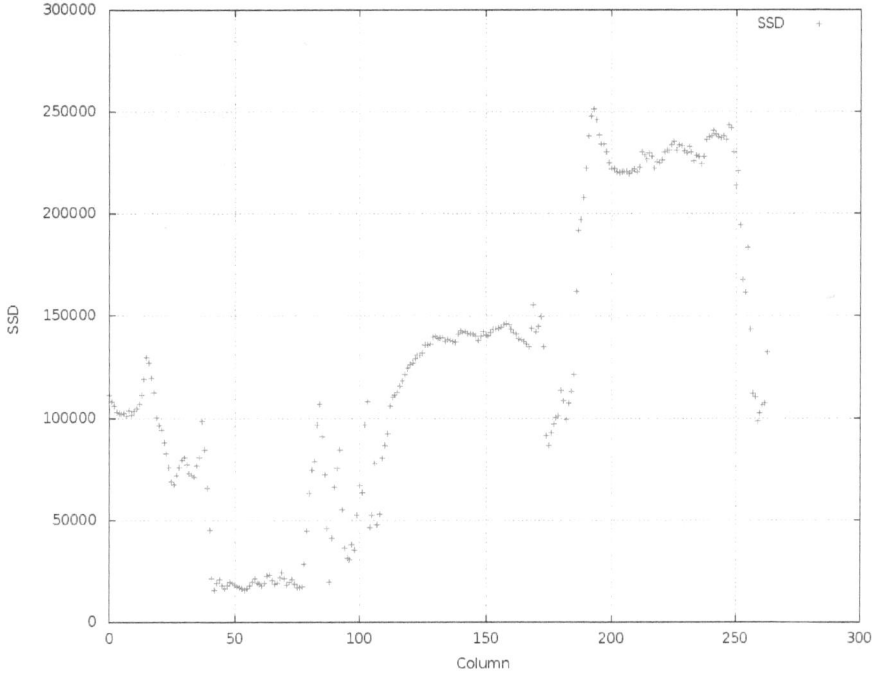

Figure 3.10: Experimental dissimilarity results of L_2^2-Norm measure.

3.3.5 Depth calculation

The projection matrix basically transforms a $3D$ point of homogeneous coordinates into a $2D$ point of homogeneous coordinates. In general terms, the coordinates of the image could be calculated as $(x/w, y/w)$ through the dot product of next expression:

$$\begin{pmatrix} x \\ y \\ w \end{pmatrix} = \mathbf{P}_{l,r} \begin{pmatrix} X \\ Y \\ Z \\ 1 \end{pmatrix} \tag{3.29}$$

conversely, the re-projection matrix $\mathbf{Q}$ is the mapping representation from the disparity map onto the depth information dataset. Therefore, by knowing a 2D homogeneous point and the disparity value that is associated to such point, then another 2D point is re-projected into the 3D space.

Figure 3.11:   A stereo pair (left and centre), with its disparity values image.

$$
\begin{pmatrix} X \\ Y \\ Z \\ W \end{pmatrix} = \mathbf{Q} \begin{pmatrix} x \\ y \\ d \\ 1 \end{pmatrix}
\tag{3.30}
$$

where the squared transformation matrix $\mathbf{Q}$ is defined by

$$
\mathbf{Q} = \begin{pmatrix} 1 & 0 & 0 & -c_x \\ 0 & 1 & 0 & -c_y \\ 0 & 0 & 1 & f \\ 0 & 0 & -1/T_x & (c_x - c'_x)/T_x \end{pmatrix}
\tag{3.31}
$$

Therefore, the $3D$ coordinates of the global system are expressed dividing the three coordinates values by the factor W ($X/W, Y/W, Z/W$).

3.3.6 Stereo data map building

A laboratory experiment to build a map using stereo data was developed. The experiment consisted of tele-operation of a mobile platform with a stereo sensor on-board. The robot basically was controlled to travel a distance significant enough to build a dense map (see 2.9), about $9m$ in length. Along such a distance 116 stereo pairs were acquired. The system specifications are: an image sensor: CCD 1/4 in progressive scan, Sony ICX-098BQ with effective pixels $659(H) \times 494(V)$. Image size: 640×480, 320×240, 160×120. Data pPath: YUV ($4:1:1$, $4:2:2$, $4:4:4$), RGB 24 bits, mono 8 bits. Cell size: $5.6\mu m \times 5.6\mu m$. Frame rate: $30, 15, 7.5, 3.75$. Focal length of lens $4.3mm$.

Furthermore, the data registration process needs a deterministic model of the robot's posture, which is defined by the posture vector $\xi_t = (x, y, \theta)^\top$ calculated recursively.

$$f_s = \frac{2\pi r}{R},$$

$$\Delta_s = f_s \left(\frac{N_r + N_l}{2} \right)$$

hence,

$$x(t) = x_0 + \Delta_s \cdot \cos \left(\theta_0 + \left(\frac{2 (N_r - N_l)}{b} \right) \right) \tag{3.32}$$

$$y(t) = y_0 + \Delta_s \cdot \sin \left(\theta_0 + \left(\frac{2 (N_r - N_l)}{b} \right) \right) \tag{3.33}$$

$$\theta(t) = \theta_0 + \left(\frac{2 (N_r - N_l)}{b} \right) \tag{3.34}$$

Data registration consists of the union of numerous clouds of points that must be correctly aligned at each robot's pose overtime. The information provided by the cloud of points is homogenised with respect to an inertial system that is consistent with the robot's motion, throughout linear roto-translation operations.

$$\mathbf{p}_r = \mathbf{R} (\alpha, \beta, \gamma) \, \mathbf{p}_v \tag{3.35}$$

and

$$\mathbf{p}_I = \mathbf{R} (\theta) \, \mathbf{p}_r + \xi \tag{3.36}$$

A 3D map of the Robotics Lab was built online by combining sensor data arising from diferetn types of sensors[16,17]. The experimental results are depicted in figure 3.12, top view and isometric view.

3.4 Divergent trinocular stereo

The term "radial" describes an arrangement of cameras placed circularly w.r.t. a common convergence origin point. The proposed radial multi-stereo system consists of three cameras radially distributed where the image planes are slightly overlapped. The geometric scheme of a radial multi-stereo system (see figure 3.13), the relationship among the common convergence

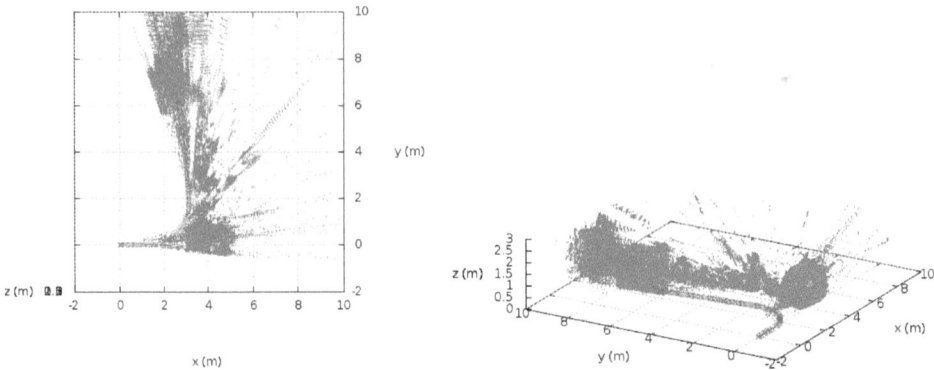

Figure 3.12:   Top view of a real environment Cartesian global map built (left). An isometric view of a real environment global map built (right).

centre, camera B, and camera C, an isosceles triangle is formed (see 1.1).   Let us call β the angle between cameras B, and C. Therefore, the angle ϕ is determined by summing the inner triangle's angles, by $\beta + \phi = \pi$ and $\phi = \pi - \beta$. Thus, $\frac{\phi}{2} = \frac{\pi}{2} - \frac{\beta}{2}$. By applying the sine's law, $\overline{BC}$ is calculated, which is the linear distance from camera B to camera C. Likewise, 1 is the distance from O to any camera. According to figure 3.13, the next relationship is stated,

$$\frac{\overline{BC}}{\sin \beta} = \frac{l}{\sin \left(\frac{\phi}{2} \right)} \qquad (3.37)$$

by substituting the angle in the right-side term, and by dropping-off the distance of interest,

$$\overline{BC} = \frac{l \sin \beta}{\sin \left(\frac{\pi}{2} - \frac{\beta}{2} \right)} \qquad (3.38)$$

The angle from the optical axis and the ray of projection of P at focal point of the camera B and camera C is calculated. x_B is the x-coordinate of the feature at the plane of camera B. x_C is the x-coordinate of the feature at the plane of camera C. And f is the focal length of the camera.

$$\theta_B = \tan^{-1} \left(\frac{x_B}{f} \right) ; \qquad \theta_C = \tan^{-1} \left(\frac{x_C}{f} \right) \qquad (3.39)$$

Besides, the complementary angles models are stated by

$$\angle B = \frac{\pi}{2} - \theta_B + \frac{\beta}{2}; \qquad \angle C = \frac{\pi}{2} - \theta_C + \frac{\beta}{2}; \qquad \angle P = \theta_B + \theta_C - \beta \qquad (3.40)$$

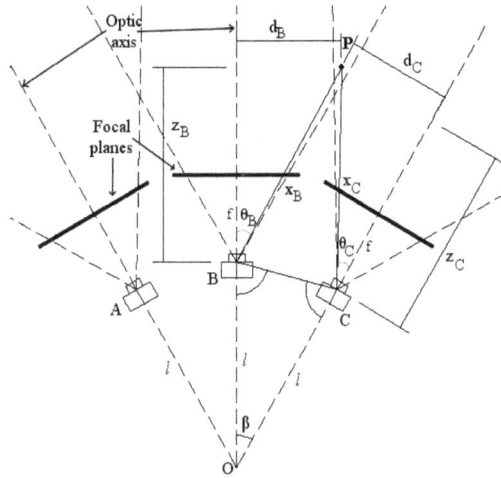

Figure 3.13: Proposed geometric model of a divergent trinocular stereo vision system.

To estimate the range from cameras B and C w.r.t. point P, the linear distance from each camera in the radial system (B and C), to the point P is calculated by sine's law by,

$$\frac{\overline{BC}}{\sin \angle P} = \frac{\overline{CP}}{\sin \angle B}; \quad \text{and dropping-off} \angle \overline{CP} \quad \overline{CP} = \frac{\overline{BC} \sin \angle B}{\sin \angle P} \tag{3.41}$$

Hence, the model to express depth information is given by $z_B = \overline{BP} \cos \theta_B$. Thus, by substituting $\overline{BP}$ and θ_B, the model is more specified,

Proposition 3.4.1 (The divergent stereo depth model.). *The depth component of an arbitrary point projected on camera B is*

$$z_B = \left(\frac{\left(\frac{l \sin \beta}{\sin\left(\frac{\pi}{2} - \frac{\beta}{2}\right)} \right) \sin \left(\frac{\pi}{2} - \theta_C + \frac{\beta}{2} \right)}{\sin \left(\theta_B + \theta_C - \beta \right)} \right) \cos \left(\tan^{-1} \left(\frac{x_B}{f} \right) \right) \tag{3.42}$$

the range of **P** *w.r.t. camera C is* $z_C = \overline{CP} \cos \theta_C$,

$$z_C = \left(\frac{\left(\frac{l \sin \beta}{\sin\left(\frac{\pi}{2} - \frac{\beta}{2}\right)} \right) \sin \left(\frac{\pi}{2} - \theta_B + \frac{\beta}{2} \right)}{\sin \left(\theta_B + \theta_C - \beta \right)} \right) \cos \left(\tan^{-1} \left(\frac{x_C}{f} \right) \right) \tag{3.43}$$

In addition, with the depth models z_B and z_C, then the real X-coordinates from the cameras B (d_B) and C (d_C) w.r.t. point P, can be estimated. Therefore, $d_B = z_B \tan \theta_B$, and

$$d_B = z_B \tan\left(\tan^{-1}\left(\frac{x_B}{f}\right)\right); \qquad d_B = \frac{z_B x_B}{f} \tag{3.44}$$

$$d_C = z_C \tan\left(\tan^{-1}\left(\frac{x_C}{f}\right)\right); \qquad d_C = \frac{z_C x_C}{f} \tag{3.45}$$

Similarly, algebraic deduction is given for Y-Component and real Y-coordinates w.r.t point P using cameras B (h_B) and C (h_C) are given by,

$$\frac{h_B}{y_B} = \frac{z_B}{f}; \quad h_B = \frac{z_B y_B}{f}; \quad \frac{h_C}{y_C} = \frac{z_C}{f}; \quad h_C = \frac{z_C y_C}{f} \tag{3.46}$$

Therefore, the next corollary is stated:

Corollary 3.4.2 (Divergent geometric model). *Divergent model, camera C w.r.t. camera A*

$$\mathbf{C}_{B,C}^R(t) = \begin{pmatrix} \dfrac{\left(\dfrac{l\sin\beta}{\sin\left(\frac{\pi}{2}-\frac{\beta}{2}\right)}\right)\sin\left(\frac{\pi}{2}-\theta_B+\frac{\beta}{2}\right)}{\sin(\theta_B+\theta_C-\beta)}\cos(\tan^{-1}(\frac{x_C}{f}))x_C}{f} \\[3em] \left(\dfrac{\left(\dfrac{l\sin\beta}{\sin\left(\frac{\pi}{2}-\frac{\beta}{2}\right)}\right)\sin\left(\frac{\pi}{2}-\theta_B+\frac{\beta}{2}\right)}{\sin(\theta_B+\theta_C-\beta)}\right)\cos\left(\tan^{-1}\left(\frac{x_C}{f}\right)\right) \\[3em] 0 \end{pmatrix} \tag{3.47}$$

likewise, for camera A, depth models are defined by

$$\mathbf{A}_{A,B}^R(t) = \begin{pmatrix} \dfrac{\left(\dfrac{l\sin\beta}{\sin\left(\frac{\pi}{2}-\frac{\beta}{2}\right)}\right)\sin\left(\frac{\pi}{2}-\theta_B+\frac{\beta}{2}\right)}{\sin(\theta_A+\theta_B-\beta)}\cos(\tan^{-1}(\frac{x_A}{f}))x_A}{f} \\[3em] \left(\dfrac{\left(\dfrac{l\sin\beta}{\sin\left(\frac{\pi}{2}-\frac{\beta}{2}\right)}\right)\sin\left(\frac{\pi}{2}-\theta_B+\frac{\beta}{2}\right)}{\sin(\theta_A+\theta_B-\beta)}\right)\cos\left(\tan^{-1}\left(\frac{x_A}{f}\right)\right) \\[3em] \varphi_A \end{pmatrix} \tag{3.48}$$

3.5 Estimation of robot's trajectory

3.5.1 Robot pose model

Our study involves the deployment of different non-holonomic robotic platforms[18,19]. For instance, the robot namely "Popeye" depicted at figure 3.14 is an example of a real robotic platform modelled for state estimation. At figure right-sided, the robot's kinematic parameters are shown. The robot's speed Cartesian components X and Y are defined as two asynchronous speeds (differential drive model). The position and motions of a robotic platform are modelled relaying on its kinematic restrictions, because they mathematically describe the geometry of movement of the robot in its surroundings.

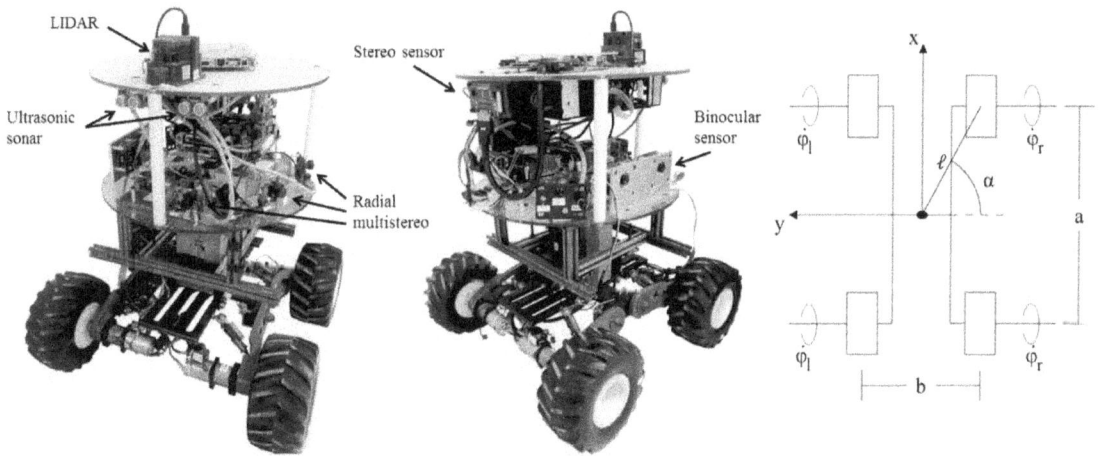

Figure 3.14: Left and center: Wheeled mobile robot "Popeye" with a sensor suite on-board; right: platform kinematic model configured as dual differential drive.

For a robot of dual asynchronous velocities deploying 4-wheel the first derivative posture model is given by,

$$\dot{x}_R(t) = v \cos \theta(t) \tag{3.49}$$

$$\dot{y}_R(t) = v \sin \theta(t) \tag{3.50}$$

and robot's orientation,

$$\dot{\theta} = \arctan \left(\frac{\dot{y}_R(t)}{\dot{x}_R(t)} \right) \tag{3.51}$$

The control vector of the system by **u** is defined with instantaneous tangential velocity, and angular speed,

$$\mathbf{u} = \begin{pmatrix} v_t \\ \omega_t \end{pmatrix} \tag{3.52}$$

The instantaneous velocity v_t model is approximated by an averaged velocity of the lateral asynchronous active wheels speeds,

$$v(t) = \frac{r}{2} \left(\dot{\varphi}_r + \dot{\varphi}_l \right) \tag{3.53}$$

The wheels' radius magnitude r is ideally the same for all wheels. The wheel instantaneous angular velocities are defined by $\dot{\varphi}_r$ (right-sided wheels), and $\dot{\varphi}_l$ (left-sided wheels). In addition, the robot's angular velocity is a direct function of the wheels rotational speeds difference. Its angular speed behaviour is consequently described by the differential magnitude of wheels speed[19]. Thus, the robot's global behaviour is given by its differential velocity is defined by,

$$\hat{v}(t) = v_r - v_l$$

The transversal differential speed component of the robot w.r.t. its geometric centre (ideally located at the centre of mass) is inferred by,

$$\omega(t) = \frac{\hat{v}(t) \cos(\alpha)}{\ell} \tag{3.54}$$

According to figure 3.14-right, ℓ is the distance between the robot's ideal centre of mass, and any wheel's contact point, with constant angle α. Algebraically substituting factors in order to state a new expression in terms of transversal and longitudinal metrics,

$$\ell = \frac{\sqrt{a^2 + b^2}}{2}$$

Then, the new equation to describe the robot's angular velocity in terms of contact point metrics, a and b, is given by,

$$\omega(t) = \frac{2br \left(\dot{\varphi}_r - \dot{\varphi}_l \right)}{a^2 + b^2} \tag{3.55}$$

Hence, the robot's pose to register non-stationary multi-sensor data is given by the expression stated in the next mathematical proposition [18],

Proposition 3.5.1 (Recursive robot's posture).

$$\begin{pmatrix} x \\ y \\ \theta \end{pmatrix} = \begin{pmatrix} x_0 \\ y_0 \\ \theta_0 \end{pmatrix} + \begin{pmatrix} \int_t \left(v_0 + \int_t \dot{v}_t dt\right) \cos\left(\theta_0 + \int_t \left(\omega_0 + \int_t \dot{\omega}_t dt\right) dt\right) dt \\ \int_t \left(v_0 + \int_t \dot{v}_t dt\right) \sin\left(\theta_0 + \int_t \left(\omega_0 + \int_t \dot{\omega}_t dt\right) dt\right) dt \\ \int_t \left(\omega_0 + \int_t \dot{\omega}_t dt\right) dt \end{pmatrix} \tag{3.56}$$

3.5.2 State estimation

The Kalman filter is a probabilistic method based on the Bayes' rule to improve the estimate of a state based on the measurements by considering uncertainty models of both, the robot and sensors. It is a recursive linear estimator based on Bayes' rule. This filter calculates successively a state based on measurements over the time, it generates a predicted state and correct that state based on the measurements [20]. The Kalman filter is typically used in tracking, location and navigation. The Kalman filter requires a kinematic model in order to predict the robot's posture. The instantaneous tangential and angular speed are given by (3.57) and (3.58).

The instantaneous velocity model is approximated by a mean of asynchronous active wheels (3.57), see please section 2.1.

$$v_t = \frac{\pi r}{\Delta_t R} (N_r + N_l) \tag{3.57}$$

and angular speed is modelled by,

$$\omega_t = \frac{4\pi b r (N_r - N_l)}{\Delta_t R (a^2 + b^2)} \tag{3.58}$$

where a, b are the transversal and longitudinal distances between the wheels' contact point. r is the wheel radius. N_r and N_l are the erncoder pulses from the right and the left wheels. R is the encoder's resolution factor. Therefore, the robot pose model is defined by,

$$f\left(\mathbf{x}_{t-1}, \mathbf{u}_t, \mathbf{w}_t\right) = \begin{pmatrix} x_t \\ y_t \\ \theta_t \end{pmatrix} = \begin{pmatrix} x_{t-1} + v_t \Delta_t \cdot \cos\left(\theta_{t-1} + \omega_t \Delta_t\right) + w_x \\ y_{t-1} + v_t \Delta_t \cdot \sin\left(\theta_{t-1} + \omega_t \Delta_t\right) + w_y \\ \theta_{t-1} + \left(\omega_t \cdot \Delta_t\right) + w_\theta \end{pmatrix} \tag{3.59}$$

Similarly, where x_{t-1}, y_{t-1}, and θ_{t-1} are the previous posture coordinates. v_t and ω_t are the tangential and the angular speeds respectively. w_x, w_y, and w_θ are the process noise, being considered statistically independent with normal distributions of mean zero, and known variance. Therefore, the process and observation models are stated by (3.60),

$$\mathbf{x}_k = \mathbf{A}\mathbf{x}_{k-1} + \mathbf{B}\mathbf{u}_{k-1} + \mathbf{w}_{k-1} \tag{3.60}$$

$$\mathbf{z}_k = \mathbf{H}\mathbf{x}_k + \mathbf{v}_k \tag{3.61}$$

where $\mathbf{x}_k$ is the state vector, $\mathbf{u}_k$ is the control vector, and $\mathbf{z}_k$ is the observation vector. $\mathbf{w}_{k-1}$ and $\mathbf{v}_k$ are process noise and measurement noise respectively. $\mathbf{A}$, $\mathbf{B}$ and $\mathbf{H}$ are transition matrices. The Kalman filter has two steps. The first step is the prediction equation (3.62) used to predict the state based on the values of previous state, and the present control input. Likewise, the covariance matrix is calculated.

$$\hat{\mathbf{x}}_k^- = \mathbf{A}\mathbf{x}_{k-1} + \mathbf{B}\mathbf{u}_k \tag{3.62}$$

$$\mathbf{P}_k^- = \mathbf{A}\mathbf{P}_{k-1}\mathbf{A}^\top + \mathbf{Q}_k \tag{3.63}$$

The second step is the correction equation (3.64), where the Kalman gain is calculated in order to correct the predicted state. Likewise, the covariance matrix is calculated for the next iteration.

$$\mathbf{K}_k = \mathbf{P}_k^- \mathbf{H}^\top \left(\mathbf{H}\mathbf{P}_k^- \mathbf{H}^\top + \mathbf{R}_k\right)^{-1} \tag{3.64}$$

$$\hat{\mathbf{x}}_k = \hat{\mathbf{x}}_k^- + \mathbf{K}_k \left(\mathbf{z}_k - \mathbf{H}\hat{\mathbf{x}}_k^-\right) \tag{3.65}$$

$$\mathbf{P}_k = \left(\mathbf{I} - \mathbf{K}_k\mathbf{H}_k\right)\mathbf{P}_k^- \tag{3.66}$$

If the process and/or observation model are non-linear (3.67), then the extended Kalman filter (EKF) is applied,

$$\mathbf{x}_k = f\left(\mathbf{x}_{k-1}, \mathbf{u}_k, \mathbf{w}_k\right) \tag{3.67}$$

$$\mathbf{z}_k = h\left(\mathbf{x}_k, \mathbf{v}_k\right) \tag{3.68}$$

where $\mathbf{x}_k$ is the state vector, $\mathbf{u}_k$ is the control vector, and $\mathbf{z}_k$ is the observation vector. Likewise, $\mathbf{w}_k$ and $\mathbf{v}_k$ are process noise and observation noise, respectively. The Taylor' expansion series make a linear approximation of the function f from the valued function and the slope of f, which is obtained from the partial derivative (3.69).

$$f'\left(u_t, x_{t-1}\right) = \frac{\partial f\left(u_t, x_{t-1}\right)}{\partial x_{t-1}} \tag{3.69}$$

Linearised models of the process and observation are shown by (3.70) and (3.71),

$$f(\mathbf{x}) = f\left(\hat{\mathbf{x}}\right) + \underbrace{f'\left(\hat{\mathbf{x}}\right)}_{=A}\left(\mathbf{x} - \hat{\mathbf{x}}\right) \tag{3.70}$$

and

$$h(\mathbf{x}) = h\left(\hat{\mathbf{x}}\right) + \underbrace{h'\left(\hat{\mathbf{x}}\right)}_{=H}\left(\mathbf{x} - \hat{\mathbf{x}}\right) \tag{3.71}$$

The steps for prediction (3.72) and correction (3.74) for EKF are adapted to the linearised models of the process and observation.

$$\hat{\mathbf{x}}_k^- = f\left(\mathbf{x}_{k-1}, \mathbf{u}_{k-1}, 0\right) \tag{3.72}$$

$$\mathbf{P}_k^- = \mathbf{A}_k \mathbf{P}_{k-1} \mathbf{A}_k^\top + \mathbf{W}_k \mathbf{Q}_{k-1} \mathbf{W}_k^\top \tag{3.73}$$

and

$$\mathbf{K}_k = \mathbf{P}_k^- \mathbf{H}_k^\top \left(\mathbf{H}_k P_k^- \mathbf{H}_k^\top + V_k \mathbf{R}_k \mathbf{V}_k^\top\right)^{-1} \tag{3.74}$$

$$\hat{\mathbf{x}}_k = \hat{\mathbf{x}}_k^- + \mathbf{K}_k \left(\mathbf{z}_k - h\left(\overline{\mathbf{x}}_k, 0\right)\right) \tag{3.75}$$

$$\mathbf{P}_k = \left(\mathbf{I} - \mathbf{K}_k \mathbf{H}_k\right) \mathbf{P}_k^- \tag{3.76}$$

The matrices and vectors used in the EKF are described next. The state vector $\mathbf{x}$ consists of the robot's posture components x, y, θ. Likewise, the control vector consists of the linear and angular velocities. The observation vector $\mathbf{z}$ consists of the values x and y obtained from odometry sensors. And, process noise vector, and measurement noise vectors are defined.

$$\mathbf{x} = \begin{pmatrix} x_t \\ y_t \\ \theta_t \end{pmatrix} ; \qquad \mathbf{u}_k = \begin{pmatrix} v \\ \omega \end{pmatrix} ; \qquad \mathbf{z}_k = \begin{pmatrix} x \\ y \end{pmatrix} ; \qquad \mathbf{w}_k = \begin{pmatrix} w_x \\ w_y \\ w_\theta \end{pmatrix} ; \qquad \mathbf{v}_k = \begin{pmatrix} v_x \\ v_y \end{pmatrix} \tag{3.77}$$

Matrix $\mathbf{A}$ is the Jacobian matrix of state function with respect to state vector.

$$\mathbf{A} = \frac{\partial f}{\partial \mathbf{x}_{k-1}} = \begin{pmatrix} \dfrac{\partial x_k}{\partial x_{k-1}} & \dfrac{\partial x_k}{\partial y_{k-1}} & \dfrac{\partial x_k}{\partial \theta_{k-1}} \\ \dfrac{\partial y_k}{\partial y_k} & \dfrac{\partial y_k}{\partial y_k} & \dfrac{\partial y_k}{\partial y_k} \\ \dfrac{\partial x_{k-1}}{\partial \theta_k} & \dfrac{\partial y_{k-1}}{\partial \theta_k} & \dfrac{\partial \theta_{k-1}}{\partial \theta_k} \\ \dfrac{\partial x_{k-1}}{\partial \theta_{k-1}} & \dfrac{\partial y_{k-1}}{\partial \theta_{k-1}} & \dfrac{\partial \theta_{k-1}}{\partial \theta_{k-1}} \end{pmatrix} \tag{3.78}$$

Modelling a linearised approach of the robot's displacement, the Jacobian is written by

$$\mathbf{A} = \begin{pmatrix} 1 & 0 & -v_t \Delta_t \sin\left(\theta_{t-1} + \omega_t \Delta_t\right) \\ 0 & 1 & v_t \Delta_t \cos\left(\theta_{t-1} + \omega_t \Delta_t\right) \\ 0 & 0 & \Delta_t \end{pmatrix} \tag{3.79}$$

Observation function represents how state vector and control vector modify the measurement.

$$\mathbf{h}\left(\mathbf{x}_{k-1}, \mathbf{u}_k, \mathbf{v}_k\right) = \begin{pmatrix} x_{k-1} + v_k \Delta_t \cdot \cos\left(\theta_{k-1} + \omega_k \Delta_t\right) + v_x \\ y_{k-1} + v_k \Delta_t \cdot \sin\left(\theta_{k-1} + \omega_k \Delta_t\right) + v_y \end{pmatrix} \tag{3.80}$$

Matrix $\mathbf{H}$ is the Jacobian matrix of observation function w.r.t. the state vector.

$$\mathbf{H} = \frac{\partial \mathbf{h}}{\partial \mathbf{x}} = \begin{pmatrix} \dfrac{\partial x_k}{\partial x_{k-1}} & \dfrac{\partial x_k}{\partial y_{k-1}} & \dfrac{\partial x_k}{\partial \theta_{k-1}} \\ \dfrac{\partial y_k}{\partial x_{k-1}} & \dfrac{\partial y_k}{\partial y_k} & \dfrac{\partial y_k}{\partial \theta_{k-1}} \end{pmatrix} \tag{3.81}$$

Modelling a linearised approach of the observation matrix, it is written by

$$\mathbf{H} = \begin{pmatrix} 1 & 0 & -v_k \Delta_t \sin\left(\theta_{k-1} + \omega_k \Delta_t\right) \\ 0 & 1 & v_k \Delta_t \cos\left(\theta_{k-1} + \omega_k \Delta_t\right) \end{pmatrix} \tag{3.82}$$

The covariance matrix $\mathbf{P}$ consists on initial values of variance for each state variable. The elements of this matrix are non-stationary.

$$\mathbf{P}_0 = \begin{pmatrix} \sigma_x^2 & 0 & 0 \\ 0 & \sigma_y^2 & 0 \\ 0 & 0 & \sigma_\theta^2 \end{pmatrix} \tag{3.83}$$

The process noise covariance matrix,

$$\mathbf{Q}_k = \begin{pmatrix} \sigma_x^2 & 0 & 0 \\ 0 & \sigma_y^2 & 0 \\ 0 & 0 & \sigma_\theta^2 \end{pmatrix} \tag{3.84}$$

The observation noise covariance matrix,

$$\mathbf{R}_k = \begin{pmatrix} \sigma_x^2 & 0 \\ 0 & \sigma_y^2 \end{pmatrix} \tag{3.85}$$

Although, both matrices are time-varying, however they are consider constant. Likewise, $\mathbf{W}_k$ is the Jacobian matrix of process model w.r.t. the process noise vector.

$$\mathbf{W}_k = \frac{\partial f}{\partial \mathbf{w}_k} = \begin{pmatrix} \frac{\partial x_k}{\partial w_x} & \frac{\partial x_k}{\partial w_y} & \frac{\partial x_k}{\partial w_\theta} \\ \frac{\partial y_k}{\partial w_x} & \frac{\partial y_k}{\partial w_y} & \frac{\partial y_k}{\partial w_\theta} \\ \frac{\partial \theta_k}{\partial w_x} & \frac{\partial \theta_k}{\partial w_y} & \frac{\partial \theta_k}{\partial w_\theta} \end{pmatrix} \tag{3.86}$$

V_k is the Jacobian matrix of observation model w.r.t. the measurement noise vector.

$$\mathbf{V}_k = \frac{\partial h}{\partial \mathbf{v}_k} = \begin{pmatrix} \frac{\partial x_k}{\partial v_x} & \frac{\partial x_k}{\partial v_y} \\ \frac{\partial y_k}{\partial v_x} & \frac{\partial y_k}{\partial v_y} \end{pmatrix} \tag{3.87}$$

Uncertainty of x and y were obtained experimentally, consisting of moving the robot linearly. The error was calculated from the difference of ideal and the real ending positions. From the experiment, $\mu_x = 0.457$ and $\mu_y = 0.569$ were measured. The Extended Kalman filter was implemented in the mobile robot to track its location indoor with experimental results shown in figure 3.15. Values of the covariance matrix $\mathbf{P}$ overtime are depicted.

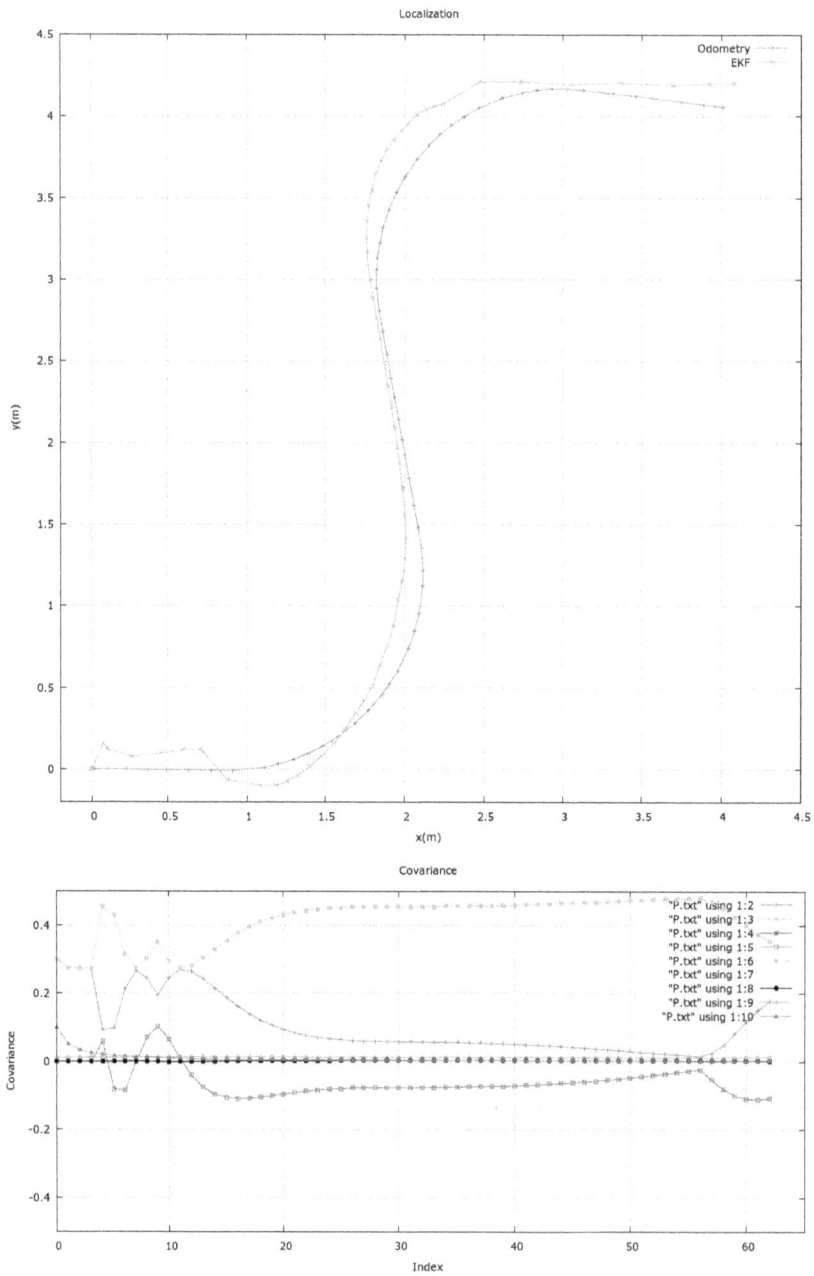

Figure 3.15: Sensor observation and Kalman estimation of the robot's position (above). Numeric calculated values of the covariance matrix **P** during experiment (below).

Bibliography

[1] Rivero-Juarez J., Martinez-Garcia E.A., Torres-Mendez A., Mohan R.E., *3D Heterogeneous Multi-sensor Global Registration*, Procedia Engineering, vol.64, pp.1552–1561, 2013, doi:10.1016/j.proeng.2013.09.237.

[2] Khaleghi, B., Khamis, A., Karray, F.O., Razavi SnN., *Multisensor data fusion: A review of the state-of-the-art*, Information Fusion, 2012.

[3] Luo, R.C., *Multisensor Fusion and Integration: A Review on Approaches and Its Applications in Mechatronics*, IEEE Transactions on Industrial Informatics, vol.8(1), pp. 49-60, 2012.

[4] Mitchell, H. B., *Multi-sensor Data Fusion: An introduction*, Springer. 2007.

[5] Martínez-García, E. A., Yoshida, T., Ohya, A. and Yuta, S., *Multi-robot Communication Architecture for Human-guiding*, Journal of Engineering Manufacture Part B, Vol. 219(1), pp. 183-190, 2005.

[6] Berná-Martínez, J., Maciá-Pérez, F., Ramos-Morillo, H. and Gilart-Iglesias, V., *Distributed Robotic Architecture based on Smart Services*, IEEE Intl. Conf. on Industrial Informatics, pp. 480-485, 2006.

[7] Siciliano, B. and Khatib, O., *Springer Handbook of Robotics*, Springer, 2008.

[8] Songmin J., Wei C., Xiuzhi L., Jinhui F., Jinbo S., *Mobile robot Bayesian map building based on laser ranging and stereovision*, IEEE Intl. Conf. on Computer Science and Automation Engineering, vol.3, pp.308–312, 2011.

[9] Carlone, L. and Bona, B., *On registration of uncertain three-dimensional vectors with application to robotics*, In Proceedings of the IFAC World Congress, 2011.

[10] Calafiore, G. and Bona, B., *Constrained optimal fitting of three-dimensional vector patterns*, IEEE Transactions on Robotics and Automation, 14(5):838-844, 1998.

[11] Cyganek B., Siebert J.P., *An introduction to 2D computer vision techniques and algorithms*, Wiley, 2001, isbn 978-0-470-01704-3.

[12] Ma Y., Soatto S., Košecká J., Sastry S.S., *An invitation to 3-S vision*, Springer, 2010, isbn 978-1-4419-1846-8.

[13] Fauregas O., *Three-dimensionadl computer vision: a geometric approach*, MIT, 1993, isbn 0-262-06158-9.

[14] Shapiro, L.G. and Stockman, G.C., *Computer Vision*, Prentice Hall, 2001.

[15] Duda R.O., Hart P.E., Stork D.G., *Pattern Classification*, 2nd Edition, Wiley-Interscience 2001, isbn 0-471-05669-3.

[16] Moghadam, P.; Wijesoma, W.S.; Dong J. F., *Improving path planning and mapping based on stereo vision and lidar*, 10th Intl. Conf. on Control, Automation, Robotics and Vision, pp.384–389, Dec. 2008.

[17] Thompson, S., Kagami, S., *Stereo vision and sonar sensor based view registration for 2.5 dimensional map generation*, IEEE/RSJ Intl. Conf. on Intelligent Robots and Systems, vol.4, pp.3444–3449, 2004.

[18] Martínez-García, E. A., Torres-Córdoba, R., *4WD skid-steer trajectory control of a rover with spring-based suspension*, Intelligent Robotics and Applications, Springer, LNCS, Vol. 6425, 2010.

[19] Lerín, E., Martínez-García, E. A., Mohan, R.E., *Design of an All-Terrain Autonomous Holonomic Rover*, 8th Intl. Conf. on Intelligent Unmanned Systems, 2012.

[20] Welch, G. and Gary, Bishop, G., *An Introduction to the Kalman Filter*, Technical Report, University of North Carolina at Chapel Hill, 1995.

Chapter 4

VISUAL INVARIANT DESCRIPTORS

Edgar A. Martínez-García

Laboratorio de Robótica, Institute of Engineering and Technology
Universidad Autónoma de Ciudad Juárez, Mexico.

Visual landmark tracking represents a major problem due to occlusion, illumination variations and affine transformations between subsequent images. A desired goal for mobile robot applications is to increase stability during invariants detection in order to minimize tracking errors. Instead of using expensive and sophisticated sensor devices, a trend in mobile robotics is using cheaper, passive, and widely available sensors operated by complex algorithms. Visual sensors, for instance, involve several perception problems such as infeasible lighting and uncontrolled illumination conditions, which prevail in mobile robot scenarios. Invariant descriptors provide relative steady geometric parameters of regions for recognition problems. Regional descriptors must be highly accurate in terms of their locations to associate correctly sets of landmarks coming from different regions, and accomplish precise geometric triangulation between two successive image frames[1]. Two issues must highly be reliable, the matching algorithm and landmarks position accuracy. If invariant descriptors locations are stable, then projective geometry algorithms will provide highly accurate robot displacement calculations. Reason of failure of the matching algorithm may be caused by missing landmarks due to lighting noise and/or occlusion; even if landmarks are matched correctly, there exists the possibility of variations of descriptors' position, yet if sensed from the same observing location. The approach is based on a feature-space analysis, although different approaches are reported[2].

Unlike other approaches [2-5], this chapter focuses on solving the problem of feature descriptors instability while tracking landmarks unlike other approaches [6]. In order to understand how the instability descriptors problem evolves in different moments and invariant moments, we carried out a comparative study on different algorithms to learn how to increase a steady state. Given the value of stability in the MSER algorithm, we can minimize errors specially where landmarks observation yields false positives and negatives.

4.1 Feature points detection

A quantitative analysis of stability tracking of invariant descriptors is presented. Two *feature extraction* algorithms are compared: the Harris Corner Detection and the Fast Corner Detection. Corner detectors are the introductory algorithmic processes of almost any regional invariant descriptor algorithm. Two of the most popular corner detectors to compare are the Harris Corner Detection (HCD) [7], and the Fast Corner Detection (FCD) algorithm [8]. The HCD has been a popular interest point detector due to its strong invariance to rotation, scale, illumination variation, and image noise. A predecessor of the HCD was presented by Moravec [7]. The HCD is based on the local auto-correlation function of a signal; where the local auto-correlation function measures the local changes of the signal with patches shifted by a small amount in different directions [9]. The discreteness refers to the shifting of the patches. Given a shift $(\Delta x, \Delta y)$ and a point $\mathbf{p}(x, y)$, the auto-correlation function is defined by

$$c(x, y) = \sum_{W} [\mathbf{I}(x_i, y_i) - I(x_i + \Delta x, y_i + \Delta y)]^2, \tag{4.1}$$

where $\mathbf{I}$ denotes the image function and (x_i, y_i) are the points in the window W centred on (x, y). For the case of the FCD, it classifies a pixel $\mathbf{p}$ as a corner, if there exists a set of n neighbouring pixels which intensities are all brighter than the intensity of the candidate pixel I_p plus a threshold t_h, otherwise, all pixels are darker than $I_p t$.

With similar setting parameter for the FCD and the HCD, the FCD usually detects a greater number of feature points than the HCD. Actually, the number of features are nearly duplicated as detector's threshold decreases at constant rate. FCD yields more density of points dispersion than the HCD; while the HCD detects points basically without redundancy (non cumulative corners around the same region) due to its statistical mechanism based on associating a regional

Figure 4.1: Left: FCD features detection. Right: HCD features detection.

covariance and its central value (a corner). HCD has a more enhanced level of suppression detection than FCD. However, the FCD detects clouds of feature points specially within a morphologically homogeneous regions. HCD is more sensitive to regional changes than FCD, being less immune than FCD and with an increased number of false positives/negatives. The HCD resulted faster than FCD in our on-board computers configuration regardless similar thresholds configuration in both algorithms (figure 4.2), but HCD yields to a more globally scattered dispersion of feature points, although less redundant than FCD (see figure 4.1). Figure 4.2 depicts a comparative plot of speed computation between FCD and HCD, using the same thresholds.

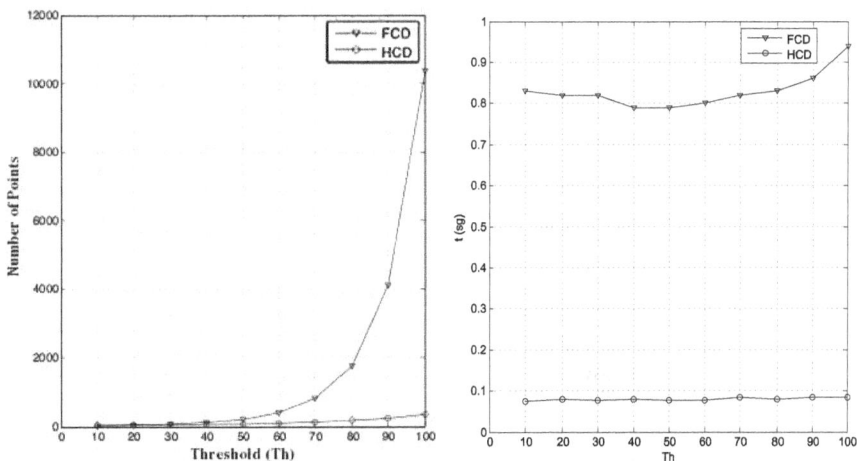

Figure 4.2: Left: FCD/HCD number of keypoints; right: computing time vs. selected Th.

It was established an average number of features, in our examples $n = 20$. A qualitative analysis is presented in table 4.1; the HCD/FCD reliabilities when facing faults are presented. In general, both corner detection algorithms (CDAs) have similar reliability, specially without

Table 4.1: Corners detection reliability, 20 features.

Time Frame	F. Positives		F. Negatives	
	HCD	FCD	HCD	FCD
t_0	5	3	3	2
t_1	4	2	3	3
t_2	2	1	2	3
t_3	3	2	4	3
t_4	3	2	3	2
t_5	4	2	4	2
t_6	3	3	2	3
t_7	4	2	4	3
t_8	5	2	3	4
t_9	5	3	4	3
t_{10}	4	4	4	2
Reliability	81%	88%	84%	86%

previous image enhancement.The resulting feature points from the CDAs are then analysed by a local invariant detector[10]. The CDA's parameters are adjusted accordingly before applying the invariant detection algorithm.

4.2 Stability analysis of invariant descriptors

A comparative study of three known *invariants* algorithms (SIFT, MSER, Quick SHIFT) is carried out to track multiple regional descriptors correlated among consecutive video frames. These algorithms were compared on faults tolerance and computational complexity. Feature-points usually lie on high-contrast regions of the image, such as edges. An important charac- teristic of these feature-points is that the relative positions between them in the original scene should not change from one image to another. Affine invariant feature descriptors are nor- mally computed by sampling the original (grey-scale) image in an invariant frame defined from each detected feature. Experiments were carried out with three popular invariants detection algorithms: the SIFT[11], the Quick SHIFT[12], and the MSER[13]. The SIFT algorithm extracts distinctive features from images in gray scale. It recognizes the same trait among different views of objects[11], extracted features are invariant to image scale and rotation (figure 4.3).

Figure 4.3: Visual local invariants: QShift, SIFT, and MSER (left to right).

The Quick SHIFT algorithm implements a mode seeking algorithm to form a tree of links to divide an image into a set of super-pixels around the nearest neighbour. This algorithm may not be seen as a invariant detection algorithm, but we found that centroids associated to each superpixel region may represent the invariants at different frames as it increases an estimate of the density. It increases an estimate of the density. With respect to the MSER method, the regions are defined solely by an extreme property of the intensity function in the region, and on its outer boundary (see bottom image in figure 4.3). We may refer to the pixels below a threshold, as black; and to those above or equal, as white. Throughout a sequence of threshold images $I(t)$ and $I(t-1)$, with corresponding threshold t_h, we would see first a white image. Subsequently, black spots corresponding with local intensity minima will appear and grow. At some point, regions corresponding with two local minima will end merging. Thus, the last image will be black. The set of all connected components of all frames are the set of all maximal regions; minimal regions could be obtained by inverting the intensity of I and running the same process.

In addition, computational complexity for all invariant algorithms is depicted in figure 4.4, whereas in Table 4.2, a faults tolerance study is summarized. The proposed EMSER algorithm has a computational complexity close to MSER (figure 4.4) but a superior performance with respect to the other algorithms in terms of reliability. Stability of invariant descriptors in this context pursuits to reduce false positives/negatives during detection.

By applying invariants, given the degree of stability in the MSER algorithm, we can minimize errors based on analysis in feature-space instead of the orthogonal sensor image, although different approaches are reported [2]. Errors can be minimized especially where a landmark might significantly be similar to others, causing confusion in the robot. Based on those findings, in-

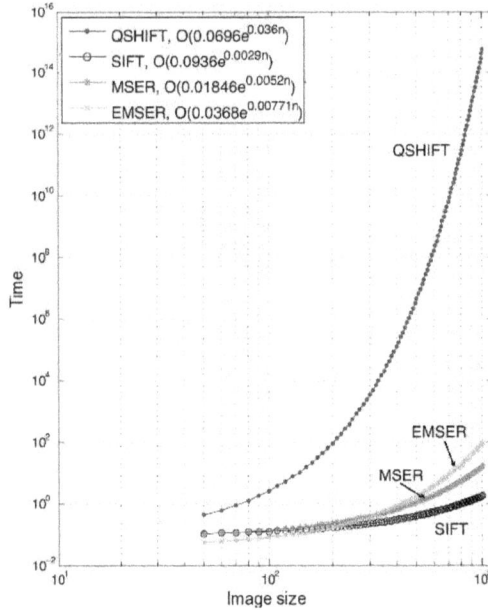

Figure 4.4: Invariant algorithms arithmetic-logic complexity ($\mathbf{I}_{n \times n}$).

creasing stability of invariant descriptors is desirable, which in turn means to reduce false positive/negatives during detection. By a preprocessing algorithm, selecting natural environmental features can be accomplished automatically. These landmarks are then transformed into invariant regional descriptors.

4.3   Image preprocessing

Image preprocessing is a set of algorithms to improve the quality, and the number of salient features-points detected. The proposed methods is described as the following order.

1. Automatic image contrast enhancement.

2. Image sharpening.

   (a) Adaptive edge detection.

   (b) Correlation-based isotropic filtering kernel.

   (c) Sharpening.

Table 4.2: Faults tolerance, 20 descriptors, 11 frames.

Frame	False Positives			False Negatives		
time	sift	mser	emser	sift	mser	emser
t_0	12	4	3	8	5	3
t_1	11	6	2	6	4	2
t_2	10	7	2	7	5	1
t_3	12	7	3	8	5	3
t_4	12	8	1	7	4	2
t_5	11	6	1	6	5	3
t_6	10	7	3	6	6	4
t_7	12	8	2	8	5	3
t_8	9	9	4	6	4	2
t_9	8	8	2	6	4	3
t_{10}	10	7	3	7	4	1
	47 %	65 %	88 %	66 %	77 %	88 %

4.3.1 Automatic contrast enhancement

Because of outdoors environmental scene structure has much greater dynamic range, lighting conditions vary enormously for a given visual sensor on-board the robot. In addition, low-contrast images may result from poor illumination, lack of dynamic range in the imaging sensor, or wrong setting of a lens aperture during image acquisition. Contrast enhancement increases the total contrast of an image by making light color lighter, and dark colours darker simultaneously in image $\mathbf{I}_C = histeq(\mathbf{I}_G)$, where $\mathbf{I}_G$ is the input image and $\mathbf{I}_C$ is the enhanced contrast image. By improving the image contrast, saliency and definition of regions corner-like are enhanced. We apply histogram equalization to the acquired images in order to recover the lost contrast by remapping the brightness values or more evenly distribute its brightness values. An example of contrast enhancement is shown in figure 4.5-A).

The histogram $h(r_k) = n_k$, r_k: the level of grey, and n_k the num. of pixels. Thus, the normalization for $k = 0, 1, \ldots, L - 1$

$$p(r_k) = \frac{n_k}{n_t} \tag{4.2}$$

then, the transformation $s = T(r)$ for $0 \leq r \leq 1$,

$$p_s(s) = pr(r)|d_r/d_s| \tag{4.3}$$

and

$$s = T(r) = \int_0^r P_r(w) r m dw \tag{4.4}$$

4.3.2 Image sharpening

Adaptive edge detection The sharpening process is to detect the edges of the scene in the enhanced contrast image. The edges are line regions used to increase the numeric value of corners points located at lines-crossing. For practicability we deployed the Canny algorithm because we can adjust the maximal suppression factor (edge thickness). Only most salient feature-points are extracted by automatically selecting corners that mutually lie collinear over crossing edges (vertical-horizontal cross points). The resulting edge image I_E is deployed subsequently (see figure 4.5-B)).

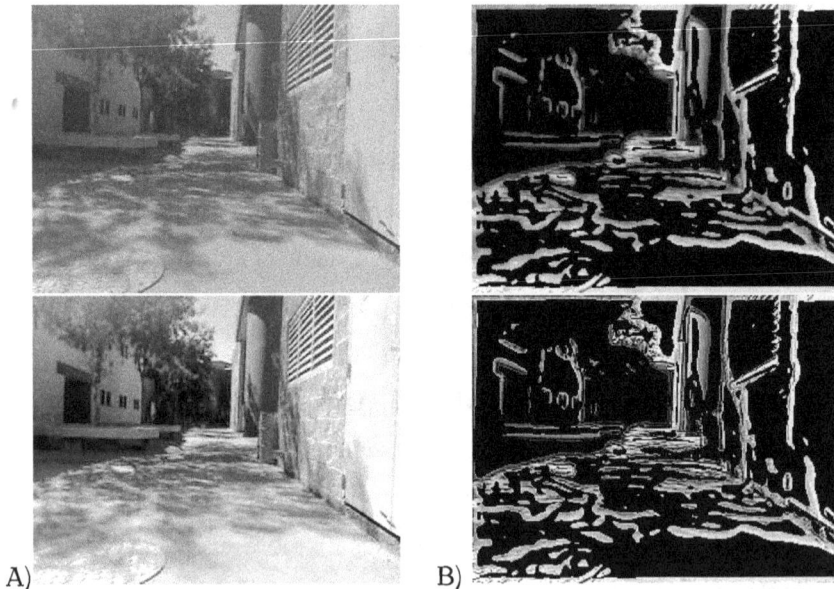

Figure 4.5: a) raw and contrast enhancement images; b) Edge detection.

Correlation-based isotropic filtering kernel Under natural light conditions, having a good definition of corners is a critical issue for the sake of key-features stability detection. We establish an non-sharp kernel through the negative of a Laplace filter. The Laplacian is a 2D isotropic measure of the 2nd spatial derivative, which in an image processing context is equally applied in all directions. The purpose of this is to highlight regions of rapid intensity change particularly for edge detection. The Laplacian ∇^2 or $\mathbf{H}$ of an image is given by:

$$\nabla^2 = \frac{\partial^2 \mathbf{I}_C}{\partial x^2} + \frac{\partial^2 \mathbf{I}_C}{\partial y^2}, \tag{4.5}$$

where $\mathbf{I}_C$ is the enhanced contrast image. This is calculated by using a convolution filter which approximates a second derivative kernel in the definition of Laplace. Thus, it can be calculated by using standard convolution methods. Nevertheless, by approximating a second derivative measurement on the image, it results very sensitive to noise. The Laplacian is often applied to an image that has first been smoothed approaching a Gauss behaviour filter to reduce noise sensitivity.

As the convolution operation is associative, we convolve the Gaussian smoothing filter with the Laplace filter first, and then convolve this hybrid filter with the image to achieve the required result, and only one convolution needs to be performed at run-time on the image.

$$LoG(x,y) = -\frac{1}{\pi\sigma^2}\left(1 - \frac{c^2 + r^2}{2\sigma^2}\right) e^{-\frac{c^2 + r^2}{2\sigma^2}}. \tag{4.6}$$

The 2D Laplacian of Gaussian function centered on zero has discrete form in an image processing context of the form:

$$\mathbf{H} = \frac{1}{\alpha + 1} \cdot \begin{pmatrix} -\alpha & \alpha - a & -\alpha \\ \alpha - a & \alpha - b & \alpha - a \\ -\alpha & \alpha - a & -\alpha \end{pmatrix}, \tag{4.7}$$

where factor $0 \le \alpha \le 1$, for the results shown in this manuscript, was set to $\alpha = \{0, 0.1\}$ as the maximal sharpening factor, under natural lighting conditions.

Sharpening The sharpening process is enhanced by introducing an adaptive function $\beta(\mathbf{I}_E)$ that highlights only the most prominent edges (referring to $\mathbf{I}_E$ where edges had been detected). All edges greater than 50% of their intensity values in $\mathbf{I}_E$ are filtered. Likely, most prominent edges associate crossing of vertical and horizontal edges (best corners featuring the scene). In the vicinity of a change in intensity, the LoG response will be positive on the darker side, and negative on the lighter side. A reasonably sharp edge between two regions of uniform but different intensities, the LoG response will be zero at a long distance from the edge; positive just to one side of the edge; negative just to the other side of the edge; zero at some point in between, on the edge itself. The enhancement function $\beta(\mathbf{I}_E)$ defined by,

$$\beta(\mathbf{I}_E(i,j)) = ((i \cdot j)^{-1} \sum_i \sum_j \mathbf{I}_E(i,j)) \leq \mathbf{I}_E(i,j). \tag{4.8}$$

In addition,

$$\mathbf{J}_E = \begin{cases} L, & \beta(\mathbf{I}_E) \cdot \mathbf{I}_C(i,j) \\ 0, & otherwise \end{cases} \tag{4.9}$$

where L is the maximum grey level.

Corollary 4.3.1. *The sharpen image $\mathbf{I}_S$ is given by the following kernel,*

$$\mathbf{I}_S(t) = \mathbf{I}_S(t-1) + \alpha \cdot \beta(\mathbf{I}_E) \cdot (\mathbf{I}_E(t-1) - \mathbf{H} \otimes \mathbf{J}_E). \tag{4.10}$$

The unsharp image $\mathbf{I}_S$ enhances edges via a procedure which subtracts an unsharp version of an image from the contrast-enhanced image.

4.4   Feature-points (corners) detection

The sharpened image $\mathbf{I}_S$ is now the input image to any corner-detection algorithm. Although, we analysed both the Fast Corner Detector (FCD), and the Harris Corner Detector (HCD), depicted results shown in figure 4.6 correspond to FCD, applied to a raw image (above), as well as to its contrast-sharpening enhanced version (below). The preprocessed version shows prominent results with respect the original raw image. In addition, figure 4.7-B) shows the resulting salient feature points on the sharpened image by *corner_detector*($\mathbf{I}_S$).

4.5   Adaptive dilation factor

The dilatation factor d_f is a concept introduced in this chapter. The factor d_f is used to easy the ability of the MSER algorithm to create connected components. The dilatation factor allow

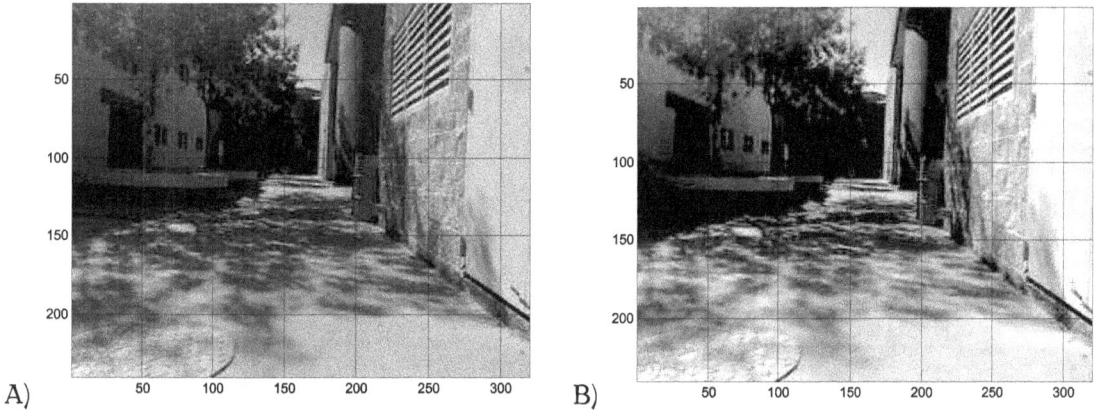

Figure 4.6: Feature-points detection. a) raw image; b) enhanced image (contrast and sharpening).

us to adaptively (based on the dispersion of key-features) dilate at a rate of the factor d_f,

$$d_f = \sqrt[2]{\frac{100}{i_n \times j_m} \sum_i (\mathbf{p} - \bar{\mu}) \cdot \mathbf{s}^{-1}}, \qquad (4.11)$$

where $\mathbf{s}$ is the vector of statistical variances of the key-feature points; $\mathbf{p}$ is the vector location of a key-feature point; and $\bar{\mu}$ is the mean-vector of key-feature points. As a result, figure 4.7 depicts a unique map of regions formed by the set of key-points dilated at a factor d_f. The original information data of the scene is no longer required, and MSER is applied in a very reduced computational search space, called *stable binary regions*.

4.6 Descriptors in Binary Stable Regions

The Binary Stable Region is a binary image compounded only of multiple dilated circular descriptors that by overlapped areas will form new regions featuring the scene. Thus, the real image scene is no longer required because binary regions are the scene descriptor themselves, with a reduced search space. Only binary regions are now uniquely predominant as candidates for analysis by the MSER algorithm to extract a small number of covariant regions (stable components) highlighted by circles to describe local invariants.

Figure 4.7 depicts some experimental results in the binary stable regions, as well as descriptors projected over the raw image. Further study can be found in [13], which is a precedent theoretical work that explain how we arrive with our algorithm to improve the invariant's

Figure 4.7:   a) key-points at cross-edges, and adaptively dilated feature-points; b) Maximally stable regions detection (bounded by circles).

Cartesian stability for robots localization. Our endeavour to alleviate the instability behaviour is not by applying MSER directly to the sensor observation (image), but only to their binary stable regions (figure 4.7 binary stable regions). We call this strategy *enhanced* MSER (EMSER). In addition, it is worth mentioning that these results improved reliability in data association.

4.7   Optical flow data association

The implemented optical flow algorithm is based on a gradient method[14]. It correlates visual environmental landmarks, which as the robot moves, such natural landmarks or key-features are displaced overtime, as depicted in figure 4.8-A), and those key-points are used as visual feedback. Optical flow provides the apparent motion in a visual scene I caused by the relative motion between the inertial frame of the robot and the scene's landmarks[15] as depicted in figure 4.8-B.

$$\mathbf{I}(x, y, \theta) = \mathbf{I}(x + \Delta x, y + \Delta y, t + \Delta t). \tag{4.12}$$

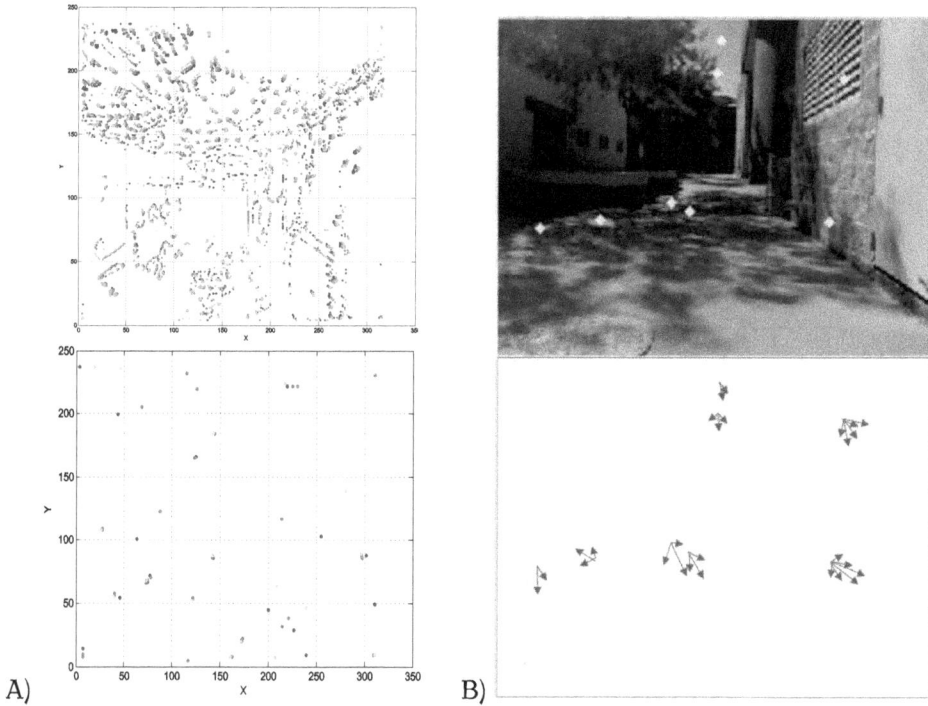

Figure 4.8: a) keypoints overlapped in time; b) optical flow from invariant keypoints.

By assuming relatively small robot's displacements (sensor suite on-board), the image constraints at $I(x, y, t)$, and changes overtime Δx and Δy in coordinates XY in time Δt as defined by,

$$\frac{\partial I}{\partial x} \frac{\Delta x}{\Delta t} + \frac{\partial I}{\partial y} \frac{\Delta y}{\Delta t} + \frac{\partial I}{\partial t} \frac{\Delta t}{\Delta t} = 0. \tag{4.13}$$

The partial derivatives of the pixels of interest will result as the optical flow of pixels (x^f, y^f) by

$$\frac{\partial I}{\partial x} \dot{x}^f + \frac{\partial I}{\partial y} \dot{y}^f + \frac{\partial I}{\partial t} = 0, \tag{4.14}$$

where $\dot{x}^f$ and $\dot{y}^f$ are the velocity component of each invariant descriptors detected in images $I(t)$ and $I(t-1)$. Both speed components allow to know the angle and magnitude of the velocity vector (figure 4.8-b)). For every key-point detected within the vision sensor field of view, its apparent motion is analysed by their optical flow feature vector. The optical flow feature vector is the input vector argument for the data association process, defined next.

$$
\mathbf{f} = \begin{pmatrix} \dot{x}^f \\ \dot{y}^f \\ \theta^f \\ v^f \\ c \\ r \end{pmatrix} = \begin{pmatrix} (\partial I/\partial x)\dot{x}^f \\ (\partial I/\partial y)\dot{y}^f \\ \arctan(\dot{y}^f/\dot{x}^f) \\ \sqrt[2]{(\dot{x}^f)^2 + (\dot{y}^f)^2} \\ c \\ r \end{pmatrix}, \tag{4.15}
$$

where θ^f is the direction of optic flow vector, v^f is the velocity vector magnitude; c and r are the column and row respectively of the invariant feature within the focal plane. With a group of optical flow feature vectors $\mathbf{f}$, the data association function $\delta_i(t)$ in eq.(4.16) correlates all invariants velocity.

Proposition 4.7.1. *The argument that minimizes the function $\delta_i(t)$ is the most similar feature vector within a multidimensional space,*

$$
\arg \min_{\mathbf{f}_j(t)} \delta_i^f(t) = \| \mathbf{f}_i(t) - (\mathbf{f}_j(t))_{j=1}^k \|. \tag{4.16}
$$

Data association of the kth landmark is illustrated in figure 4.9. Dots and empty-circles that overlap together indicate successful landmarks data association of their inertial frames w.r.t. to the robot's displacement. The detected invariant descriptors are extracted from the scene image and placed onto an empty image frame (white background) at same spatial coordinates. Thus, the optical flow process is assured to be performed only to detected invariants.

Table 4.3) show results on data association between two successive images at robot speeds of about 0.3m/s. The original data association index as well as the closer descriptor are recorded, and a specific threshold is used to determine the number of matching features. This can be seen in the sequences of images in figure 4.9.

4.8   Local invariants consideration for robot motion

For the sake of visual odometry, the locations of these descriptors are geometrically triangulated to infer the robot's displacement, as illustrated by figure 4.10. Detecting each descriptor's

Frame	Correlated
t_0	10
t_1	10
t_2	10
t_3	9
t_4	10
t_5	10
t_6	10
t_7	10
t_8	8
t_9	9
t_{10}	9
Reliability	95%

Table 4.3:   Data Association, 10 landmarks, 11 frames.

key-point with no perturbation in the same scene's configuration regional location is desired; otherwise robot's positioning is not accurate. Descriptors stability is important for numerous mobile robot applications with visual feedback (i.e. positioning control, trajectory tracking, visual odometry, robot pose estimation). Descriptors are advantageously used as natural environmental landmarks by geometrically triangulating their locations.

Detection of key-points with no perturbations (same regional location of the scene overtime) is a desired condition [5].

Proposition 4.8.1. *The key-point coordinates* $\dot{\mathbf{x}} = f(x)$, $\mathbf{x} = (r,c)^T$ *for* $f : D \to \Re^n$ *is stable, if the equilibrium point* $\mathbf{x} = \mathbf{x}_e$ *($\mathbf{x}_e$ ideal location), and*

$$f(\mathbf{x}_e) = 0; \quad \|\mathbf{x}_0 - \mathbf{x}_e\| < \delta$$

and

$$\lim_{t \to \infty} \mathbf{x}(t) = \mathbf{x}_e$$

such that,

$$\delta = (\Delta c^2 + \Delta r^2)^{1/2}$$

in which δ is a marginal error due to instability of the visual perception process.

Figure 4.9: Data association results with the proposed method.

The key-points location is ideally inferred, if the robot's motion is robustly controlled through the control vector $\mathbf{u} = (v, \omega)^T$, where v_t and ω_t are the robot's instantaneous linear and yaw velocities. There for, estimation of robot's displacement w.r.t. the visual landmark is modelled by the linearised state equation,

$$\dot{\mathbf{x}} = \mathbf{Ax} + \mathbf{Bu} \qquad (4.17)$$

Thus, $x = x_e$ for $\dot{x} = f(x) = \mathbf{Ax}$, which is asymptotically stable, it is both stable and convergent. The model for the descriptor location $\mathbf{x} = \mathbf{x}_0 + g(\dot{x}, \dot{y}, t)$ arises from a kinematic robot's motion model $g(\cdot)$, which is not treated in this chapter because of its extensive nature and lack of space.

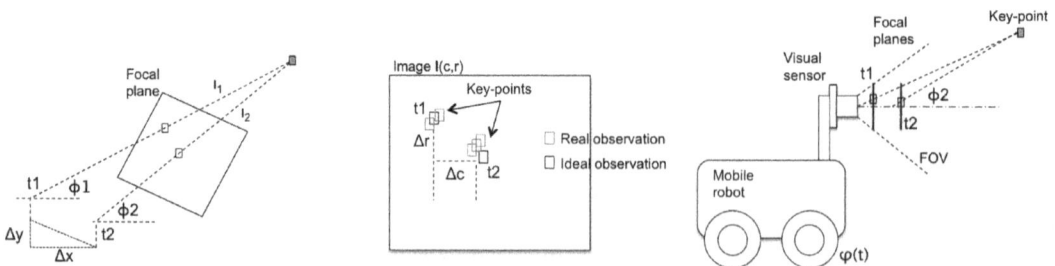

Figure 4.10: Stability invariants for robot motion.

However, expected key-point in focal plane is inferred by

$$
\begin{pmatrix} \Delta c \\ \Delta r \end{pmatrix} = f \begin{pmatrix} \tan(\arccos(\frac{x_1}{l_1})) - \tan(\arccos(\frac{x_2}{l_2})) \\ \tan(\arccos(\frac{y_1}{k_1})) - \tan(\arccos(\frac{y_2}{k_2})) \end{pmatrix}
\tag{4.18}
$$

where f is the focal length of the visual sensor; $(\Delta c, \Delta r)$ are the expected location of the descriptor in focal plane according to the actual robot's motion; (x_i, y_i) is the Euclidian position of the descriptor; l_i and k_i are the line distances between robot and descriptor in Cartesian space (see figure 4.10). Perturbations will affect the stable equilibrium point if overpassing the magnitude of δ.

In addition, x_i and y_i (robot position w.r.t. key-feature) arise from the deterministic robot's motion model that defines its small displacement $\Delta \xi(t)$

$$
\Delta \xi(t) = \begin{pmatrix} x_2 \\ y_2 \end{pmatrix} - \int_{t_1}^{t_2} v(t) \begin{pmatrix} \cos(\int_t \omega(t)dt) \\ \sin(\int_t \omega(t)dt) \end{pmatrix} dt
\tag{4.19}
$$

The input control vector $\mathbf{u}(t) = (v, \omega)^T$ defines the robot's linear and angular velocities for any k number of wheels with actual speed $\dot{\varphi}_i(t)$ of constant radius r,

$$
v(t) = \frac{r}{k} \sum_i \dot{\varphi}_i(t); \quad \omega(t) - v(t)/h
\tag{4.20}
$$

where h is the robot's body radius (width from geometric centre to any wheel's contact point). For an ideal model, we might say that Δc, Δr are deterministic variables in terms of $\Delta \xi(v(t), \omega(t), g(x, y), t)$.

Bibliography

[1] López G., Guerrero J.J., Sagués C., *Visual control of vehicles using two-view geometry*, Mechatronics, Vol. 20, pp. 315-325, 2010, doi: 10.1016/j.mechatronics.2010.01.005.

[2] Hayet J.B., Lerasle F., Devy M., *A visual landmark framework for mobile robot navigation*, Image and vision computing, Vol. 25, pp. 1341-1351, 2007, doi: 10.1016/j.imavis.2006.08.006.

[3] Ramisa A., Tapus A., Aldavert D., Toledo R., Lopez R., *Robust vision-based robot localization using combinations of local feature region detectors*, Autonomous Robots, Vol. 27(4), pp. 373-385, 2009, doi: 10.1007/s10514-009-9136-9.

[4] Kunze L., Lingemann K., Nuchter A., Hertzberg J., *Salient visual features to help close the loop in 6D SLAM*, 5th Intl. Conf. on Computer Vision Systems, Bielefeld Germany, 2007.

[5] Lowe D., Little J., *Vision-based mobile robot localization and mapping using scale-invariant features*, IEEE ICRA, Vol. 2, pp. 2051-2058, 2001.

[6] Bayramoğlu E., Andersen N.A., Poulsen N.K.,Andersen J.C., Ravn O., *Mobile robot navigation in a corridor using visual odometry*, Intl. Conf. on Adv. Robotics, pp. 1-6, 2009.

[7] Harris C., Stephens M., *A combined corner and edge detector*, Proc. 4th Alvey Vision Conf., pp. 147-151, 1988, doi: 10.5244/C.2.23.

[8] Trajkovic M., Hedley M., *Fast corner detection*, Image and Vision Computing, 16(2), pp. 75–87, 1998, doi: dx.doi.org10.1016S0262-8856(97)00056-5.

[9] Košecká J., Li F., *Vision based topological Markov localization*, IEEE ICRA, pp. 1481-1486, 2004, doi: dx.doi.org10.1109robot.2004.1308033.

[10] Flusser J., Suk T., Zitová B., *Moments and moment invariant in pattern recognition*, Wiley, 2009, doi: 10.1002/9780470684757.

[11] Lowe, D., *Distintive image features from scale invariant keypoints*, University of BC Canada, 2004.

[12] Vedaldi A., Soatto J., *Quick shift and kernel methods for mode seeking*, In Proc. of the European Conf. on Computer Vision, 2008, doi: 10.1007/978-3-540-88693-8_52.

[13] Fellow R.K., Zhang C., Bronstein A., Bronstein M., *Are MSER features really interesting?*, IEEE Trans. on Patt. Analysis and Mach, Intel, 2011.

[14] Mccarthy C., Barnes N., *Performance of Optical Flow Techniques for Indoor Navigation with a Mobile Robot*, IEEE Intl. Conf. on Robotics & Automation, pp. 5093-5098, 2004, doi: dx.doi.org/10.1109/robot.2004.1302525.

[15] Martínez-García E.A., Torres-Mendez L.A., Mohan R.E., *Multi-Legged Robot Dynamics Navigation Model with Optical Flow*, Intl. Journal of Intelligent Unmanned Systems, Vol. 2, Iss. 2, pp. 121-139, 2014, Emerald, doi: 10.1108/IJIUS-04-2014-0003.

Part II

Robot Navigation and Planning

Chapter 5

MULTI-LEGGED OBSTACLE AVOIDANCE

Cesar García Sariñaga and Edgar A. Martínez García

Laboratorio de Robótica, Institute of Engineering and Technology
Universidad Autónoma de Ciudad Juárez, Mexico.

In this chapter the kinematics, and the navigation model for obstacle avoidance of a six-leg (hexapod) robot are discussed. A navigation model controls the course of a mobile robot from a starting position towards a goal destination. The kinematics describes the geometry of motion of a body (i.e. limbs, links, a joint, a walker) regardless the causes that produced such motion. Therefore, a robotic navigation model entirely depends on the kinematic models in order to describe the set of local Cartesian goals that the mobile robot must reach. The present approach is on combining the optic flow information observed from the obstacles, with a decision engine comprised of the image motion analysis, and a kinematic control law. The optical flow describes the apparent motion of a body w.r.t. the observer location (i.e. the robot's visual sensor on board), and such description consists of the velocity components within a local inertial frame. In the present scheme, a decision engine is a robot's algorithm that considers information descriptors that are used to take decisions on how to avoid collisions online against static and dynamic obstacles. The decision engine considers motion information from the actual image frames to feedback a linear state equation control law about the environment's near obstacles. The inverse and direct kinematic analysis of a leg is presented separately to approach an algebraic solution that obtains the Jacobian matrix of the limbs. The limb's inverse kinematic solves the independent variables that control the system, the rotational joints; given

that the workspace variables are known. Inversely, the forward kinematics obtains the solution
of the unknown workspace Cartesian variables, given that the independent control variables are
known. The Jacobians are deployed to build the input vector model $\mathbf{u} = (v, \omega)^\top$ of a linearised
displacement state feedback equation control $\dot{\mathbf{x}} = \mathbf{A} \cdot \mathbf{x} + \mathbf{B} \cdot \mathbf{u}$. The function of the state control
is to provide a fast response navigation reaction for avoiding obstacles. The Jacobian matrices
are the functions in terms of the workspace state variables derived w.r.t. the all independent
control variables. Thus, the solution of the Jacobians are directly involved with the input vector.
Finally, the hexapod robot is a statically stable multi-legged robot comprised of six limbs able
to walk over rough terrains, developing a variety of gait configurations.

5.1   Limb forward kinematics

Figure 5.1 depicts the robotic platform body, and the limb's kinematic. The platform is an AH3-
R hexapod model, which is a radially symmetric multi-legged walker with 3 DOF for each leg.
A general view of the mechanical structure is shown in figure 5.1. The limb inertial space E
is defined with Cartesian origin at the base of the first joint ϕ_0, following the next two joints
ϕ_1 and ϕ_2, and links called l_0, l_1, and l_2. The leg's contact point is modelled by the position
vector $^{\prime\prime}$, which defines the instantaneous position usually stepping the ground surface.

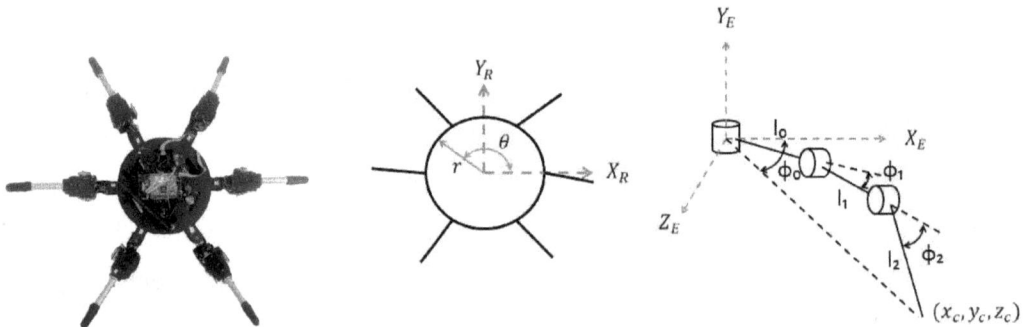

Figure 5.1:   The AH3-R hexapod: body and leg kinematics.

The extremity position vector $\mathbf{p} = (x_{ce}, y_{ce}, z_{ce})^\top$ is locally described w.r.t. limb's base Cartesian coordinate system; analysing independently each workspace variable, the x component is
defined by

$$x_{ce} = \cos(\phi_0)(l_0 + l_1 \cos(\phi_1) + l_2 \cos(\phi_1 + \phi_2)) \tag{5.1}$$

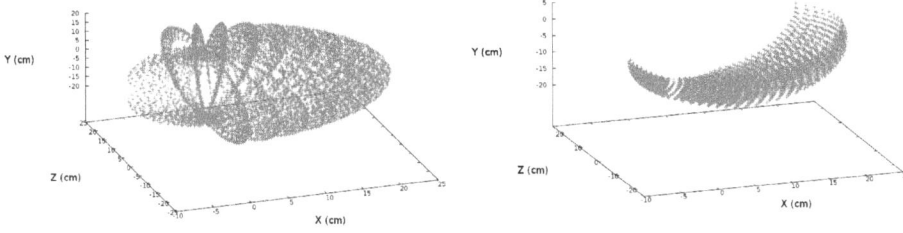

Figure 5.2: Leg whole workspace, and step-limit workspace.

for the position projected along the y axis,

$$y_{ce} = l_1 \sin(\phi_1) + l_2 \sin(\phi_1 + \phi_2) \tag{5.2}$$

as well as for the component along the z axis.

$$z_{ce} = \sin(\phi_0)(l_0 + l_1 \cos(\phi_1) + l_2 \cos(\phi_1 + \phi_2)) \tag{5.3}$$

With the equations (5.1), (5.2) and (5.3), the plots in figure 5.2 are produced. The left-hand side plot shows the limb's workspace; being the workspace each three-dimension Cartesian point that the limb's contact point is able to reach. Metric scale is represented with the real physical size of the links $l_0 = 0.04$, $l_1 = 0.06$ and $l_2 = 0.14$ (m). Likewise, the right-hand side plot depicts the angular independent variables with numeric limits to approach the AH3-R joints' movements to one-step in three-dimension space.

Equations (5.1)-(5.3) represent the forward kinematics using an algebraic approach to analyse the limb's kinematics. With this forward position equations and by knowing the control independent articular variables, the workspace variables are obtained. Nevertheless, it is our interest to get an algebraic inverse solution. Unlike geometric approaches, such as using the sine and cosine laws in order to solve for the limb's angles, our approach is rather flexible to mathematically model in case of reconfiguring or making physical changes to the limbs.

In this algebraic context, to solve for the inverse kinematic equations of the position model that is denoted by the vector $\mathbf{p} \in \mathbb{R}^3$, it is proposed to algebraically expand, and then derive w.r.t. time as to express the new mathematical relation in terms of velocities. Thus, being the expressions in terms of velocities, it is easier to drop-off the angular speeds vector since the robot AH3-R software technology solely allows to move the joints by using the angular velocities $\dot{\phi}_i$ and the angular positions ϕ_i. Therefore, by algebraically expanding the position equations, the next expression for the component x is produced,

$$x_{ce} = l_0 \cos(\phi_0) + l_1 \cos(\phi_0) \cos(\phi_1) + l_2 \cos(\phi_0) \cos(\phi_1 + \phi2) \tag{5.4}$$

However, for the y component (5.3), it does not require any further algebraic expansion. And for the component z, the following expression is yielded,

$$z_{ce} = l_0 \sin(\phi_0) + l_1 \sin(\phi_0) \cos(\phi_1) + l_2 \sin(\phi_0) \cos(\phi_1 + \phi2) \tag{5.5}$$

5.1.1 First order derivatives

The present approach states the first order derivatives w.r.t. time for the position components. The partial derivatives of each workspace variable are defined as functions of the three independent angular variables.

$$dx_{ce} = \frac{\partial x_{ce}}{\partial \phi_0} d\phi_0 + \frac{\partial x_{ce}}{\partial \phi_1} d\phi_1 + \frac{\partial x_{ce}}{\partial \phi_2} d\phi_2 \tag{5.6}$$

for the component y,

$$dy_{ce} = \frac{\partial y_{ce}}{\partial \phi_0} d\phi_0 + \frac{\partial y_{ce}}{\partial \phi_1} d\phi_1 + \frac{\partial y_{ce}}{\partial \phi_2} d\phi_2 \tag{5.7}$$

and for the component z.

$$dz_{ce} = \frac{\partial z_{ce}}{\partial \phi_0} d\phi_0 + \frac{\partial z_{ce}}{\partial \phi_1} d\phi_1 + \frac{\partial z_{ce}}{\partial \phi_2} d\phi_2 \tag{5.8}$$

The three previous equations if arranged in the matrix form, then the Jacobian matrix of derivatives is obtained. Hence, previous definitions are solved by developing the first order derivative w.r.t. time, for the component $\dot{x}$,

$$\dot{x}_{ce} = -l_0 \sin(\phi_0)\dot{\phi}_0 + l_1[-\sin(\phi_0) \cos(\phi_1)\dot{\phi}_0 - \cos(\phi_0)sin(\phi_1)\dot{\phi}_1]$$
$$+ l_2[-\sin(\phi_0) \cos(\phi_1 + \phi2)\dot{\phi}_0 - \cos(\phi_0) \sin(\phi_1 + \phi2)(\dot{\phi}_1 + \dot{\phi}_2)] \tag{5.9}$$

for component $\dot{y}$ we have,

$$\dot{y}_{ce} = l_1 \cos(\phi_1)\dot{\phi}_1 + l_2 cos(\phi_1 + \phi_2)(\dot{\phi}_1 + \dot{\phi}_2) \tag{5.10}$$

and finally for the component $\dot{z}$,

$$\dot{z}_{ce} = l_0 \cos(\phi_0)\dot{\phi}_0 + l_1[cos(\phi_0)cos(\phi_1)\dot{\phi}_0 - \sin(\phi_0)\sin(\phi_1)\dot{\phi}_1]$$
$$+ l_2[\cos(\phi_0)\cos(\phi_1 + \phi_2)\dot{\phi}_0 - \sin(\phi_0)\sin(\phi_1 + \phi_2)(\dot{\phi}_1 + \dot{\phi}_2)] \tag{5.11}$$

Since our interest is on solving for the angular derivatives, we firstly expand algebraically the same equations,

$$\dot{x}_{ce} = -l_0 \sin(\phi_0)\dot{\phi}_0 - l_1 \cos(\phi_0)sin(\phi_1)\dot{\phi}_1 - l_1 \sin(\phi_0)\cos(\phi_1)\dot{\phi}_0$$
$$- l_2 \sin(\phi_0)\cos(\phi_1 + \phi_2)\dot{\phi}_0 - l_2 \cos(\phi_0)\sin(\phi_1 + \phi_2)\dot{\phi}_1 - l_2 \cos(\phi_0)\sin(\phi_1 + \phi_2)\dot{\phi}_2 \tag{5.12}$$

for the velocity $\dot{y}$,

$$\dot{y}_{ce} = l_1 \cos(\phi_1)\dot{\phi}_1 l_2 cos(\phi_1 + \phi_2)\dot{\phi}_1 + l_2 cos(\phi_1 + \phi_2)\dot{\phi}_2 \tag{5.13}$$

and for the velocity $\dot{z}$,

$$\dot{z}_{ce} = l_0 \cos(\phi_0)\dot{\phi}_0 + l_1 \cos(\phi_0)cos(\phi_1)\dot{\phi}_0 - l_1 \sin(\phi_0)\sin(\phi_1)\dot{\phi}_1$$
$$+ l_2 \cos(\phi_0)\cos(\phi_1 + \phi_2)\dot{\phi}_0 - l_2 \sin(\phi_0)\sin(\phi_1 + \phi_2)\dot{\phi}_1 - l_2 \sin(\phi_0)\sin(\phi_1 + \phi_2)\dot{\phi}_2 \tag{5.14}$$

Thus, in order to factorise the higher order derivatives, it is reduced algebraically in the following manner,

$$\dot{x}_{ce} = -\sin(\phi_0)[l_0 + l_1 \cos(\phi_1) + \cos(\phi_1 + \phi_2)]\dot{\phi}_0$$
$$- \cos(\phi_0)[l_1 sin(\phi_1) + l_2 \sin(\phi_1 + \phi_2)]\dot{\phi}_1$$
$$- l_2 \cos(\phi_0)\sin(\phi_1 + \phi_2)\dot{\phi}_2 \tag{5.15}$$

likewise, for the velocity $\dot{y}$,

$$\dot{y}_{ce} = [l_1 \cos(\phi_1) + l_2 cos(\phi_1 + \phi_2)]\dot{\phi}_1 + l_2 cos(\phi_1 + \phi_2)\dot{\phi}_2 \tag{5.16}$$

and for the velocity $\dot{z}$.

$$
\begin{aligned}
\dot{z}_{ce} = \; & \cos(\phi_0)[l_0 + l_1 cos(\phi_1) + l_2 \cos(\phi_1 + \phi_2)]\dot{\phi}_0 \\
& -sin(\phi_0)[l_1 \sin(\phi_1) + l_2 \sin(\phi_1 + \phi_2)]\dot{\phi}_1 \\
& -l_2 \sin(\phi_0) \sin(\phi_1 + \phi_2)\dot{\phi}_2
\end{aligned}
\tag{5.17}
$$

Stating (5.15)-(5.17) in the matrix form, the second derivatives are factorised and the Jacobian matrix $J(\Phi) = J(\phi_0, \phi_1, \phi_2)$ is established. Therefore, the squared Jacobian matrix in the general form is defined by the next expression (5.18).

$$
J(\phi_0, \phi_1, \phi_2) =
\begin{pmatrix}
\frac{\partial x_c}{\partial \phi_0} & \frac{\partial x_c}{\partial \phi_1} & \frac{\partial x_c}{\partial \phi_2} \\
\frac{\partial y_c}{\partial \phi_0} & \frac{\partial y_c}{\partial \phi_1} & \frac{\partial y_c}{\partial \phi_2} \\
\frac{\partial z_c}{\partial \phi_0} & \frac{\partial z_c}{\partial \phi_1} & \frac{\partial z_c}{\partial \phi_2}
\end{pmatrix}
\tag{5.18}
$$

Thus, since the Jacobian matrix expression terms are too long, then for purpose of practicality they are substituted by new abbreviated expressions. Likewise, the trigonometric functions sin and cos are substituted by the letters s and c, respectively. The values of the variables along the first row are,

$$
a = -s_0(l_0 + l_1c_1 + l_2c_{12}); \qquad b = -c_0(l_1s_1 + l_2s_{12}); \qquad c = -l_2c_0s_{12}
$$

for the second row,

$$
d = 0; \qquad e = l_1c_1 + l_2c_{12}); \qquad f = l_2c_{12}
$$

and the third row,

$$
g = c_0(l_0 + l_1c_1 + l_2c_{12}); \qquad h = -s_0(l_1s_1 + l_2s_{12}); \qquad i = -l_2s_0s_{12}
$$

The Jacobian matrix is then simplified,

$$
J =
\begin{pmatrix}
a & b & c \\
d & e & f \\
g & h & i
\end{pmatrix}
\tag{5.19}
$$

Therefore, forward kinematics equation of the whole limb is presented in its matrix linear form as expressed by the equation (5.20); where $\dot{\Phi}$ is the vector of angular velocities, with vector components $\dot{\Phi} = (\dot{\phi}_0, \dot{\phi}_1, \dot{\phi}_2)^\top$.

$$\dot{p} = J \cdot \dot{\Phi} \tag{5.20}$$

The limb's contact point Cartesian positions provided by the forward kinematics of equation (5.3) are depicted by the plots of figure 5.3.

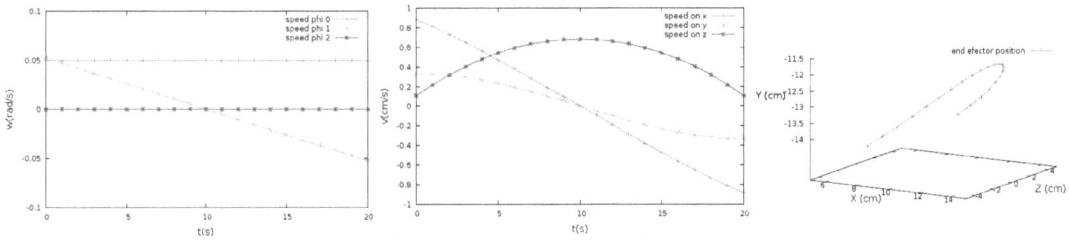

Figure 5.3:   Forward kinematics numerical simulations: Angular speeds(left); Cartesian speeds (center); and limb's contact point positions (right).

5.2   Limb inverse kinematics

The limb's inverse kinematic solution allows to obtain the joints control variables as functions of the known workspace variables. The inverse kinematic solution is useful to directly control the kinematic structure joints, knowing where it is desired to move the limb's contact point.

From expression (5.20), the vector $\dot{\Phi}$ is solved to represents the inverse equation form in expression (5.22). This, by multiplying the inverse Jacobian in both sides of the expression:

$$J^{-1} \cdot \dot{p} = J^{-1} \cdot (J \cdot \Phi) \tag{5.21}$$

and then the identity matrix I is the product of $J^{-1} \cdot J$, hence

$$\dot{\Phi} = J^{-1} \cdot \dot{p} \tag{5.22}$$

5.2.1 Singular matrix inversion

To solve the linear equation (5.22), the method of the inverse matrix is applied to the Jacobian matrix. This algebraic process is developed since the Jacobian is a squared matrix, and therefore by obtaining its non zero determinant a solution is possible. The linear algebraic method was already discussed in chapter 1.2.3.

$$\mathbf{J}^{-1} = \frac{1}{|\mathbf{J}|} \text{Adj}(\mathbf{J})^{\top} \tag{5.23}$$

Its algebraic form is obtained with the cofactors matrix $\text{Adj}(\mathbf{A})$

$$\text{Adj}(\mathbf{J}) = \begin{pmatrix} J_{11} & J_{12} & J_{13} \\ J_{21} & J_{22} & J_{23} \\ J_{31} & J_{32} & J_{33} \end{pmatrix} \tag{5.24}$$

where each cofactor term for the 3×3 matrix is defined by,

$$J_{11} = (-1)^2 \begin{vmatrix} e & f \\ h & i \end{vmatrix} ; \quad J_{12} = (-1)^3 \begin{vmatrix} d & f \\ g & i \end{vmatrix} ; \quad J_{13} = (-1)^4 \begin{vmatrix} d & e \\ g & h \end{vmatrix}$$

$$J_{21} = (-1)^3 \begin{vmatrix} b & c \\ h & i \end{vmatrix} ; \quad J_{22} = (-1)^4 \begin{vmatrix} a & c \\ g & i \end{vmatrix} ; \quad J_{23} = (-1)^5 \begin{vmatrix} a & b \\ g & h \end{vmatrix}$$

$$J_{31} = (-1)^4 \begin{vmatrix} b & c \\ e & f \end{vmatrix} ; \quad J_{32} = (-1)^5 \begin{vmatrix} a & c \\ d & f \end{vmatrix} ; \quad J_{33} = (-1)^6 \begin{vmatrix} a & b \\ d & e \end{vmatrix}$$

By defining the terms $u = l_0 + l_1 c_1 + l_2 c_{12}$ and $v = l_1 c_1 + l_2 c_{12}$, the cofactors are represented by the next expressions,

$$J_{11} = -l_1 l_2 s_0 s_2; \quad J_{12} = l_2 c_0 c_{12} u; \quad J_{13} = -c_0 u v$$

$$J_{21} = 0; \quad J_{22} = l_2 s_{12} u (c_0^2 + s_0^2); \quad J_{23} = -u(l_1 s_1 + l_2 s_{12})(s_0^2 + c_0^2)$$

$$J_{11} = l_1 l_2 c_0 s_2; \qquad J_{12} = l_2 s_0 c_{12} u; \qquad J_{13} = -s_0 u v$$

In addition the determinant of the Jacobian matrix det($\mathbf{A}$) is obtained by the rule of Sarrus (see chapter 1.2.1),

$$\det(\mathbf{J}) = aei + bfg + cdh - gec - hfa - idb \qquad (5.25)$$

hence,

$$\det(\mathbf{J}) = l_1 l_2 (l_0 s_2 - l_2 s_1 + l_1 c_1 s_2 + l_2 s_1 c_2 + l_2 c_1 c_2 s_2) \qquad (5.26)$$

and by using the equation (5.23) the inverse Jacobian is obtained, which is depicted by the following expression (5.27),

$$\mathbf{J}^{-1} = \begin{pmatrix} j & k & l \\ m & n & o \\ p & q & r \end{pmatrix} \qquad (5.27)$$

where the terms of the inverse matrix $\mathbf{J}^{-1}$ are expressed as given by the following expressions. For the first row, the matrix factors are,

$$j = \frac{-s_0}{u}; \qquad k = 0; \qquad l = \frac{c_0}{u}$$

the matrix second row terms,

$$m = \frac{c_0 c_{12}}{l_1 s_2}; \qquad n = \frac{s_{12}}{l_1 s_2}; \qquad o = \frac{s_0 c_{12}}{l_1 s_2}$$

and the matrix third row terms,

$$p = \frac{-c_0 v}{l_1 l_2 s_2}; \qquad q = -\frac{l_1 s_1 + l_2 s_2}{l_1 l_2 s_2}; \qquad r = \frac{-s_0 v}{l_1 l_2 s_2}$$

5.2.2   Inverse matrix by Cramer theorem

The Jacobian inverse matrix was obtained by using the Cramer's rule (5.28) as a second choice. This method was already discussed in chapter 1.2.4. Let us state the next expression,

$$\phi_i = \frac{\det(\mathbf{A}_i(\dot{\mathbf{p}}))}{\det(\mathbf{A})} \qquad (5.28)$$

where $\mathbf{A}_i(\dot{\mathbf{p}})$ is the matrix formed by replacing the column i for the vector $\dot{\mathbf{p}}$ such that,

$$\det(\mathbf{A_1}) = \begin{vmatrix} \dot{x} & b & c \\ \dot{y} & e & f \\ \dot{z} & h & i \end{vmatrix} ; \qquad \det(\mathbf{A_2}) = \begin{vmatrix} a & \dot{x} & c \\ d & \dot{y} & f \\ g & \dot{z} & i \end{vmatrix} ; \qquad \det(\mathbf{A_3}) = \begin{vmatrix} a & b & \dot{x} \\ d & e & \dot{y} \\ g & h & \dot{z} \end{vmatrix} \qquad (5.29)$$

the determinant of each matrix is given by the next expressions,

$$\det(\mathbf{A_1}) = l_1 l_2 s_2 (\dot{z} c_0 - \dot{x} s_0) \qquad (5.30)$$

and,

$$\det(\mathbf{A_2}) = l_2 u (\dot{y} s_{12} + \dot{x} c_0 c_{12} + \dot{z} s_0 c_{12}) \qquad (5.31)$$

and,

$$\det(\mathbf{A_3}) = -u(\dot{y} l_2 c_0^2 s_{12} + \dot{y} l_2 s_0^2 s_{12} + \dot{y} l_1 c_0^2 s_1 + \dot{y} l_1 s_0^2 s_1$$
$$+ \dot{x} l_2 c_0 c_{12} + \dot{z} l_2 s_0 c_{12} + \dot{x} l_1 c_0 c_1 + \dot{z} l_1 s_0 c_1) \qquad (5.32)$$

Furthermore, the inverse solution of the joints' angular speed are the results of the next expressions,

$$\dot{\phi}_1 = \frac{\dot{z} c_0 - \dot{x} s_0}{u} \qquad (5.33)$$

for $\dot{\phi}_2$,

$$\dot{\phi}_2 = \frac{\dot{y} c_1 s_2 + \dot{y} s_1 c_2 + \dot{x} c_0 c_1 c_2 + \dot{z} s_0 c_1 c_2 - \dot{x} c_0 s_1 s_2 - \dot{z} s_0 s_1 s_2}{l_1 s_2} \qquad (5.34)$$

and for $\dot{\phi}_3$,

$$\dot{\phi}_3 = -(\dot{y} l_1 s_1 + \dot{x} l_1 c_0 c_1 + \dot{y} l_2 c_1 s_2 + \dot{y} l_2 s_1 c_2 + \dot{z} l_1 s_0 c_1$$
$$+ \dot{x} l_2 c_0 c_1 c_2 + \dot{z} l_2 s_0 c_1 c_2 - \dot{x} l_2 c_0 s_1 s_2 - \dot{z} l_2 s_0 s_1 s_2)/l_1 l_2 s_2 \qquad (5.35)$$

Thus, by algebraically arranging and reducing the equations (5.33), (5.34) and (5.35), the same inverse matrix result $\mathbf{J}^{-1}$ is obtained as the previous section equation (5.27). Numerical simulations of the inverse kinematics model (5.22) yields the numerical results depicted in figure 5.4.

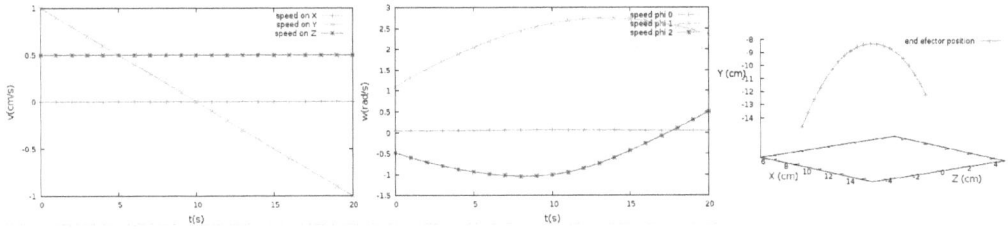

Figure 5.4:   Inverse kinematics numerical simulations: Cartesian speeds (left); angular speeds (center); and limb's contact point positions (right).

5.3   Robot's posture kinematic model

The global robot's locomotion is contributed by all limbs' movement configuration[4,5]. Depending on the limbs angles of phase and their synchronous motion, the robot's angular and linear velocity (figure 5.5), both are impacted. The models in the previous sections were obtained from the leg inertial systems, based on those will get the kinematic control models for the robot.   Previous sections analysis considered solely the limb kinematics, hence the limbs

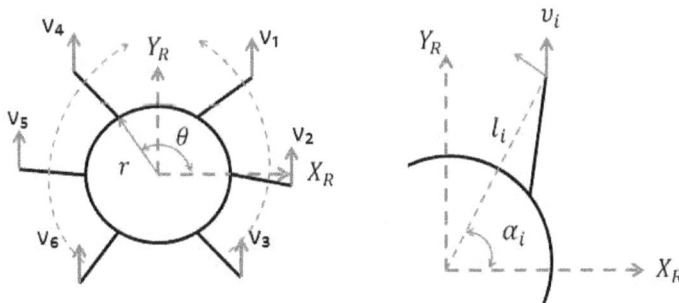

Figure 5.5:   Cartesian limb's kinematic in robot's fixed inertial space.

speed vector might be transformed into the robot's inertial frame system by,

$$\dot{\mathbf{p}}_i^R = \mathbf{R}(\gamma_i) \cdot \dot{\mathbf{p}}_i + \mathbf{t}_i \tag{5.36}$$

where the index i represents the leg number; γ_i is the z-axis rotation angle for each leg, and the vector $\mathbf{t}_i$ is the position of each limb around the robot's body. Thus, $\dot{\mathbf{p}}_i^R$ is the limb's speed vector w.r.t. the robot's centroid (see figure 5.5). Hence, the following proposition is stated:

Proposition 5.3.1 (robot's linear velocity). *The robot's instantaneous linear velocity is modelled as an averaged limbs' Cartesian speed, such that*

$$v = \frac{1}{6} \sum_{i=1}^{6} v_i = \frac{1}{6} \sum_{i=1}^{6} \sqrt[2]{\dot{x}_i^2 + \dot{y}_i^2 + \dot{z}_i^2} \tag{5.37}$$

In fact, as $\sqrt[2]{\dot{x}_i^2 + \dot{y}_i^2 + \dot{z}_i^2} = \|\dot{\mathbf{p}}_i^R\|$ and by substituting that direct kinematic model to represent each leg's velocity in terms of the Jacobians, it is also assumed that $\|\dot{\mathbf{p}}_i^R\| = \|\mathbf{J}_i^R \cdot \dot{\boldsymbol{\Phi}}_i^R\|$. Hereafter, the super index ^{R} is omitted and we will assume the the limbs contact points are in represented in robot's fixed inertial frame². Thus, the robotic platform with averaged absolute velocity is defined by,

$$v = \frac{1}{6} \sum_{i=1}^{6} \|\mathbf{J}_i \cdot \dot{\boldsymbol{\Phi}}_i\| \tag{5.38}$$

Similarly, by approaching a differential velocity approach, from a top view the right-hand side legs have positive sense of motion, while the left-hand side have negative values, thus,

$$\hat{v} = \sum_{i=1}^{3} v_i - \sum_{j=4}^{6} v_j \tag{5.39}$$

By establishing a model for the robots angular velocity ω_t w.r.t. an averaged radius ℓ obtained by the robot's geometric centre, and the limbs contact point positions, thus the instantaneous angular velocity $\omega_t = \hat{v}/\ell$. Where,

$$\ell = \frac{1}{6} \sum_i l_i \tag{5.40}$$

Each limb distance to the robot's centroid l_i has an angle α_i, and an instantaneous average value that represent all angles,

$$\hat{\alpha}_t = \frac{1}{6} \sum_i \alpha_i \tag{5.41}$$

Given that, each α_i can be obtained by,

$$\alpha_i = \arccos \left(\frac{x_{cr}}{\|x_{cr} + y_{cr}\|} \right) \tag{5.42}$$

Where the positions of the contact points w.r.t. the robot inertial space are,

$$x = r \cos(\theta_i) + (x^2 + z^2)^{1/2} \cos(\theta + \phi_{i0}) \tag{5.43}$$

and,

$$y = r \sin(\theta_i) + (x^2 + z^2)^{1/2} \sin(\theta + \phi_{i0}) \tag{5.44}$$

Therefore, it follows that, the transversal velocity component is stated as the numerator of the following equation,

$$\omega = \frac{\hat{v} \cos(\hat{\alpha}_t)}{\ell} \tag{5.45}$$

And by substituting the whole terms in order to have the complete model equation (5.46),

Proposition 5.3.2 (robot's yaw speed). *The robot's angular velocity is inversely proportional to the distance of the limbs' contact point w.r.t. the robot's centroid, and directly proportional to the transversal Cartesian component of the limbs' differential speed $\hat{v}$, equation* (5.39).

$$\omega = \frac{\frac{1}{6} \sum_{i=1}^{6} ||\mathbf{J} \cdot \dot{\boldsymbol{\Phi}}_i|| \cos(\alpha_i)}{\frac{1}{6} \sum_{i=1}^{6} (r + (x_i^2 + y_i^2 + z_i^2)^{1/2})} \tag{5.46}$$

Now the input vector to control the robotic platform in terms of linear and angular speeds is defined by equation (5.47), which is comprised of the equations (5.38) and (5.46).

$$\mathbf{u} = \begin{pmatrix} v, & \omega \end{pmatrix}^{\top} \tag{5.47}$$

5.4 Optical flow analysis

The optical flow vectors represent the apparent motion of the objects in the scene w.r.t. the visual sensor perspective, where the sensor is the local observer [$-$]. Thus, it is of interest to measure the optical flow of the feature points that are invariant to scale and rotation (SIFT) detected in the scene (feature extraction). In order to measure the optical flow, the image $I(c, r, t_1)$ is sensed and it is established data correspondence with the next consecutive image $I(c, r, t_2)$.

$$\mathbf{I}(c_f + udt, r_f + vdt, t + dt) = \mathbf{I}(c_f, r_f, t) \tag{5.48}$$

Previous expression represents the small differential values of columns c_f, and rows r_f w.r.t. dt between two successive image frames. We refer to feature points (SIFT) with valid correspondences in order infer their optical flow speed components.

$$\frac{\partial \mathbf{I}}{\partial c}\frac{dc_f}{dt} + \frac{\partial \mathbf{I}}{\partial r}\frac{dr_f}{dt} + \frac{\partial \mathbf{I}}{\partial t} = 0 \tag{5.49}$$

For the key points correspondence, we present the nearest-neighbour method that minimizes differences among the feature vectors. Each SIFT key-point[10,11] detected with its optic flow components are comprised of the features vector $\mathbf{f} \in \mathbb{R}^6$ defined next,

$$\mathbf{f} = (\dot{c}_f, \dot{r}_f, \theta_f, v_f, c_f, r_f)^\top \tag{5.50}$$

or as the following definition,

Definition 5.4.1 (feature vector). The feature vector is $\mathbf{f} \in \mathbb{R}^6$, where a given $\mathbf{f}_i(t)$ is any feature vector detected in image frame at actual time t

$$\mathbf{f} = \left(\frac{\partial \mathbf{I}}{\partial c}\frac{dc_f}{dt}, \frac{\partial \mathbf{I}}{\partial r}\frac{dr_f}{dt}, \arctan\left(\frac{\dot{r}_f}{\dot{c}_f}\right), \sqrt[2]{(\dot{c}_f)^2 + (\dot{r}_f)^2}, c_f, r_f \right)^\top \tag{5.51}$$

Figure 5.6 illustrates the optic-flow-based feature vectors $\mathbf{f}$, and depicts a set of correlated key points measured during experimental navigation[12]. Such feature points are detected by deploying the SIFT algorithm because of its suitability and reliability for the present application. Furthermore, the descriptors obtained in the present context are the optical flow of the SIFT points because those poses invariance of scale and rotation.

Therefore, one of the main interest of this data association approach is to establish approximated displacement speeds of the legged robot w.r.t. to the approaching objects[13]. This information allows to infer how the robot may quickly avoid collisions by directly involving such motion information in the robot's navigation equations.

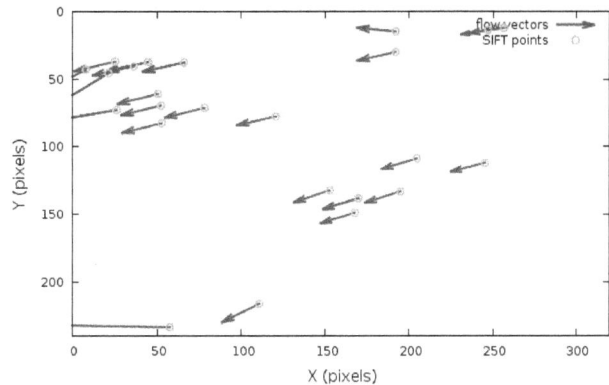

Figure 5.6:   Optic flow of invariant features (SIFT). Real scene (left); optic flow vectors extraction (right).

5.5   Navigation control

For the sake of the robot's safe navigation, the critical areas of consideration to detect obstacles moving towards a collision generally converge in the visual feature centre (figure 5.7). The area of convergence is depicted as a circular region that is of critical interest in order to detect the collisions that might occur as the nearest objects empirically scoped by the radius r_e. The extrinsic parameters depicted in figure 5.7 are explained in the subsequent paragraphs of this section, provided as postulates that comprise an inference engine for safe navigation.

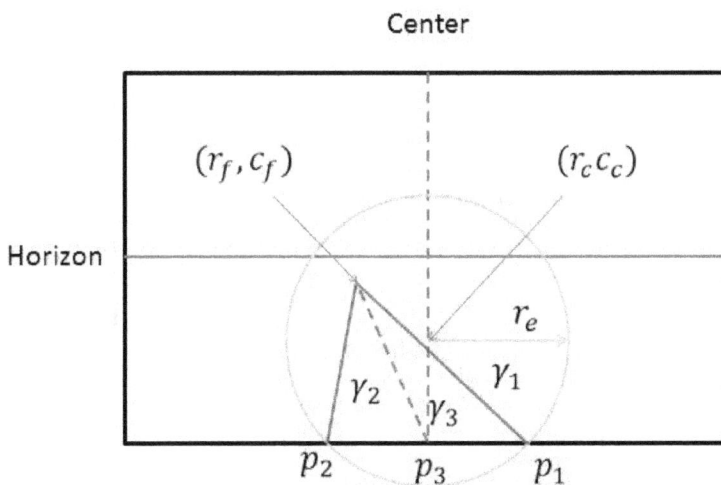

Figure 5.7:   Image frame with area of critical interest of a motion feature point.

The optic flow vectors falling within the encircled area are deployed as elements that contribute to take decisions in order to yield changes of speeds for evasion, according to the next coordinates criterion,

Postulate 5.5.1 (evasion distance). *When the distance in focal plane of a feature vector is less than the magnitude r_e, the optic flow feature vector representing an obstacle in the local scene is in close proximity to the robot.*

$$\sqrt{(r_f - r_c)^2 + (c_f - c_c)^2} < r_e \qquad (5.52)$$

Besides, the optic flow vectors with direction to the robot are defined by the angles criterion

Postulate 5.5.2 (direction of collision). *The obstacle angle of direction θ_f is leading towards the robot, if the following criterion occurs.*

$$\gamma_1 < \theta_f < \gamma_2 \qquad (5.53)$$

Furthermore, in order to determine the magnitudes of speed displacements, the inequality (5.54) is a criterion to discriminate whether or not a motion feature vector $\mathbf{f}$ in image frame $t - 1$ w.r.t. its correlated one $\mathbf{g}$ in image frame t produces a relevant motion toward a collision.

Postulate 5.5.3. *The magnitude of motion $\mathbf{f}$ at time $t - 1$, and at time t called $\mathbf{g}$ of an object, indicates whether a relevant motion toward a collision has been produced or not.*

$$\sum_i \left(\sum_j \|\mathbf{f}_i - \mathbf{g}_j\| \right) \lessgtr \epsilon_f \qquad (5.54)$$

With equations (5.52)-(5.54) it is possible to establish a condition, which indicates that there is a feature motion vector within an emergency proximity area,

Postulate 5.5.4 (near object detection). *There exist is a near object in possibility of collision, when the feature motion vector is within an emergency proximity area, and there is a considerable speed magnitude greater than a speed threshold ϵ_f.*

$$\left(\sqrt{(r_f - r_c)^2 + (c_f - c_c)^2} < r_e \right) \wedge \left(\sum_i \sum_j \| \mathbf{f}_i - \mathbf{g}_j \| > \epsilon_f \right) \tag{5.55}$$

Similarly, it also possible to state a condition criterion to establish that there exist a free collision path in front of the robot by the next postulate,

Postulate 5.5.5 (free-collision path). *There is not any object collision when any optic feature vector is out of the scope r_e, and angle θ_f is out of criteria γ_1, γ_2.*

$$\left(\sqrt{(r_f - r_c)^2 + (c_f - c_c)^2} > r_e \right) \wedge (\gamma_1 > \theta_f > \gamma_2) \tag{5.56}$$

Likewise, it is possible to determine whether or not there are obstacles toward the direction the robot is moving along, by using the next criterion

Postulate 5.5.6 (potential collision detection). *There are obstacles within the critical circular area, and in close direction to the robot.*

$$(\gamma_1 < \theta_f < \gamma_2) \wedge \left(\sum_i \sum_j \| \mathbf{f}_i - \mathbf{g}_j \| > \epsilon_f \right) \tag{5.57}$$

In addition, in order to know along which side of the robot the feature vectors are moving, the angle γ_3 is analysed. The obstacles moving to the right side of the robot the next criterion is used,

Postulate 5.5.7 (right-hand side object avoidance). *An obstacle is right-hand side avoided, if this criterion occurs:*

$$(\gamma_3 < \theta_f < \gamma_1) \wedge \left(\sum_i \sum_j \|\mathbf{f}_i - \mathbf{g}_j\| \lesseqgtr \epsilon_f \right) \tag{5.58}$$

Similarly, analysing when the feature vectors are moving to the left side of the robot,

Postulate 5.5.8 (left-hand side object avoidance). *An obstacle is left-hand side avoided, if this criterion occurs:*

$$(\gamma_2 < \theta_f < \gamma_3) \wedge \left(\sum_i \sum_j \|\mathbf{f}_i - \mathbf{g}_j\| \lesseqgtr \epsilon_f \right) \tag{5.59}$$

Therefore, from previous definitions of the avoidance criteria, it follows to propose the robot's velocity models that will provide suitable navigation behaviours[14]. Starting from a first kinematic definition $a\,dt = dv$, let us complete the integrals for each differential dt and dv respectively, thus

$$a \int_{t_1}^{t_2} dt = \int_{v_1}^{v_2} dv \tag{5.60}$$

hence by performing the integrations in both sides of the equation,

$$a(t_2 - t_1) = v_2 - v_1$$

the recursive model for v_{t+1} is stated by

$$v_{t+1} = v_t + a(t_2 - t_1) \tag{5.61}$$

Similarly, the kinematic equations $a = dv/dt$, and $v = ds/dt$ are true for our model, and both are set for a second kinematic model,

$$v\,dv = a\,ds \tag{5.62}$$

thus, the change of velocity w.r.t. the change of position is described by the next expression,

$$\int_{v_1}^{v_2} v\, dv = a \int_{s_1}^{s_2} ds \tag{5.63}$$

and by solving the defined integrals,

$$\frac{v_2^2 - v_1^2}{2} = a(s_2 - s_1) \tag{5.64}$$

Hence, exchanging the sub-index $v_{1,2}$ and $s_{1,2}$ by the time counters, a non linear recursive solution for the robot's instantaneous velocity is stated by the following expression,

$$v_{t+1}^2 = v_t^2 + 2a(s_{t+1} - s_t) \tag{5.65}$$

Therefore, by combining the expressions of the optic-flow feature vectors stated as criteria from previous postulates, the new control input velocity models are proposed as theorems. The velocity model theorems comprise the inference engine, which is fundamental for the control vector $\hat{\mathbf{u}} = (v, \omega)^\top$ that combines the acting-sensing models in terms of the linear and angular velocities. Therefore, the robot's instantaneous linear velocity behaviour is given by the following expression,

Theorem 5.5.9 (linear velocity model). *The robot's acting-sensing behaviour of its linear velocity model is stated by the present inference engine:*

$$v_{t+1} = \begin{cases} \sqrt{v_t^2 + 2a(s_{t+1} - s_t)}, & \left(\sqrt{(r_f - r_c)^2 + (c_f - c_c)^2} > r_e\right) \wedge (\gamma_1 > \theta_f > \gamma_2) \\[3mm] v_t - a(t_2 - t_1), & (\gamma_1 < \theta_f < \gamma_2) \wedge \left(\sum_i \sum_j \|f_i - g_j\| < \epsilon_f\right) \\[3mm] 0, & \left(\sqrt{(r_f - r_c)^2 + (c_f - c_c)^2} < r_e\right) \wedge \left(\sum_i \sum_j \|f_i - g_j\| < \epsilon_f\right) \end{cases} \tag{5.66}$$

Likewise, the robot's acting-sensing behaviour of its angular velocity is stated by the next theorem,

Theorem 5.5.10 (angular velocity model). *The robot's acting-sensing behaviour of its angular velocity model is stated by the present inference engine:*

$$\omega_{t+1} = \begin{cases} \omega_t + \ddot{\theta}(t_2 - t_1), & (\gamma_3 < \theta_f < \gamma_1) \wedge \left(\sum_i \sum_j ||f_i - g_j|| < \epsilon_f\right) \\[2ex] \omega_t - \ddot{\theta}(t_2 - t_1), & (\gamma_2 < \theta_f < \gamma_3) \wedge \left(\sum_i \sum_j ||f_i - g_j|| < \epsilon_f\right) \\[2ex] 0, & \left(\sqrt{(r_s - r_c)^2 + (c_s - c_c)^2} > r_e\right) \wedge (\gamma_1 > \theta_f > \gamma_2) \end{cases}$$

(5.67)

Thus, a state space linear equation that controls the robot motion in global inertial frame is defined by

$$\dot{\mathbf{x}} = \mathbf{A}\mathbf{x} + \mathbf{B}\mathbf{u}$$

(5.68)

Then, by defining the state vector $\mathbf{x} = (x_g, y_g, \theta_g)^{\top}$, it is also presented the state transition and input matrices to complete the state equation

$$\begin{pmatrix} \dot{x}_g \\ \dot{y}_g \\ \dot{\theta}_g \end{pmatrix} = \begin{pmatrix} \frac{1}{t} & 0 & 0 \\ 0 & \frac{1}{t} & 0 \\ 0 & 0 & \frac{1}{t} \end{pmatrix} \begin{pmatrix} x_g \\ y_g \\ \theta_g \end{pmatrix} + \begin{pmatrix} \cos\theta_g & 0 \\ \sin\theta_g & 0 \\ 0 & 1 \end{pmatrix} \begin{pmatrix} v \\ \omega \end{pmatrix}$$

(5.69)

The feedback robot's displacement [16] arising from the optical flow observations are used to control the navigation model, it deploys the position information provided by the optical flow vectors. With the flow vector $(c_f, r_f)^{\top}$ on the focal plane [17], the angular factor f_c is obtained from the horizontal angle of view of the visual sensor by

$$f_C = c_f \frac{\varphi_h}{C}$$

(5.70)

likewise, for a numerical factor related to the vertical angle is given by

$$f_R = r_f \frac{\varphi_v}{R}$$

(5.71)

The optical parameters of the sensor are depicted in figure 5.8. In addition, the sensor height position is provided in metric units, such parameter is used in next equation to obtain metric data about the feature points sensed in world-space w.r.t. the robot's local inertial system,

$$x_{rc} = \pm \frac{h}{\cos(\varphi - \varphi_v/2 + \Delta\varphi_v)} \tan(f_C) \tag{5.72}$$

likewise, for the position along the y metric component,

$$y_{rc} = h \tan(\varphi - \varphi_v/2 + f_R) \tag{5.73}$$

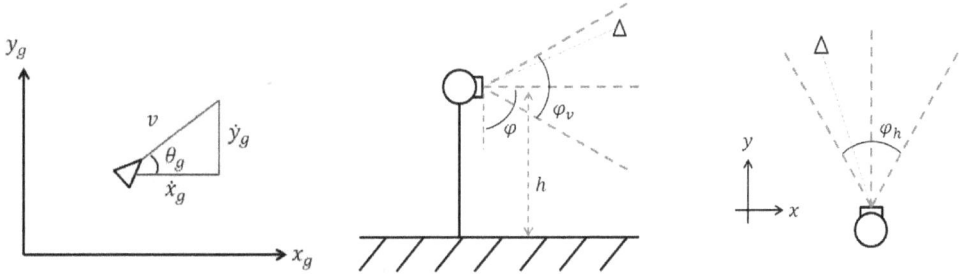

Figure 5.8: Robot's plane, and sensor coordinate systems.

From previous metric definitions, the feature points metric locations[18] are used to infer the robot's yaw variation in times $t - 1$ and t by

$$\varphi_v = \arctan\left(\frac{y_1(t-1) - y_2(t-1)}{x_1(t-1) - x_2(t-1)}\right) - \arctan\left(\frac{y_1(t) - y_2(t)}{x_1(t) - x_2(t)}\right) \tag{5.74}$$

It follows that, a type of visual odometry is to estimate the robot's displacement, as the position[19] variates incrementally over time, and is transformed onto previous observation inertial frame by the Euler orthogonal matrix $\mathbf{R}(\varphi_v)$,

$$\begin{pmatrix} \Delta x \\ \Delta y \end{pmatrix} = \mathbf{R}(\varphi_v) \cdot \left(\begin{pmatrix} x_1 \\ y_1 \end{pmatrix} - \begin{pmatrix} x_2 \\ y_2 \end{pmatrix} \right) \tag{5.75}$$

Therefore, the navigation control law that projects the velocity vector in $t + 1$ is defined in proposition (5.5.11),

Proposition 5.5.11 (positioning control law). *The reference model based positioning mo-tion control is proposed by*

$$\mathbf{v}_{t+1} = \dot{\boldsymbol{\xi}}_t + \int_t \frac{1}{\tau}(\hat{v}\mathbf{u} - \mathbf{v}_r)dt \qquad (5.76)$$

where the state vector is defined by $\dot{\boldsymbol{\xi}} = (\dot{x}, \dot{y})^\top$, and it comes from the state equation in actual time t. The relaxation time τ is the time required to reach a velocity change. And $\hat{v}$ is a reference velocity model described by,

$$\dot{s} = v + at + 3a\omega t^2 \qquad (5.77)$$

Hence, given a reference velocity, the unit vector $\mathbf{u}$ only provides the directions of the reference velocity vector,

$$\mathbf{u} = \frac{\mathbf{x}_{t+1} - \mathbf{x}_t}{||\mathbf{x}_{t+1} - \mathbf{x}_t||}; \qquad \mathbf{x} = \begin{pmatrix} x \\ y \end{pmatrix} \qquad (5.78)$$

In addition, by substituting the visual feedback planar displacement equation (5.75) in the pro-posed navigation control law [11], the observed speed $\mathbf{v}_r$ is obtained:

$$\mathbf{v}_{t+1} = \begin{pmatrix} \dot{x}_g \\ \dot{y}_g \end{pmatrix} + \frac{1}{\tau} \int_{t_1}^{t_2} \left[(v + at + 3a\omega t^2)\left(\frac{\mathbf{x}_{t+1} - \mathbf{x}_t}{||\mathbf{x}_{t+1} - \mathbf{x}_t||}\right) - \frac{1}{t}\left(\varphi_v\begin{pmatrix} x_1 - x_2 \\ y_1 - y_2 \end{pmatrix}\right) \right] dt \qquad (5.79)$$

Bibliography

[1] Görner, M. and Stelzer, A., *A leg proprioception based 6 DOF odometry for statically Go stable walking robots*, Autonomous Robots, Vol. 34(4), pp. 311–326, 2013, doi: dx.doi.org/10.1007/s10514-013-9326-3.

[2] Kahlouche, S. and Achour, K., *Optical flow based obstacle avoidance*, Intl. Journal of Advanced Robotic System, Vol. 4(1), pp. 13–16, 2007.

[3] Lin, P.C., Komsuoglu, H. and Koditschek, D.E., *A leg configuration measurement system for full body posture estimates in a hexapod robot*, IEEE Trans. Robotics and Automation, Vol. 21(3), pp. 411–422, 2005, doi: dx.doi.org/10.1109/TRO.2004.840898.

[4] Cobano, J.A., Estremera, J. and Gonzalez de Santos, P., *Location of legged robots in outdoor environments*, Robotics and Autonomous Systems, Vol. 56(9), pp. 751–761, 2008, doi: dx.doi.org/10.1016/j.robot.2007.12.003.

[5] Gassmann, B., Zacharias, F., Zollner, J.M. and Dillmann, R., *Localization of walking robots*, IEEE Intl. Conf. On Robotics and Automation, Barcelona, pp. 1471–1476, 2005.

[6] Roy, S.S., Singh, A.K. and Pratihar, D.K., *Analysis of six-legged walking robots*, 14th Conf. Machines and Mechanisms, Durgapur, Dec 2009, pp. 259-265, 2009.

[7] Giachetti, A., Campani, M. and Torre, V., *The use of optical flow for road navigation*, IEEE Trans. on Robotics and Automation, Vol. 14(1), pp. 34–48, 1998.

[8] Sorensen, D.K., Smukala, V., Ovinis, M. and Lee, S., *On-line optical flow feedback for mobile robot localization/navigation*, IEEE/RSJ IROS, Vol. 2, pp. 1246-1251, 2003.

[9] Surya, P.N., Csonka, P.J., Kenneth, J. and Waldron, K.J., *Optical flow aided motion estimation for legged locomotion*, IEEE/RSJ Intl. Conf. in Robotics Systems, Beijing, pp. 1738-1743, 2006.

[10] Martínez-García, E.A. and Torres-Mendez, L.A., *Regional invariant descriptors in mobile robotics part II: enhanced detection and matching*, IEEE Andean Region Intl. Conf. Andescon, Cuenca Ecuador, pp. 91–94, 2012.

[11] Martinez-Garcia E.A., Torres-Méndez L.A., and Mohan R.E., *Multi-Legged Robot Dynamics Navigation Model with Optical Flow*, Intl. Journal of Intelligent Unmanned Systems, Vol.2(2), pp.121–139, Emerald, 2014.

[12] Mccarthy, C. and Barnes, N., *Performance of optical flow techniques for indoor navigation with a mobile robot*, IEEE Intl. Conf. on Robotics & Automation, pp. 5093-5098, 2004.

[13] Milushev, M., Petrov, P. and Boumbarov, O., *Active target tracking from a six legged robot*, IEEE 4th Intl. Conf. Intelligent Systems, pp. 4–61, 2008.

[14] Wawrzynaski, P., Mozaryn, J. and Klimaszewski, J., *Robust velocity estimation for legged robot using on-board sensors data fusion*, 18th Intl. Conf. on Methods and Models in Automation and Robotics, Miedzyzdroje, pp. 717–722, 2013.

[15] Martínez-García, E., Torres, D., Ortega, A., Zamora, A., Torres-Cordoba, R. and Martínez-Villafane, A, *Modeling dynamics and navigation control of an explorer hexapod*, Autonomous Robots: Control, Sensing and Perception, Cuvillier Verlag, pp. 82–115, 2011.

[16] Belter, D. and Skrzypczynski, P., *Precise self-localization of a walking robot on rough terrain using parallel tracking and mapping*, Industrial Robot: An Intl. Journal, vol.40(3), pp. 229–237, 2013.

[17] Samperio, R. and Hu, H., *Real-time landmark modelling for visual-guided walking robots*, Int. J. Computer Applications in Technology, Vol. 41 Nos 3/4, pp. 253-261, 2011.

[18] Perez-Sala, X., Angulo, C. and Escalera, S., *Biologically inspired path execution using SURF flow in robot navigation*, in Joan, C., Ignacio, R. and Gonzalo, J. (Eds), Lecture Notes in Computer Science, Vol. 6692, June, pp. 581–588, 2011 doi: 0302-9743, 10.1007/978-3-642-21498-1_76.

[19] Stelzer, A., Hirschmuller, H. and Gorner, M., *Stereo-vision-based navigation of a six-legged walking robot in unknown rough terrain*, Intl. Journal of Robotics Research, Vol. 31(4), pp. 381-402, 2012, doi: dx.doi.org/10.1177/0278364911435161.

Chapter 6

Navigation using exponential

derivatives

Edgar A. Martínez García

Laboratorio de Robótica, Institute of Engineering and Technology
Universidad Autónoma de Ciudad Juárez, Mexico.

In this chapter, an autonomous navigation algorithm for wheeled mobile robots (WMR) for operating in dynamic environments (indoors or structured outdoors), and based on field vectors [1-3] is formulated. The planning scheme is of critical importance for autonomous navigational tasks in complex dynamic environments. To avoid potential crashes, reliable planning algorithms must be computationally efficient while considering important WMR and motion dynamic effects. The focus concerns a model for autonomous navigation with capabilities that help to avoid collisions. This chapter presents a framework that includes the kinematics and motion dynamics model in continuous-time merged with a general model to solve the motion-planning problem. The model approaches a general velocity-based motion framework that models causes and effects of motion. The speed control considers the functional form of motors rotational speed rate, and the robot's size to determine the vehicle yaw speed, and with such basis the actual and posterior position vectors are formulated. The combined scheme allows any forward kinematics, and allows weighting factors to yield motion effects from multiple sensed features.

Figure 6.1: Multiple wheeled mobile robots using the proposed navigation scheme.

The topic of this chapter consist of given a WMR μ with fixed inertial frame $\mathbf{x}_t^\mu = (x, y)^T$, and heading to θ_t, it should pass through a sequence of local goals at $\mathbf{x}_t$ with direction $\mathbf{m}_t$ each. And eventually the robot must reach a global goal destination γ. The Cartesian distance between two points is generally defined by the norm of their geometric difference $\|\vec{\delta}_{\mu\alpha}\| = \|\mathbf{x}_t^\mu - \mathbf{x}_t^\alpha\|$ (distance between robot μ and obstacle α_t). Goals are established to exert attractive accelerative fields $\mathbf{F}_t^\gamma$ which easily conduct the robot. Detected obstacles α exert repulsive accelerative fields $\mathbf{F}_t^\alpha$. Both types of fields in combination form an enriched directional map.

6.1   General navigation model

The navigation models begins by introducing a general control equation to govern the robot's speed (6.1). This relationship keeps the robot around a safe ideal velocity v^o, while navigating along the direction θ_t. The factor η is the gain value that if adjusted, defines the control convergence of the speed.

Proposition 6.1.1 (controlled velocity). $\mathbf{v}_t$ *is proportionally controlled by the feedback error w.r.t. the reference* $\mathbf{v}_t^o$*, and magnitude limit h,*

$$\mathbf{v}_t = (\hat{\mathbf{v}}_{t-1} + \eta(\mathbf{v}_t^o - \hat{\mathbf{v}}_t))h\left(\frac{v^{max}}{\|\hat{\mathbf{v}}_t\|}\right) \tag{6.1}$$

The velocity vector $\mathbf{v}_t$ is proportionally adjusted using the feedback velocity error w.r.t. a reference speed, which in particular happens when the magnitude reduction function $h(\cdot) = 1$. The non-stationary ideal velocity is denoted by $\mathbf{v}^o = v^o(\cos(\theta_t), \sin(\theta_t))^\top$. When an unexpected collision suddenly occurs, very short periods of time are generally taken. As a result, the model (6.1) controls the velocity peaks exceeding maximal allowable velocities when the function value is in the range $0 < h(\cdot) < 1$, as modelled by equation (6.2).

Definition 6.1.2 (velocity limit). The limit conditions factor h allow three behaviors of $\hat{\mathbf{v}}_t$,

$$h\left(v^{max}, \|\hat{\mathbf{v}}_t\|\right) = \begin{cases} 0, & \|\hat{\mathbf{v}}_t\| = 0 \\ 1, & 0 < \|\hat{\mathbf{v}}_t\| \leq v^{max} \\ \frac{v^{max}}{\|\hat{\mathbf{v}}_t\|}, & v^{max} < \|\hat{\mathbf{v}}_t\| \end{cases} \tag{6.2}$$

The velocity model (6.1) recursively controls the real velocity fluctuating around v^o, and removes divergent magnitudes overpassing the maximal allowable velocity. The real velocity vector $\hat{\mathbf{v}}_t$ at actual time t is defined in (6.3). The real velocity model in this context involves the causes and effects of motion, as well as random fluctuations perturbing acceleration components. The real velocity vector $\hat{\mathbf{v}}_t$ is expressed in terms of two global accelerative components that yield $\frac{d\hat{\mathbf{v}}}{dt}$ as in equation (6.3). One term is the directional field vector $\mathbf{F}_t = (f_x, f_y)^\top$ that expresses the internal and external causes of motion by Newton's 2nd law of motion $\mathbf{F}/m$ with $m = 1$. The second term $\mathbf{a}_t = (a_x, a_y)^\top$ is the general accelerative behaviour for any inertial system (global $\mathbf{a}^I$, or local $\mathbf{a}^R$).

$$\frac{d\hat{\mathbf{v}}}{dt} = \mathbf{F}_t - \mathbf{a}_t^I \tag{6.3}$$

The real motion model $\hat{\mathbf{v}}_t$ expresses the robot's global behaviour yielded by external causes of motion. We encompass three external causes of motion (sensors are deployed in the process for detection), *next desired goals* at sight, *obstacles position*, and *final goal destinations*. The equation (6.4) models the directional fields described in terms of global accelerations $\mathbf{F}_t$. Causes of motion: internal ($\mathbf{F}_t^o$), and external ($\mathbf{F}_t^\alpha$ and $\mathbf{F}_t^\gamma$).

$$\mathbf{F}_t = \mathbf{F}_t^o + \sum_\alpha \mathbf{F}_t^\alpha + \sum_\gamma \mathbf{F}_t^\gamma \tag{6.4}$$

Since the robot's navigation depends on sensor observations, only sensor data feature are used as regions of interest to exert weighted navigation functions. Each acceleration is defined with an adaptive numeric weight $w(\mathbf{m}_t, \mathbf{f}_t)$ yielded by the bearing location of the targets (local goal destination, or obstacles) within the sensors field of view as defined in equation (6.5).

Proposition 6.1.3 (Vector field).

$$\mathbf{F}_t = \mathbf{F}_t^o + \sum_\alpha w(\mathbf{m}_t, -\mathbf{f}_t^\alpha)\mathbf{f}_t^\alpha + \sum_\gamma w(\mathbf{m}_t, \mathbf{f}_t^\gamma)\mathbf{f}_t^\gamma \qquad (6.5)$$

The repulsive and attractive behaviour, which affect the robot's behaviour accentuate the magnitudes of the motion functions given in expressions (6.6) and (6.7), where $\vec{\delta}_\mu = \mathbf{x}_t^\alpha - \mathbf{x}_t^\mu$ is a distance vector between the positions of a goal/obstacle and the actual robot μ. $\mathbf{m}_t = (\mathbf{x}_{t+1}^\mu - \mathbf{x}_t^\mu)/\|\mathbf{x}_{t+1}^\mu - \mathbf{x}_t^\mu)\|$ is a unit vector expressing the direction towards a next desired location $\mathbf{x}_{t+1}$. Thus, the weighting factor w_t will affect the repulsive acceleration behaviour according to,

Definition 6.1.4.

$$\mathbf{F}_t^\alpha(\mathbf{m}, \vec{\delta}_{\mu\alpha}) = w(\mathbf{m}_t, -\mathbf{f}_t^\alpha)\mathbf{f}(\vec{\delta}_{\mu\alpha}) \qquad (6.6)$$

similarly the weighting factor will affect the attractive acceleration by,

$$\mathbf{F}_t^\alpha(\mathbf{m}, \vec{\delta}_{\mu\gamma}) = w(\mathbf{m}_t, \mathbf{f}_t^\gamma)\mathbf{f}(\vec{\delta}_{\mu\gamma}) \qquad (6.7)$$

The influence of the weight w_t depends on the sensing direction ϕ_t of a goal/obstacle and the actual acceleration $\mathbf{f}$, as defined by the equation (12.2). If the actual orientation of the vector acceleration $\mathbf{f}$ is about the same as the actual desired orientation $\mathbf{m}_t$, then no change of direction is required for the robot. It is expected that the orientation of the goal/obstacle sensed at bearing ϕ_t is approximately along the direction of the next desired position. But, if the orientations of vectors $\mathbf{m}_t$ and ϕ_t are different, then the component $\mathbf{f}_t \cos\phi$ must be decreased by the yaw changes.

$$w_t = \begin{cases} 1, & \mathbf{m}_t \cdot \mathbf{f}_t \geq \|\mathbf{f}_t\| \cos(\phi_t^i) \\ \lambda_t, & \text{otherwise} \end{cases} \qquad (6.8)$$

The influence of rotations that the robot must carry out is given by an influence term λ_t which is an average of the fusion of all multi-sensory observations. As we established that the robot is instrumented with s_n different sensor devices i. Thus, λ_t is valued within the range $0 \leq \lambda \leq 1$ based on an effective angle of view ϕ_t,

$$\lambda_t = \sin\left(\frac{1}{s_n} \sum_i \frac{\phi_t^i}{\pi}\right) \tag{6.9}$$

Where s_n is the total number of sensors involved in the perception of the objective (goal/obstacle), and $(\phi_t)_i^{s_n}$ are the angles at which each sensor i detected the same objective. Expression (6.9) defines a greater numeric weight to objectives located nearly along the longitudinal robot's axis (fixed-frame, defined at $\pi/2$). The sensing modality for environment mapping is by deploying a laser range finder. The important features, which the robot is able to perceive are very critical because on this issue, the robot defines the numeric weighting factors to impact significantly the navigation functions.

6.2 Inertial frames

The definition of the robot's motion is described in local and global Cartesian frames to represent the accelerations map (figure 6.5). Such scheme is useful to model accelerative motion behaviour denoted by $\mathbf{a}_t$, already described by equation (6.3) to describe part of the real acceleration $\frac{d\dot{\theta}}{dt}$. Let us consider the linear velocity components of a robot with averaged velocity v_t. By defining the velocity vector $\mathbf{v}_t^R = (v_x, v_y)^T$ in the robot fixed-frame, the components XY represent the 2D plane of motion and is given by the expression equation (6.10),

$$\mathbf{v}_t^R = v_t \begin{pmatrix} \cos(\theta_t) \\ \sin(\theta_t) \end{pmatrix} \tag{6.10}$$

Where θ_t is the robot's angle of motion w.r.t. robot's initial posture. By transforming the original robot fixed-frame using a transformation matrix $\mathbf{R}$ with rotation angle ψ between the robot's frame, and the global system. The new expression for the global frame becomes as expressed by equation (6.28), which is the velocity behaviour without wheels kinematic constraints. The Euler rotation matrix $\mathbf{R}$ and its inverse are critical in most mathematical definitions of this manuscript. Because of $\mathbf{R}$ is a non-singular matrix according to $\mathbf{R}^{-1}\mathbf{R} = \mathbf{I}$ or $\mathbf{R}\mathbf{R}^{-1} = \mathbf{I}$, and since $\mathbf{R}$ is an orthogonal matrix, hence $\mathbf{R}^{-1} = \mathbf{R}^T$. Thus, let us demonstrate it (see 1.2.3),

$$\mathbf{R}(\psi_t) = \begin{pmatrix} \cos\psi_t & -\sin\psi_t \\ \sin\psi_t & \cos\psi_t \end{pmatrix} ; \qquad \mathbf{R}^{-1}(\psi_t) = \begin{pmatrix} \cos\psi_t & \sin\psi_t \\ -\sin\psi_t & \cos\psi_t \end{pmatrix} \tag{6.11}$$

If $\cos\psi_t \cos\psi_t - (-\sin\psi_t)\sin\psi_t \neq 0$, then $\mathbf{R}$ is invertible, if only if $\det\mathbf{R} \neq 0$, where $\det\mathbf{R} = \cos\psi_t \cos\psi_t - (-\sin\psi_t \sin\psi_t)$.

$$\mathbf{R}^{-1} = \frac{1}{\det\mathbf{R}} \begin{pmatrix} \cos\psi_t & -(-\sin\psi_t) \\ -(\sin\psi_t) & \cos\psi_t \end{pmatrix} \tag{6.12}$$

and,

$$\mathbf{R}^{-1} = \frac{1}{\cos\psi_t \cos\psi_t - (-\sin\psi_t)\sin\psi_t} \begin{pmatrix} \cos\psi_t & -(-\sin\psi_t) \\ -(\sin\psi_t) & \cos\psi_t \end{pmatrix} \tag{6.13}$$

Thus, from previous definitions, $\mathbf{R}$ as well as $\mathbf{R}^{-1}$ will be used to describe motion in both inertial frames accordingly. Two ways to formulate an equation for $\mathbf{a}_t^R$ are presented. Firstly, with linear matrix algebra (see 1.2), inversely transform the acceleration into the robot coordinate framework by,

$$\mathbf{a}_t^R = \mathbf{R}_Z^{-1}(\psi_t)\mathbf{a}_t^I \tag{6.14}$$

Thus, substituting terms in previous equations,

$$\mathbf{a}_t^R = \begin{pmatrix} \cos(\psi_t) & \sin(\psi_t) \\ -\sin(\psi_t) & \cos(\psi_t) \end{pmatrix} \left(v_t(\dot\theta_t + \dot\psi_t)\begin{pmatrix} -\sin(\theta_t + \psi_t) \\ \cos(\theta_t + \psi_t) \end{pmatrix} + \dot v_t \begin{pmatrix} \cos(\theta_t + \psi_t) \\ \sin(\theta_t + \psi_t) \end{pmatrix} \right) \tag{6.15}$$

Algebraically expanding,

$$\begin{aligned} \mathbf{a}_t^R = v_t(\dot\theta_t + \dot\psi_t) &\begin{pmatrix} -\cos(\psi_t)\sin(\psi_t + \theta_t) + \sin(\psi_t)\cos(\psi_t + \theta_t) \\ \sin(\psi_t)\sin(\psi_t + \theta_t) + \cos(\psi_t)\cos(\psi_t + \theta_t) \end{pmatrix} \\ + \dot v_t &\begin{pmatrix} \cos(\psi_t)\cos(\theta_t + \psi_t) + \sin(\psi_t)\sin(\theta_t + \psi_t) \\ -\sin(\psi_t)\cos(\psi_t + \theta_t) + \cos(\psi_t)\sin(\psi_t + \theta_t) \end{pmatrix} \end{aligned} \tag{6.16}$$

By substituting trigonometric identities, our expression is simplifyied,

$$\mathbf{a}_t^R = v_t(\dot{\psi}_t + \dot{\theta}_t) \begin{pmatrix} \sin(\psi_t - (\psi_t + \theta_t)) \\ \cos(\psi_t - (\psi_t + \theta_t)) \end{pmatrix} + \dot{v}_t \begin{pmatrix} \cos(\psi_t - (-\psi_t + \theta_t)) \\ \sin(\psi_t + (-\psi_t + \theta_t)) \end{pmatrix} \tag{6.17}$$

Thus,

$$\mathbf{a}_t^R = v_t(\dot{\psi}_t + \dot{\theta}_t) \begin{pmatrix} \sin(-\theta_t) \\ \cos(-\theta_t) \end{pmatrix} + \dot{v}_t \begin{pmatrix} \cos(-\theta_t) \\ \sin(-\theta_t) \end{pmatrix} \tag{6.18}$$

Using the identities $\sin(-\theta_t) = -\sin(\theta_t)$ and $\cos(-\theta_t) = \cos(\theta_t)$. Thus, without lost of generality, the resulting simplified mathematical expression is now written as in equation (6.19), where in the first term, $\dot{\psi}_t$ is still existing, nevertheless it does not yield any impact because within the robot's motion frame ψ_t is always zero.

$$\mathbf{a}_t^R = v_t\dot{\theta} \begin{pmatrix} -\sin(\theta_t) \\ \cos(\theta_t) \end{pmatrix} + \dot{v}_t \begin{pmatrix} \cos(\theta_t) \\ \sin(\theta_t) \end{pmatrix} \tag{6.19}$$

Secondly, another way to find a functional form for $\mathbf{a}_t^R$ is from equation $\mathbf{v}_t^R = \mathbf{R}^{-1}\mathbf{v}_t^I$ and its derivative is as it follows,

$$\mathbf{a}_t^R = \dot{\mathbf{v}}_t^R = \mathbf{R}^{-1}\dot{\mathbf{v}}_t^I + \dot{\mathbf{R}}^{-1}\mathbf{v}_t^I \tag{6.20}$$

By algebraically developing the second term in the right side of previous equation,

$$\mathbf{a}_t^R = \mathbf{R}^{-1}\mathbf{a}_t^I + \begin{pmatrix} -\sin\psi_t & \cos\psi_t \\ -\cos\psi_t & -\sin\psi_t \end{pmatrix} \dot{\psi}_t v_t \begin{pmatrix} \cos(\psi_t + \theta_t) \\ \sin(\psi_t + \theta_t) \end{pmatrix} \tag{6.21}$$

arranging terms and signs

$$\mathbf{a}_t^R = \mathbf{R}^{-1}\mathbf{a}_t^I - \dot{\psi}_t v_t \begin{pmatrix} \sin\psi_t & -\cos\psi_t \\ \cos\psi_t & \sin\psi_t \end{pmatrix} \begin{pmatrix} \cos(\theta_t + \psi_t) \\ \sin(\theta_t + \psi_t) \end{pmatrix} \tag{6.22}$$

thus,

$$\mathbf{a}_t^R = \mathbf{R}^{-1}\mathbf{a}_t^I - \dot{\psi}_t v_t \begin{pmatrix} \sin\psi_t \cos(\theta_t + \psi_t) - \cos\psi_t \sin(\theta_t + \psi_t) \\ \cos\psi \cos(\theta_t + \psi_t) + \sin\psi_t \sin(\theta + \psi_t) \end{pmatrix} \tag{6.23}$$

$$\mathbf{a}_t^R = \mathbf{R}^{-1}\mathbf{a}_t^I - \dot{\psi}_t v_t \begin{pmatrix} \sin(\psi_t - \theta_t - \psi_t) \\ \cos(\psi_t - \theta_t - \psi_t) \end{pmatrix} \tag{6.24}$$

and

$$\mathbf{a}_t^R = \mathbf{R}^{-1}\mathbf{a}_t^I - \dot{\psi}_t v_t \begin{pmatrix} -\sin\theta_t \\ \cos\theta_t \end{pmatrix} \tag{6.25}$$

Now, developing the first term of right-side of equation,

$$\mathbf{a}_t^I = v_t(\dot{\theta}_t) \begin{pmatrix} -\sin\theta_t \\ \cos\theta \end{pmatrix} + \dot{v}_t \begin{pmatrix} \cos\theta_t \\ \sin\theta_t \end{pmatrix} - \dot{\psi}_{t t} v_t \begin{pmatrix} -\sin\theta_t \\ \cos\theta_t \end{pmatrix} \tag{6.26}$$

Finally,

$$\mathbf{a}_t^I = v_t \dot{\theta}_t \begin{pmatrix} -\sin\theta_t \\ \cos\theta_t \end{pmatrix} + \dot{v}_t \begin{pmatrix} \cos\theta_t \\ -\sin\theta_t \end{pmatrix} \tag{6.27}$$

$$\mathbf{v}_t^I = v_t \begin{pmatrix} \cos(\theta_t + \psi_t) \\ \sin(\theta_t + \psi_t) \end{pmatrix} \tag{6.28}$$

Now, the acceleration vector in global frame is obtained by equation (6.29),

$$\mathbf{a}_t^I = v_t(\dot{\theta}_t + \dot{\psi}_t) \begin{pmatrix} -\sin(\theta_t + \psi_t) \\ \cos(\theta_t + \psi_t) \end{pmatrix} + \dot{v}_t \begin{pmatrix} \cos(\theta_t + \psi_t) \\ \sin(\theta_t + \psi_t) \end{pmatrix} \tag{6.29}$$

Inversely transforming the acceleration into the robot's coordinate framework, and without lost of generality, the resulting simplified mathematical expression is now written in equation (6.30) as the robot's local frame, hence $\dot{\psi} = 0$.

$$\mathbf{a}_t^R = v_t \dot{\theta} \begin{pmatrix} -\sin(\theta_t) \\ \cos(\theta_t) \end{pmatrix} + \dot{v}_t \begin{pmatrix} \cos(\theta_t) \\ \sin(\theta_t) \end{pmatrix} \tag{6.30}$$

Equation (6.29) is about the same as equation (6.30). In the former, the rotation frame angle ψ is being considered for transformation into the global frame. Latter expression has no rotated inertial frames, hence $\psi = 0$. Hereafter, equation (6.29) may be used as the general frame solution. Being $\psi_t \neq 0$ when global inertial Cartesian frame is required.

6.3 Navigation model derivation

Based on the Newton's law of motion, the next equation establishes that the sum of all accelerations in the system is equal to a global acceleration,

$$a_t = \frac{1}{m} \sum_i \sqrt[2]{fx_i^2 + fy_i^2} \equiv \sum_i \sqrt[2]{fx_i^2 + fy_i^2} \tag{6.31}$$

For the sake of analysis, we define an unitary mass $m = 1$; and by simplifying it, equation (6.31)
it becomes (6.32). The acceleration a_t is known as the robot's global behaviour at any inertial
frame, and F_t is defined as the descriptive equations of dynamic effects. The approximated
real acceleration model is denoted by the next equilibrium condition,

$$\mathbf{a}_t^R = \mathbf{F}_t^\mu \tag{6.32}$$

The equation (6.32) describes the boundary case of equilibrium for $\mathbf{a}_t = 0$ constrained by the
following statements,

Theorem 6.3.1. *(equilibrium conditions)*

1. *when $d\hat{\mathbf{v}}_t/dt = 0$, $\hat{\mathbf{v}}_t$ is constant.*

   (a) *therefore for equation (6.1), $\hat{\mathbf{v}}_{t-1} \equiv \hat{\mathbf{v}}_t$, and $\eta = 1$, then $\mathbf{v}_t = 0$*

2. *the condition for equation (6.1), $\hat{\mathbf{v}}_t = 0$, when the robot is initially stopped, or when
   it reached its final goal destination.*

   (a) *therefore for equation (6.1), $\hat{\mathbf{v}}_t = 0$ and $\eta = 1$, then $\mathbf{v}_t = 0$*

*when such limit case condition occurs, sum of all accelerations will meet the condition
for the equilibrium case when,*

$$\frac{d\hat{\mathbf{v}}}{dt} = \mathbf{F}_t - \mathbf{a}_t^R = 0$$

The equilibrium condition of global accelerative model is then demonstrated by algebraic
development to validate the expression (6.32).

By substituting (6.29) and (6.4) in (6.3) with $\psi_t = 0$ (both described in common inertial frame) as follows,

$$v_t \dot{\theta}_t \begin{pmatrix} -\sin \theta_t \\ \cos \theta_t \end{pmatrix} + \dot{v}_t \begin{pmatrix} \cos \theta_t \\ \sin \theta_t \end{pmatrix} = \mathbf{F}_t^o + \sum_\alpha \mathbf{F}_t^\alpha + \sum_\gamma \mathbf{F}_t^\gamma \qquad (6.33)$$

The term $\mathbf{F}_t^o$ is the robot's internal motivation with functional form in (6.34), which makes the robot move along multiple local goals. Where $\delta = \|\mathbf{x}_{t+1} - \mathbf{x}_t\|$ is the distance between the actual robot location $\mathbf{x}_t$ and the next desired goal $\mathbf{x}_{t+1}$.

$$\mathbf{F}_t^o = \frac{1}{\tau} \left(\frac{v^o}{\|\vec{\delta}_t\|} (\mathbf{x}_{t+1} - \mathbf{x}_t) - \mathbf{v}_t \right) \qquad (6.34)$$

The ideal linear speed v^o sets a desired speed in xy components. There is a vector of actual measured velocity $\mathbf{v}_t$, and a relaxation time τ that defines the time taken for speeds change.

Arranging (6.33) by dropping off the velocity measurement, the next algebraic steps are developed

$$v_t \dot{\theta}_t \begin{pmatrix} -\sin \theta_t \\ \cos \theta_t \end{pmatrix} + \dot{v}_t \begin{pmatrix} \cos \theta_t \\ \sin \theta_t \end{pmatrix} = \frac{1}{\tau} \left(\frac{v^o}{\|\vec{\delta}_t\|} (\mathbf{x}_t - \mathbf{x}_{t+1}) - \mathbf{v}_t \right) + \sum_\alpha \mathbf{F}_t^\alpha + \sum_\gamma \mathbf{F}_t^\gamma \qquad (6.35)$$

therefore,

$$\tau \left(v_t \dot{\theta}_t \begin{pmatrix} -\sin \theta_t \\ \cos \theta_t \end{pmatrix} + \dot{v}_t \begin{pmatrix} \cos \theta_t \\ \sin \theta_t \end{pmatrix} - \sum_\alpha \mathbf{F}_t^\alpha - \sum_\gamma \mathbf{F}_t^\gamma \right) = \frac{v^o}{\|\vec{\delta}_t\|} (\mathbf{x}_t - \mathbf{x}_{t+1}) - \mathbf{v}_t \qquad (6.36)$$

We are treating the usual condition where both v_t and $\dot{\theta}_t$ are approximately uniforms in small periods of time τ (with very small variations for τ), where in our context we define that $\int_t v_t \dot{\theta}_t \sin \theta_t dt \approx v_t \theta_t \sin \theta_t$, since $\sin \theta_{t-1} \approx \sin \theta_t$.

$$v_t \theta_t \begin{pmatrix} -\sin \theta_t \\ \cos \theta_t \end{pmatrix} + v_t \begin{pmatrix} \cos \theta_t \\ \sin \theta_t \end{pmatrix} - \sum_\alpha \mathbf{v}_t^\alpha - \sum_\gamma \mathbf{v}_t^\gamma = \frac{v^o}{\|\vec{\delta}_t\|} (\mathbf{x}_t - \mathbf{x}_{t+1}) - \mathbf{v}_t \qquad (6.37)$$

Hence, due to units of time of τ, left-sided equation terms changed from m/s^2 into m/s, for which there is an integrable functional form for $\hat{\mathbf{v}}_t$.

Thus, by dropping off the real approximated velocity vector,

$$\hat{\mathbf{v}}_t = \frac{v^o}{\|\vec{o}_t\|}(\mathbf{x}_t - \mathbf{x}_{t+1}) + \sum_\alpha \mathbf{v}_t^\alpha + \sum_\gamma \mathbf{v}_t^\gamma - v_t \left(\theta_t \begin{pmatrix} -\sin\theta_t \\ \cos\theta_t \end{pmatrix} - \begin{pmatrix} \cos\theta_t \\ \sin\theta_t \end{pmatrix} \right) \tag{6.38}$$

Our equation has now been solved in (6.40) as to have a model of motion that combines the robot's fixed-frame, with external dynamic constraints,

$$\hat{\mathbf{v}}_t = \frac{v^o}{\|\vec{o}_t\|}(\mathbf{x}_t - \mathbf{x}_{t+1}) + \sum_\alpha \mathbf{v}_t^\alpha + \sum_\gamma \mathbf{v}_t^\gamma - v_t \begin{pmatrix} \sin\theta_t & \cos\theta \\ -\cos\theta_t & \sin\theta_t \end{pmatrix} \cdot \begin{pmatrix} -\theta_t \\ 1 \end{pmatrix} \tag{6.39}$$

Proposition 6.3.2 (The navigation model). *A model of motion for any inertial frame.*

$$\hat{\mathbf{v}}_t = -d_\tau(\dot{\theta} + \dot{\psi}) \begin{pmatrix} -\sin\theta_t \\ \cos\theta_t \end{pmatrix} - v_t \begin{pmatrix} \cos\theta_t \\ \sin\theta_t \end{pmatrix} + \sum_\alpha \mathbf{v}_t^\alpha + \sum_\gamma \mathbf{v}_t^\gamma + \frac{v^o}{\|\vec{o}_t\|}(\mathbf{x}_{t+1} - \mathbf{x}_t) \tag{6.40}$$

Where $d_\tau = \tau v_t$, it is defined as a short robot's displacement during the relaxation period of time. The expression (6.34) will be developed in next sections, as it involves the robot's actual and posterior position vectors.

6.4 Position model

The robot position vector $\mathbf{x}_t = (x, y)^T$ is a summation of all estimated positions overtime w.r.t. a common inertial frame from its starting position up to the actual time. The position vector is calculated based on the actuators rotational kinematic model to quantify displacements. We propose a solution to deduce $\mathbf{x}_t$ based on controlling the wheels actuator. Although, the kinematic parameters are fundamentals for any planning algorithm as reported [41], we provide a general solution that only consider any robot's angular velocity ω_t equation.

$$\omega_t = \frac{v_x^\pm}{l} \tag{6.41}$$

The distance l from robot's centroid to any wheel's contact point with surface is a numeric constant value that is geometrically inferred from the robot's size, as in (6.42),

$$l = \frac{\sqrt{W^2 + L^2}}{2} \tag{6.42}$$

With W as the robot width, and L the robot's length. On such basis, it is formulated that,

$$\cos(\alpha) = \frac{W/2}{l} \tag{6.43}$$

As we are interested on the linear differential velocity projected over X-axis, we call it $v_x^{\pm}$ to satisfy the equation (6.41), we defined the following relation between the differential linear velocity and its x-component

$$\cos(\alpha) = \frac{v_x^{\pm}}{v^r - v^l} = \frac{v_x^{\pm}}{v^{\pm}}; \quad v^{\pm}\cos(\alpha) = v_x^{\pm} \tag{6.44}$$

The equations (6.43) and (6.44) are equivalent, thus expressing (6.45) and then dropping off $v_x^{\pm}$

$$\frac{v_x^{\pm}}{v^{\pm}} = \frac{W/2}{l} \tag{6.45}$$

We substitute $v_x^{\pm}$ into (6.41) and algebraically arranging we deduce the following expression,

Proposition 6.4.1. *Robot's instantaneous angular velocity model.*

$$\omega_t = \frac{(v_t^r - v_t^l)W}{l(W^2 + L^2)} = \frac{r(\dot{\varphi}_t^r - \dot{\varphi}_t^l)W}{l(W^2 + L^2)} = K(\dot{\varphi}_t^r - \dot{\varphi}_t^l) \tag{6.46}$$

For simplicity in this approach, we assume equivalents all wheels nominal radius, so that let K be defined as a constant numeric value by

$$K = \frac{rW}{l(W^2 + L^2)} \tag{6.47}$$

For the case of two-wheel (dual asynchronous velocities) instead of four-wheel, $L = 0$ and the same proposition applies.

$$\omega_t = Kg(\dot{\varphi}_t^1, \dots, \dot{\varphi}_t^k) \tag{6.48}$$

Where K is a constant, and the function $g(\cdot)$ represents the yaw rate model with wheels rotational velocities φ_t^i as input parameters (its number will depend on the type of kinematic structure). In equation (6.48), the angular velocity is directly controlled by the wheels rotation. It is worth highlighting that the framework allows integration of other kinds of kinematic constraints by changing the ω_t model accordingly, such as the Ackerman type, synchronised type, differential drive, or the platforms studied in [ ].

The approach to infer $\mathbf{x}_t$ and $\mathbf{x}_{t+1}$ is by quantifying the wheels angular displacement directly by the speed drivers. We take advantage of the control hardware (motor drivers) which works under asymptotic functions (although a non-linear motor speed curve will vary from product to product). A general relationship between actuator's angular speed $\dot{\varphi}_t$ and a digital control variable Ω_t kinematics is given by

$$\dot{\varphi}_t(\Omega) = \left(\frac{a}{1 + e^{-\Lambda\Omega-\mu}} \right) - b \tag{6.49}$$

Where a and b are constants that adjust the non-linear angular velocity behaviour curve, Λ is the constant of fast asymptotic fall, Ω is a control digital word which is associated with an angular speed given directly by a user program, and μ is the central value of the velocity curve. By solving (6.49), we integrate the equation to obtain the next expression (6.50) en terms of wheels instantaneous angle of rotation,

$$\int_a^b \dot{\varphi}(\Omega)d\Omega = \varphi(\Omega) = (a - b)\Omega + \frac{a}{\Lambda}\ln(1 + e^{\Lambda(\mu-\Omega)}) \tag{6.50}$$

We synthesize the robot's direction and deduce a formal position model equation as expressed in the vector form by (6.51) with the k rotation velocities $\dot{\varphi}_t^i$.

$$\mathbf{x}_t = \begin{pmatrix} x_0 \\ y_0 \end{pmatrix} + \int_{t_0}^{t_n} \left(v_0 + \int_{t_1}^{t_n} \dot{v}_t dt \begin{pmatrix} \cos(\theta_0 + K\int^t g(\dot{\varphi}_t^1,\ldots,\dot{\varphi}_t^k)dt) \\ \sin(\theta_0 + K\int^t g(\dot{\varphi}_t^1,\ldots,\dot{\varphi}_t^k)dt) \end{pmatrix} dt \right) dt \tag{6.51}$$

Nevertheless, the problem of robot skid/slip is overcome by combining with the method reported in [ ] that deploys an in-house made inertial unit. It works reasonable because yaw rates can directly be controlled by using low level commands. Thus, by simplifying previous expression,

$$\Phi_t = K\int_{t_1}^{t_n} g(\dot{\varphi}_t^1,\ldots,\dot{\varphi}_t^n)dt \tag{6.52}$$

Substituting (6.50) in (6.52) and algebraically solving,

$$\Phi_t = K\left((a-b)\Omega^r + \frac{a}{\Lambda}\ln(1+e^{\Lambda(\mu-\Omega^r)}) - (a-b)\Omega^l + \frac{a}{\Lambda}\ln(1+e^{\Lambda(\mu-\Omega^l)})\right) \quad (6.53)$$

then,

$$\Phi_t = K\left((a-b)(\Omega^r - \Omega^l) + \frac{a}{\Lambda}\ln\left(\frac{1+e^{\Lambda(\mu-\Omega^r)}}{1+e^{\Lambda(\mu-\Omega^l)}}\right)\right) \quad (6.54)$$

Hence, the actual position vector is written as,

$$\mathbf{x}_{t_n} = \begin{pmatrix} x_0 \\ y_0 \end{pmatrix} + \int_{t_0}^{t_n}\left(v_0 + \int_{t_1}^{t_n} \dot{v}_t dt \begin{pmatrix} \cos(\theta_0 + \Phi_t) \\ \sin(\theta_0 + \Phi_t) \end{pmatrix} dt\right) dt \quad (6.55)$$

The orientation θ_t is solved by integration w.r.t. the time interval $[t_0, t_n]$, in which wheels rotations are controlled rather than collecting absolute odometry measurements (as commonly proposed by other approaches). Thus, reformulating the robot's angle by

$$\theta_t = \theta_0 + K\int_{t_1}^{t} g(\dot{\varphi}_t^1, \ldots, \dot{\varphi}_t^n) dt \quad (6.56)$$

and solving for the instantaneous angle,

$$\theta_t = \theta_0 + \Phi_t \quad (6.57)$$

We assumed that the magnitude of the robot's angular acceleration $d\omega/dt$ at every control loop is much smaller than the magnitude of the angular velocity. Arranging the actual position vector to be implemented in terms of the robot kinematic structure,

$$\mathbf{x}_t = \begin{pmatrix} \dot{x}_{t+1} \\ \dot{y}_{t+1} \end{pmatrix}\Delta t + \frac{r}{2}g\left(\dot{\varphi}_t^r, \ldots, \dot{\varphi}_t^l\right)\Delta t \begin{pmatrix} \cos(\theta_0 + \Phi_t) \\ \sin(\theta_0 + \Phi_t) \end{pmatrix} \quad (6.58)$$

According to figure 6.2, in order to alter the robot's orientation towards a next desired destination $\mathbf{x}_{t+1}$, the motion control is based on the collection of consecutive sensor data that feedback the controller. The next desired position $\mathbf{x}_{t+1}$ defines a Cartesian objective, either attractive or repulsive, which will depend on the nature of the objective. Rather than a single Cartesian point, this objective is referred as a territorial section, or area scoping the ideal Cartesian position ()no accuracy is required.

Furthermore, the next position vector model arises from a function $\mathbf{f}'_t$ that uses the actual orientation θ_t.

$$\mathbf{x}_{t+1} = \mathbf{x}_t + \int_t w_t(\mathbf{m}_t, \mathbf{f}'_t)\mathbf{f}'_t(\delta_\mu)\mathrm{d}t \tag{6.59}$$

In this approach we alter the actual orientation θ_t by weighting the accelerative navigation function $w_t(\cdot)\mathbf{f}_t^{\gamma,\alpha}$ previously given in equation (6.6) and equation (6.7). The fundamentals of this algorithm is focused on equation (12.2) describing $\mathbf{m}_t \cdot \mathbf{f}_t$. This expression quantifies the alignment of perpendicularity between yaw rate and a desired orientation $\mathbf{m}_t$. If $\mathbf{m}_t$ and $\mathbf{f}_t$ are approximately aligned, then it means that the velocity orientation is projected along the actual desired goal and altering direction is not required. However, $\|\mathbf{f}_t\| \cos(\phi_t)$ is the acceleration

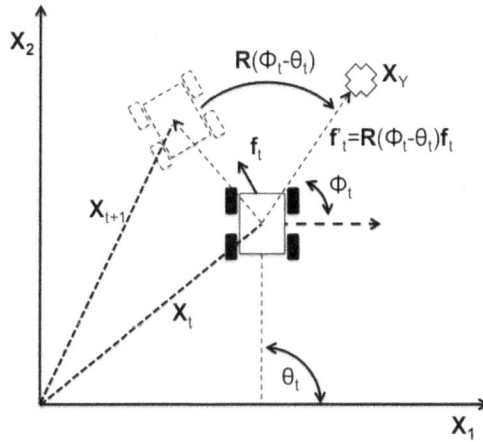

Figure 6.2: Geometric definition of $\mathbf{f}'_t$.

magnitude along the horizontal axis (common frame) respect to objective angle ϕ. A very small value of $\|\mathbf{f}_t\| \cos(\phi)$, signifies that practically no change in direction is required. If such magnitude is too large, an important correction in orientation must be established through the weighting factor λ. The objective (attractive or repulsive) is represented by sensor data features, the more sensors detect the same feature, the more certainty about the dirtection objective will improve the weighting factor λ. If ϕ is very near or along the robot heading axis (about 90°), then $\lambda = 1$ approximately (see equation (6.9). The actual accelerative force $\mathbf{f}_t$ is altered and defined as $\mathbf{f}'_t$, there is an objective angle correction $(\phi - \theta_t)$, thus, the direction of $\|\mathbf{f}_t\|$ is rotated by equation (6.60),

$$\mathbf{f}'_t = \mathbf{R}(\phi_i - \theta_t)\mathbf{f}_t \tag{6.60}$$

where $\mathbf{R}(\phi - \theta_t)$ an Euler rotation matrix that corrects the yaw. Thus, extending the expression, we now have,

$$\mathbf{f}'_t = \begin{pmatrix} f_x \cos(\phi_i - \theta) - f_y \sin(\phi_i - \theta) \\ f_x \sin(\phi_i - \theta) - f_y \cos(\phi_i - \theta) \end{pmatrix} \tag{6.61}$$

by developing in the vector form $\mathbf{f}'_t = (f_x, f_y)^T$, the new next-position vector,

$$\mathbf{x}_{t+1} = \begin{pmatrix} x_t \\ y_t \end{pmatrix} + \iint_t \mathbf{f}'_t \, d^2t = \begin{pmatrix} x_t \\ y_t \end{pmatrix} + \iint_t \begin{pmatrix} f_x \cos(\phi - \theta_t) - f_y \sin(\phi - \theta_t) \\ f_x \sin(\phi - \theta_t) - f_y \cos(\phi - \theta_t) \end{pmatrix} d^2t \tag{6.62}$$

The new desired direction w.r.t. the actual orientation is given by the vector $\mathbf{f}'_t$, which is a transformation into the global coordinate frame, since observations are locals.

6.5   Exponential derivatives

A gradient vector field assigns the direction of a function leading to each Cartesian point. The gradients can be viewed as accelerations acting on a positive sense, attracted to the negative goal. Obstacles have a positive sense that forms a repulsive acceleration leading the robot away from the obstacles. The combination of repulsive and attractive accelerations directs the robot from a starting location to the goal location while avoiding obstacles.

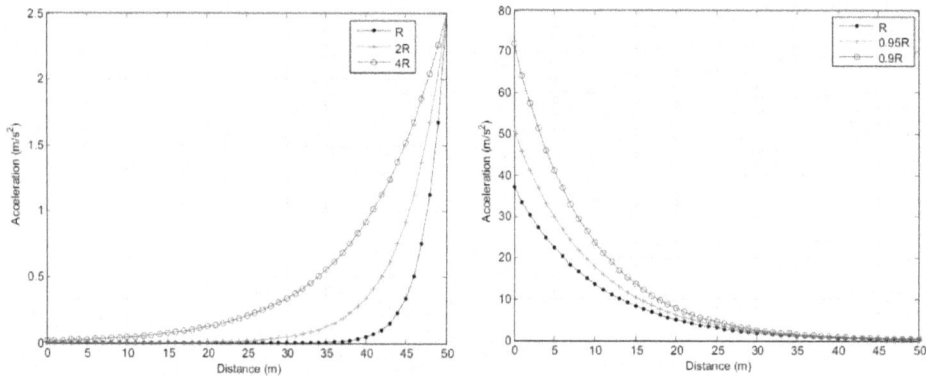

Figure 6.3: Exponential derivatives in 1D. Attractive (left); Repulsive (right).

A vector field is a set of vectors or potential functions, which are differentiable real-valued functions; where the gradient of the potential is the acceleration.

The gradient is a vector with components that point out along a direction that locally maximises the function.

Definition 6.5.1 (Repulsive exponential derivative).

$$f_t^{\alpha} = -\nabla_{\mu\alpha} u_{\alpha}^{o} R \frac{e^{\|\mathbf{x}_{\alpha}-\mathbf{x}_{\mu}\|/R}}{\|\mathbf{x}_{\alpha} - \mathbf{x}_{\mu}\|} \tag{6.63}$$

The denominator is determined by the factor $(R^{-1}\|\mathbf{x}_{\alpha} - \mathbf{x}_{\mu}\|)$ and defines the function to respond fast against situations in too close interaction with obstacles. Solving for its gradient operator, we obtain that $\mathbf{f}_t^{\alpha} \equiv \left(\frac{\partial f}{\partial x}, \frac{\partial f}{\partial x}\right)$. The equation (6.64) is a general function with the gradient operator, where u_{α}^{o} is a constant defining the acceleration amplitude, and R is a stationary value defining the asymptotic potential falling value.

$$f_t^{\alpha} = -\nabla_{\mu\alpha} u_{\alpha}^{o} R \frac{e^{\|\mathbf{x}_{\alpha}-\mathbf{x}_{\mu}\|/R}}{\|\mathbf{x}_{\alpha} - \mathbf{x}_{\mu}\|} \tag{6.64}$$

Differentiating w.r.t. $\partial f^{\alpha}/\partial x$ and $\partial f^{\alpha}/\partial y$,

$$\left(\frac{\partial f_t}{\partial x}, \frac{\partial f_t}{\partial y}\right) = \begin{pmatrix} \frac{Ru_{\alpha}^{o}}{\|(\mathbf{x}_{\alpha} - \mathbf{x}_{\mu})\|} e^{\|\mathbf{x}_{\alpha}-\mathbf{x}_{\mu}\|/R}(\frac{1}{2R})\|\mathbf{x}_{\alpha} - \mathbf{x}_{\mu}\|^{-1}2(x_{\alpha} - x_{\mu}) \\ \frac{Ru_{\alpha}^{o}}{\|(\mathbf{x}_{\alpha} - \mathbf{x}_{\mu})\|} e^{(x_{\alpha}-x_{\mu})^{1/2}/R}(\frac{1}{2R})\|\mathbf{x}_{\alpha} - \mathbf{x}_{\mu}\|^{-1}2(y_{\alpha} - y_{\mu}) \end{pmatrix} + \begin{pmatrix} e^{\|\mathbf{x}_{\alpha}-\mathbf{x}_{\mu}\|/R} u_{\alpha}^{o} R(\frac{-1}{2})\|\mathbf{x}_{\alpha} - \mathbf{x}_{\mu})\|^{-3}2(x_{\alpha} - x_{\mu}) \\ e^{\|\mathbf{x}_{\alpha}-\mathbf{x}_{\mu}\|/R} u_{\alpha}^{o} R(\frac{-1}{2})\|\mathbf{x}_{\alpha} - \mathbf{x}_{\mu}\|^{-3}2(y_{\alpha} - x_{\mu}) \end{pmatrix} \tag{6.65}$$

Arranging and ordering the terms,

$$\left(\frac{\partial f^{\alpha}}{\partial x}, \frac{\partial f^{\alpha}}{\partial y}\right) = \begin{pmatrix} \dfrac{u_{\alpha}^{o} R(x - x)e^{\|\mathbf{x}_{\alpha}-\mathbf{x}_{\mu}\|/R}}{\|\mathbf{x}_{\alpha} - \mathbf{x}_{\mu}\|^3} - \dfrac{u_{\alpha}^{o}(x - x)e^{\|\mathbf{x}_{\alpha}-\mathbf{x}_{\mu}\|/R}}{(\mathbf{x}_{\alpha} - \mathbf{x}_{\mu})} \\ \dfrac{u_{\alpha}^{o} R(y - y)e^{\|\mathbf{x}_{\alpha}-\mathbf{x}_{\mu}\|/R}}{\|\mathbf{x}_{\alpha} - \mathbf{x}_{\mu}\|^3} - \dfrac{u_{\alpha}^{o}(y - y)e^{\|\mathbf{x}_{\alpha}-\mathbf{x}_{\mu}\|/R}}{(\mathbf{x}_{\alpha} - \mathbf{x}_{\mu})} \end{pmatrix} \tag{6.66}$$

Thus, algebraically arranging and factorising common terms,

$$\left(\frac{\partial \mathbf{f}^\alpha}{\partial x}, \frac{\partial \mathbf{f}^\alpha}{\partial y} \right) = \begin{pmatrix} u_\alpha^o (x - x) e^{\|\mathbf{x}_\alpha - \mathbf{x}_\mu\|/R} \left(\dfrac{1}{(\mathbf{x}_\alpha - \mathbf{x}_\mu)} - \dfrac{R}{(\mathbf{x}_\alpha - \mathbf{x}_\mu)^{3/2}} \right) \\ u_\alpha^o (y - y) e^{\|\mathbf{x}_\alpha - \mathbf{x}_\mu\|/R} \left(\dfrac{1}{(\mathbf{x}_\alpha - \mathbf{x}_\mu)} - \dfrac{R}{(\mathbf{x}_\alpha - \mathbf{x}_\mu)^{3/2}} \right) \end{pmatrix} \tag{6.67}$$

in order to facilitate let us use notation $\mathbf{f}_t^\alpha$ instead, and let us define too $\|\delta_{\mu\alpha}\| = \|\mathbf{x}_\alpha - \mathbf{x}_\mu\|$,

$$\mathbf{f}_t^\alpha = \frac{u_\alpha^o e^{\|\mathbf{x}_\alpha - \mathbf{x}_\mu\|/R}}{\|\vec{\delta}_{\mu\alpha}\|} \begin{pmatrix} x_\alpha - x_\mu \\ y_\alpha - y_\mu \end{pmatrix} \left(\frac{1}{\|\vec{\delta}_{\mu\alpha}\|} - \frac{R}{(\mathbf{x}_\alpha - \mathbf{x}_\mu)} \right) \tag{6.68}$$

Finally, some terms of the derived equation may be substituted and simply expressed as,

$$\mathbf{f}_t^\alpha = u_\alpha^o \frac{e^{\|\vec{\delta}_{\mu\alpha}\|/R}}{\|\vec{\delta}_{\mu\alpha}\|} \left(\frac{1}{\|\vec{\delta}_{\mu\alpha}\|} - \frac{R}{\vec{\delta}_{\mu\alpha}} \right) \begin{pmatrix} x_\alpha - x_\mu \\ y_\alpha - y_\mu \end{pmatrix} \tag{6.69}$$

Previous expression if defined in terms of a velocity vector,

$$\mathbf{v}_t^\alpha = \int_t \mathbf{f}_t^\alpha dt \tag{6.70}$$

$$\mathbf{f}_t^\alpha = u_\alpha^o \frac{e^{\|\vec{\delta}_{\mu\alpha}\|/R}}{\|\vec{\delta}_{\mu\alpha}\|} \left(\frac{1}{\|\vec{\delta}_{\mu\alpha}\|} - \frac{R}{\vec{\delta}_{\mu\alpha}} \right) \begin{pmatrix} x_\alpha - x_\mu \\ y_\alpha - y_\mu \end{pmatrix} \tag{6.71}$$

Previous expression is defined in terms of velocities by (6.72), where such term will satisfy the real velocity of equation (6.40),

$$\mathbf{v}_t^\alpha = \int_t \mathbf{f}_t^\alpha dt \tag{6.72}$$

Similarly, equations controlling the robot course to a global goal destination yield motion behaviour as depicted by figure 6.3-left. The set of goals γs are defined a priori as intersection points along the full course path. The general $1D$ artificial potential equation is defined by,

Definition 6.5.2 (Attractive exponential derivative).

$$f_t^\gamma = -\nabla_{\mu\gamma} u_\gamma^o e^{-\|\vec{\delta}_{\mu\gamma}\|/R} \tag{6.73}$$

Where R represents the radius of a goal's territorial scope. The distance vector between the robot μ and the goal γ is defined by $\vec{\delta} = \mathbf{x}_t^\gamma - \mathbf{x}_t^\mu$. The constant factor u_γ^o scales the attractive accelerative forces amplitude. Solving for its gradient operator the attractive potential function general equation is defined by,

$$f_t^\gamma = -\nabla_{\mu\gamma} u^o e^{-\|\vec{\delta}_{\mu\gamma}\|/R} \tag{6.74}$$

Thus, by deriving the function w.r.t. x and y, it yields,

$$\left(\frac{\partial f_t^\gamma}{\partial x}, \frac{\partial f_t^\gamma}{\partial y} \right) = \begin{pmatrix} -u^o e^{-\|\mathbf{x}_\gamma - \mathbf{x}_\mu\|/R} (\frac{-1}{2}) \dfrac{\|\mathbf{x}_\gamma - \mathbf{x}_\mu\|}{R} 2(x_\gamma - x_\mu) \\ -u^o e^{-\|\mathbf{x}_\gamma - \mathbf{x}_\mu\|/R} (\frac{-1}{2}) \dfrac{\|\mathbf{x}_\gamma - \mathbf{x}_\mu\|}{R} 2(y_\gamma - y_\mu) \end{pmatrix} \tag{6.75}$$

simplifying the expression it now becomes

$$\mathbf{f}_t^\gamma = u^o e^{-\|\mathbf{x}_\gamma - \mathbf{x}_\mu\|/R} \begin{pmatrix} \dfrac{x_\gamma - x_\mu}{R\|\mathbf{x}_\gamma - \mathbf{x}_\mu\|} \\ \dfrac{y_\gamma - y_\mu}{R\|\mathbf{x}_\gamma - \mathbf{x}_\mu\|} \end{pmatrix} \tag{6.76}$$

Algebraically arranging, the 2D potential function becomes as follows,

$$\mathbf{f}_t^\gamma = u^o \frac{e^{-\|\vec{\delta}_{\mu\gamma}\|/R}}{R\|\vec{\delta}\|} \begin{pmatrix} x_\mu - x_\gamma \\ y_\mu - y_\gamma \end{pmatrix} \tag{6.77}$$

$$\mathbf{f}_t^\gamma = u_\gamma^o \frac{e^{-\|\vec{\delta}_{\mu\gamma}\|/R}}{R\|\vec{\delta}\|} \begin{pmatrix} x_\mu - x_\gamma \\ y_\mu - y_\gamma \end{pmatrix} \tag{6.78}$$

Finding a general solution for equation (6.40), previous expression is rather defined in terms of velocities, than accelerations (6.79), as a term that partially satisfies (6.40),

$$\mathbf{v}_t^\gamma = \int_t \mathbf{f}_t^\gamma dt \tag{6.79}$$

By combining both directional fields $\mathbf{F}_t^\alpha + \mathbf{F}_t^\gamma$, figure 6.4 illustrates an obstacle and a goal gradients interaction of robot accelerations. Experimentally, a set of known goal destinations γ^i were established and the robot was able to reach them all. The robot built up a map of accelerative interactions between attractive and repulsive directional fields (figure 6.5). Furthermore, during outdoor sensing experiments, numerous features of critical interest for system feedback

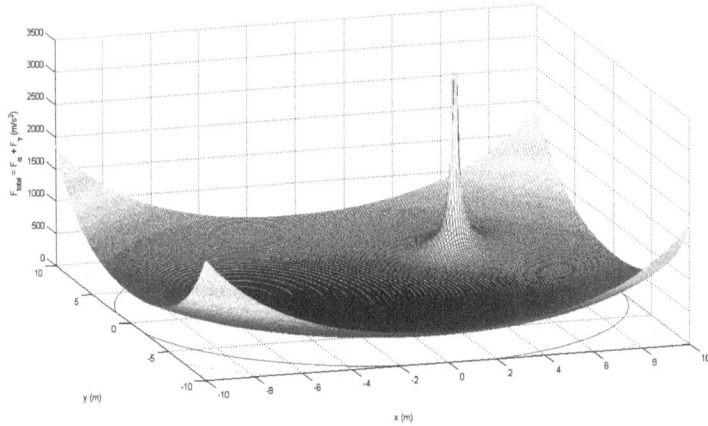

Figure 6.4: $3D$ potential fields combining $\mathbf{F}_\gamma + \mathbf{F}_\alpha$, at $< -5, 0 > \gamma$ is located, and at coordinates $(4, 5)$ an obstacle is located.

were detected. Such features are the directions to local goals $(\mathbf{m}_t)$. Each sensor observation is comprised of a high density repulsive local map, but concurrently combined with a priori attractive directional fields [12]. The resulting experimental directional fields map is depicted in figure 6.5. The nearer the obstacle, the larger becomes the accelerative potential force magnitude exerted by the proposed model.   In further experimental simulations, figure 6.6 shows how

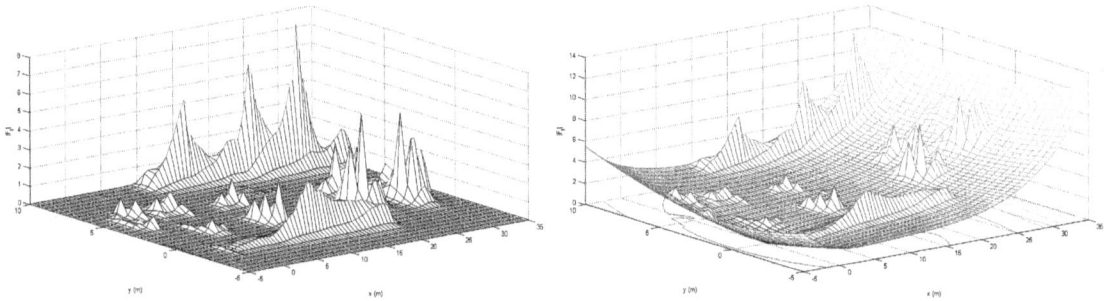

Figure 6.5:   Robot's directional fields dynamic interaction of accelerations. Left: repulsive accelerations; Right: attractive and repulsive.

functions evolve to safely avoid two obstacles. There exist two obstacles at $\mathbf{x}_\alpha = (15, 50)^T$ and $\mathbf{x}_\beta = (-2, 20)^T$. The robot parameters for this experiment are $v^o = 0.5m/s$ from 0m to 100m along the vertical Y-axis. In figure 6.6(top-left), the robot navigates vertically from $\mathbf{x}_{t_0}^\mu = (0, 0)^T$ up to $\mathbf{x}_{t_n}^\mu = (0, 100)^T$, where location of γ is denoted by a triangle. Figure 6.6 (top-right) depicts the robot's attractive accelerations towards the goal. As the distance $\|\gamma - \mu\|$ is getting shorter gradually, the rate of motion behaviour is decreasing until the robot reaches γ. In figure 6.6 (middle left-right), the acceleration components yielded by the presence of both obstacles are depicted versus the distance respect the actual robot positions. It is worth noting at figure 6.6 (down left and right) how the acceleration components evolve (dots x-component, and crosses y-component) along the y-axis. In particular at y-coordinate 20m and 50m to avoid obstacles β, and subsequently α, and then return to the desired trajectory.

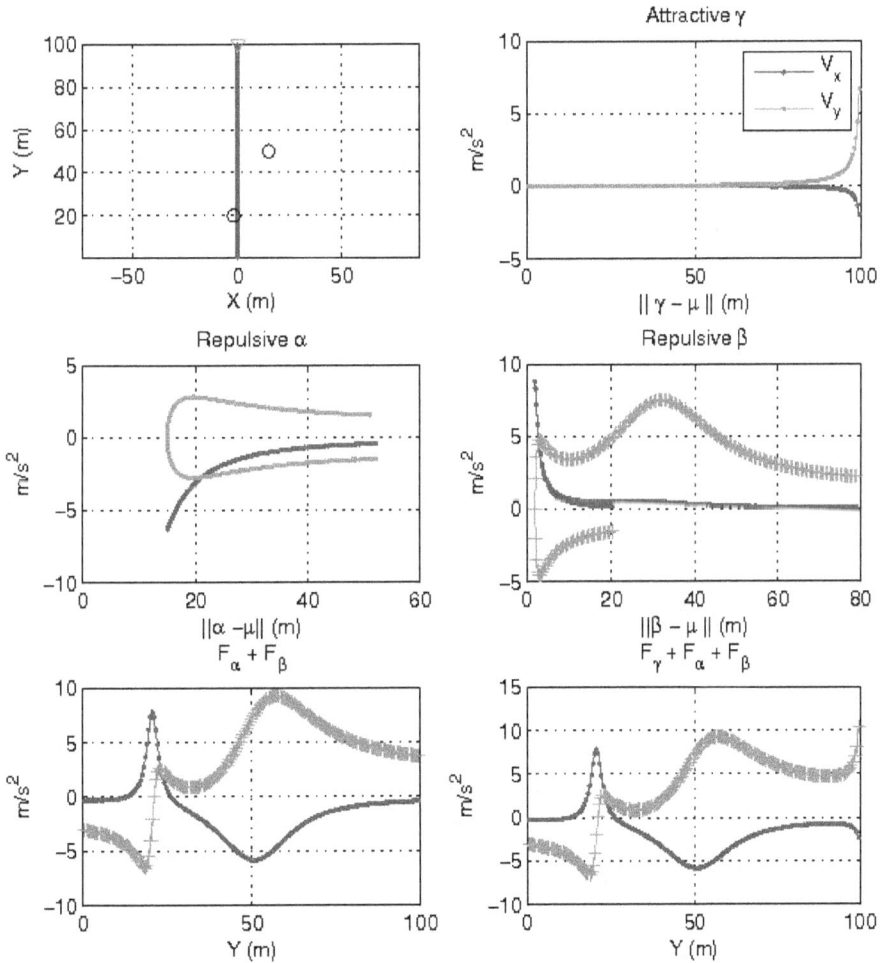

Figure 6.6:   Robot's initial position $\mathbf{x}_\mu = (0,0)^T$; goal at $\mathbf{x}_\gamma = (0,100)^T$; obstacles at $\mathbf{x}_\alpha = (15,50)^T$, and $\mathbf{x}_\beta = (-2,20)^T$.

Bibliography

[1] Mora M.C., Tornero J., *Path planning and trajectory generation using multi-rate predictive artificial potential fields*, IEEE/RSJ Intl. Conf. on Intelligent Robots and systems, France, pp. 2990–2995, 2008.

[2] Masound A.A., *Managing the dynamics of a harmonic potential field-guided robot in a cluttered environment*, IEEE Transaction on industrial electronics, Vol.56(2), pp. 488–496, 2010, doi: 10.1109/TIE.2008.2002720.

[3] Goncalves V.M., Pimienta L., Maia C.A., Dutra B.C., Pereira G.A., *Vector fields for robot navigation along time-varying curves in n-dimensions*, IEEE Trans. on Robotics, Vol. 26(4), pp. 647–659, 2010, doi: 10.1109/TRO.2010.2053077.

[4] Tychonievich L.A., Burton R.P., Tychonievich L.P., *Versatile Reactive Navigation*, IEEE/RSJ Intl. Conf. on Intelligent Robots and Systems, USA, pp. 2966–2972, Oct 2009.

[5] Kareem M., Garibeh M., Feilat E., *Dynamic motion planning for autonomous mobile robot using fuzzy potential field*, Proc. 6th Intl. Symposium on Mechatronics and its applications, UAE, Mar 2009.

[6] Baronov D., Baillieul J., *Autonomous vehicle control for ascending/descending along a potential field with two applications*, American Control Conf., USA, pp. 678–683, 2008, doi: 10.1109/ACC.2008.4586571.

[7] Ren I., Mclsaac K.A., Patel R.V., Peters T.M., *A Potential Field Model Using Generalized Sigmoid Functions*, IEEE Trans. on Systems, Man and Cybernetics–Part B, Vol.37(2), pp. 447–484, 2007, doi: 10.1109/TSMCB.2006.883866.

[8] Chunyu J., Qu Z., Pollak E., Falash M., *A New Reactive Target-tracking Control with Obstacle Avoidance in a Dynamic Environment*, American Control Conf., MO USA, pp. 3872–3877, Jun 2009.

[9] Charifa S., Bikdash M., *Comparison of geometrical, kinematic, and dynamic performance of several potential field methods*, IEEE Southeastcon, pp. 18–23, Mar 2009.

[10] Martínez-García E., Torres-Córdoba R., *4WD Skid-steer trajectory control of a rover with spring-based suspension analysis: Direct and inverse kinematic parameters solution*, Intelligent Robotics and Applications, Springer LNCS, Vol. 6425, Nov 2010.

[11] Loizou S.G., Kyriakopolous K.J., *Navigation of multiple kinematically constrained robots*, IEEE Trans. on Robotics, Vol. 24(1), pp. 221–231, 2008, doi: 10.1109/TRO.2007.912092.

[12] Huang W.H., Fajen B.R., Fink J.R., Warren W. H., *Visual navigation and obstacle avoidance using a steering potential function*, Robotics and Autonomous Systems, Vol 54, pp. 288-299, 2005, doi: 10.1016/j.robot.2005.11.004.

Chapter 7

EXPLORATION AND SEARCH ROBOT

NAVIGATION

Edgar A. Martínez García

Laboratorio de Robótica, Institute of Engineering and Technology
Universidad Autónoma de Ciudad Juárez, Mexico.

The process of a mobile robot exploring an environment to search for a goal, might be a common robotic mission in numerous applications. Assuming this process as a generalised engineering problem, we may assume that the unknown goal emits an arbitrary form of energy, and by deploying a mobile robot instrumented with a suitable sensing device, it will explore in order to found the goal. Hence, the robot's sensing device requires a calibration process to fit the data with the robot's Cartesian location. In this chapter, it is assumed that a mobile robot is instrumented with although general sensing device, but appropriate to search an objective goal. A general navigation model for exploring and searching is formulated for any type of mission, considering the sensor's measurements $\hat{E}$ minimally informative for the robot to have a good estimation of the goal location. Where $\hat{E}$ represents the general measurement of a type of energy (active or passive) by the goal destination. In next sections we will establish a mathematical approach to fit an empirical measurement model with a theoretical model to calibrate our equations assuming that the distance is a universal variable of interest for many researchers.

7.1 Fitting a theoretical-empirical model E_t-r_t

A calibration process between the instantaneous experimental measurement E_t and the theoretical distance r_t will yield a non-linear relation for generalised conditions.

Postulate 7.1.1. The actual measurement $\hat{E}_t$ has close relationship with actual distance $\|\boldsymbol{\xi}_t - \boldsymbol{\xi}_0\|$, and actual magnitude arising from an unknown destination. The general model of instantaneous distance r_t is,

$$r_t = \sqrt[2]{(x_0 - x_t)^2 + (y_0 - y_t)^2 + z_0^2} = \|\boldsymbol{\xi}_t - \boldsymbol{\xi}_0\|$$

When $\|\boldsymbol{\xi}_t - \boldsymbol{\xi}_0\| = 0$, the goal is reached. It denotes the robot at $\boldsymbol{\xi}_t$, and the unknown goal location at $\boldsymbol{\xi}_0 = (x_0, y_0, z_0)$.

The postulation is sustained by an empirical model to calibrate the theoretical model, using measurement data. The data representing distance measurements $\hat{r}$, versus a measured type of energy $\hat{E}$ is approached by a polynomial regression. The sum of quadratic residuals $s_r = \sum_i^n e^2$ is defined by the ith empirical observation of n measurements of $\hat{r}_i$ as a function of $\hat{E}_i$. The error $e = \hat{r}_i - r_i$, is given by the observation $\hat{r}_i$, and a theoretical model $r_i(E_i) = a_0 + a_1 E_i + a_2 E_i^2 + \cdots + a_k E_i^k$. Substituting a cubic model that may fit suitable for numerous physical systems (see sec. 1.6.3),

$$s_r = \sum_i^n (\hat{r}_i - r_i)^2 = \left(\hat{r}_i - a_0 + a_1 \hat{E}_i + a_2 \hat{E}_i^2 + a_3 \hat{E}_i^3\right)^2 \tag{7.1}$$

The partial derivations w.r.t. unknown parameters are led to find cubic model[1] parameters next,

$$\frac{\partial s_r}{\partial a_0} = 2\sum_i^n \left(\hat{r}_i - a_0 + a_1 \hat{E}_i + a_2 \hat{E}_i^2 + a_3 \hat{E}_i^3\right)(-1) \tag{7.2}$$

$$\frac{\partial s_r}{\partial a_1} = 2\sum_i^n \left(\hat{r}_i - a_0 + a_1 \hat{E}_i + a_2 \hat{E}_i^2 + a_3 \hat{E}_i^3\right)(-E_i) \tag{7.3}$$

---

[1]Nevertheless, the degree of the polynomial may be determined according to the practical problem.

$$\frac{\partial s_r}{\partial a_2} = 2 \sum_i^n \left(\hat{r}_i - a_0 + a_1 \hat{E}_i + a_2 \hat{E}_i^2 + a_3 \hat{E}_i^3 \right) \left(-E_i^2 \right) \tag{7.4}$$

$$\frac{\partial s_r}{\partial a_3} = 2 \sum_i^n \left(\hat{r}_i - a_0 + a_1 \hat{E}_i + a_2 \hat{E}_i^2 + a_3 \hat{E}_i^3 \right) \left(-E_i^3 \right) \tag{7.5}$$

Then, we arrange algebraically,

$$\sum_i^n \hat{r}_i = a_0(n) + a_1 \sum_i^n \hat{E}_i + a_2 \sum_i^n \hat{E}_i^2 + a_3 \sum_i^n \hat{E}_i^3 \tag{7.6}$$

$$\sum_i^n \hat{r}_i \hat{E}_i = a_0 \sum_i^n \hat{E}_i + a_1 \sum_i^n \hat{E}_i^2 + a_2 \sum_i^n \hat{E}_i^3 + a_3 \sum_i^n \hat{E}_i^4 \tag{7.7}$$

$$\sum_i^n \hat{r}_i \hat{E}_i^2 = a_0 \sum_i^n \hat{E}_i^2 + a_1 \sum_i^n \hat{E}_i^3 + a_2 \sum_i^n \hat{E}_i^4 + a_3 \sum_i^n \hat{E}_i^5 \tag{7.8}$$

$$\sum_i^n \hat{r}_i \hat{E}_i^3 = a_0 \sum_i^n \hat{E}_i^3 + a_1 \sum_i^n \hat{E}_i^4 + a_2 \sum_i^n \hat{E}_i^5 + a_3 \sum_i^n \hat{E}_i^6 \tag{7.9}$$

it follows to arrange and simplify in matrix form,

$$\begin{pmatrix} \sum_i^n \hat{r}_i \\ \sum_i^n \hat{r}_i \hat{E}_i \\ \sum_i^n \hat{r}_i \hat{E}_i^2 \\ \sum_i^n \hat{r}_i^n \hat{E}_i^3 \end{pmatrix} = \begin{pmatrix} n & \sum_i^n \hat{E}_i & \sum_i^n \hat{E}_i^2 & \sum_i^n \hat{E}_i^3 \\ \sum_i^n \hat{L}_i & \sum_i^n \hat{E}_i^2 & \sum_i^n \hat{E}_i^3 & \sum_i^n \hat{E}_i^4 \\ \sum_i^n \hat{E}_i^2 & \sum_i^n \hat{E}_i^3 & \sum_i^n \hat{E}_i^4 & \sum_i^n \hat{E}_i^5 \\ \sum_i^n \hat{E}_i^3 & \sum_i^n \hat{E}_i^4 & \sum_i^n \hat{E}_i^5 & \sum_i^n \hat{E}_i^6 \end{pmatrix} \cdot \begin{pmatrix} a_0 \\ a_1 \\ a_2 \\ a_3 \end{pmatrix} \tag{7.10}$$

As the main matrix is of squared size, we applied the theorem of Cramer to solve for the cubic polynomial coefficients (see sec. 1.2.4). Thus, the solution for coefficient a_0 is,

$$a_0 = \frac{\det \begin{pmatrix} \sum_i^n \hat{r}_i & \sum_i^n \hat{E}_i & \sum_i^n \hat{E}_i^2 & \sum_i^n \hat{E}_i^3 \\ \sum_i^n \hat{r}_i \hat{E}_i & \sum_i^n \hat{E}_i^2 & \sum_i^n \hat{E}_i^3 & \sum_i^n \hat{E}_i^4 \\ \sum_i^n \hat{r}_i \hat{E}_i^2 & \sum_i^n \hat{E}_i^3 & \sum_i^n \hat{E}_i^4 & \sum_i^n \hat{E}_i^5 \\ \sum_i^n \hat{r}_i \hat{E}_i^3 & \sum_i^n \hat{E}_i^4 & \sum_i^n \hat{E}_i^5 & \sum_i^n \hat{E}_i^6 \end{pmatrix}}{\det \begin{pmatrix} n & \sum_i^n \hat{E}_i & \sum_i^n \hat{E}_i^2 & \sum_i^n \hat{E}_i^3 \\ \sum_i^n \hat{E}_i & \sum_i^n \hat{E}_i^2 & \sum_i^n \hat{E}_i^3 & \sum_i^n \hat{E}_i^4 \\ \sum_i^n \hat{E}_i^2 & \sum_i^n \hat{E}_i^3 & \sum_i^n \hat{E}_i^4 & \sum_i^n \hat{E}_i^5 \\ \sum_i^n \hat{E}_i^3 & \sum_i^n \hat{E}_i^4 & \sum_i^n \hat{E}_i^5 & \sum_i^n \hat{E}_i^6 \end{pmatrix}} \tag{7.11}$$

the solution for coefficient a_1,

$$a_1 = \frac{\det \begin{pmatrix} n & \sum_i^n \hat{r}_i & \sum_i^n \hat{E}_i^2 & \sum_i^n \hat{E}_i^3 \\ \sum_i^n \hat{E} & \sum_i^n \hat{r}_i \hat{E}_i & \sum_i^n \hat{E}_i^3 & \sum_i^n \hat{E}_i^4 \\ \sum_i^n \hat{E}_i^2 & \sum_i^n \hat{r}_i \hat{E}_i^2 & \sum_i^n \hat{E}_i^4 & \sum_i^n \hat{E}_i^5 \\ \sum_i^n \hat{E}_i^3 & \sum_i^n \hat{r}_i \hat{E}_i^3 & \sum_i^n \hat{E}_i^5 & \sum_i^n \hat{E}_i^6 \end{pmatrix}}{\det \begin{pmatrix} n & \sum_i^n \hat{E}_i & \sum_i^n \hat{E}_i^2 & \sum_i^n \hat{E}_i^3 \\ \sum_i^n \hat{E}_i & \sum_i^n \hat{E}_i^2 & \sum_i^n \hat{E}_i^3 & \sum_i^n \hat{E}_i^4 \\ \sum_i^n \hat{E}_i^2 & \sum_i^n \hat{E}_i^3 & \sum_i^n \hat{E}_i^4 & \sum_i^n \hat{E}_i^5 \\ \sum_i^n \hat{E}_i^3 & \sum_i^n \hat{E}_i^4 & \sum_i^n \hat{E}_i^5 & \sum_i^n \hat{E}_i^6 \end{pmatrix}} \tag{7.12}$$

the solution for coefficient a_2,

$$a_2 = \frac{\det \begin{pmatrix} \sum_i^n & \sum_i^n \hat{E}_i & \sum_i^n \hat{r}_i & \sum_i^n \hat{E}_i^3 \\ \sum_i^n \hat{E}_i & \sum_i^n \hat{E}_i^2 & \sum_i^n \hat{r}_i \hat{E}_i & \sum_i^n \hat{E}_i^4 \\ \sum_i^n \hat{E}_i^2 & \sum_i^n \hat{E}_i^3 & \sum_i^n \hat{r}_i \hat{E}_i^2 & \sum_i^n \hat{E}_i^5 \\ \sum_i^n \hat{E}_i^3 & \sum_i^n \hat{E}_i^4 & \sum_i^n \hat{r}_i \hat{E}_i^3 & \sum_i^n \hat{E}_i^6 \end{pmatrix}}{\det \begin{pmatrix} n & \sum_i^n \hat{E}_i & \sum_i^n \hat{E}_i^2 & \sum_i^n \hat{E}_i^3 \\ \sum_i^n \hat{E}_i & \sum_i^n \hat{E}_i^2 & \sum_i^n \hat{E}_i^3 & \sum_i^n \hat{E}_i^4 \\ \sum_i^n \hat{E}_i^2 & \sum_i^n \hat{E}_i^3 & \sum_i^n \hat{E}_i^4 & \sum_i^n \hat{E}_i^5 \\ \sum_i^n \hat{E}_i^3 & \sum_i^n \hat{E}_i^4 & \sum_i^n \hat{E}_i^5 & \sum_i^n \hat{E}_i^6 \end{pmatrix}} \tag{7.13}$$

the solution for coefficient a_3,

$$a_3 = \frac{\det \begin{pmatrix} \sum_i^n & \sum_i^n \hat{E}_i & \sum_i^n \hat{E}_i^2 & \sum_i^n \hat{r}_i \\ \sum_i^n \hat{E}_i & \sum_i^n \hat{E}_i^2 & \sum_i^n \hat{E}_i^3 & \sum_i^n \hat{r}_i \hat{E}_i \\ \sum_i^n \hat{E}_i^2 & \sum_i^n \hat{E}_i^3 & \sum_i^n \hat{E}_i^4 & \sum_i^n \hat{r}_i \hat{E}_i^2 \\ \sum_i^n \hat{E}_i^3 & \sum_i^n \hat{E}_i^4 & \sum_i^n \hat{E}_i^5 & \sum_i^n \hat{r}_i \hat{E}_i^3 \end{pmatrix}}{\det \begin{pmatrix} n & \sum_i^n \hat{E}_i & \sum_i^n \hat{E}_i^2 & \sum_i^n \hat{E}_i^3 \\ \sum_i^n \hat{E}_i & \sum_i^n \hat{E}_i^2 & \sum_i^n \hat{E}_i^3 & \sum_i^n \hat{E}_i^4 \\ \sum_i^n \hat{E}_i^2 & \sum_i^n \hat{E}_i^3 & \sum_i^n \hat{E}_i^4 & \sum_i^n \hat{E}_i^5 \\ \sum_i^n \hat{E}_i^3 & \sum_i^n \hat{E}_i^4 & \sum_i^n \hat{E}_i^5 & \sum_i^n \hat{E}_i^6 \end{pmatrix}} \tag{7.14}$$

Thus, the new adjusted theoretical model $r(E)$ is a cubic polynomial of the general form,

$$r(E_t) = a_0 + a_1 E_t + a_2 E_t^2 + a_3 E_t^3 \qquad (7.15)$$

7.2   Model for directional derivatives

To establish a math form for the directional fields in terms of distances and potentials, the following postulate is stated,

Postulate 7.2.1. The equilibrium condition is stated as $\hat{E}_t = 0$, when $\boldsymbol{\xi}_t = \boldsymbol{\xi}_0$, through its continuous search by $-\nabla_{r_t} Q$.

$$\| - \nabla Q_t \| = \left\| \left(\frac{\partial Q(\hat{E}_t, r_t)}{\partial x} + \frac{\partial Q(\hat{E}_t, r_t)}{\partial y} \right) \right\| = 0 \qquad (7.16)$$

We apply the gradient operator to derive E_t w.r.t. x and y, so we obtain an analytical solution on how the energy of E_t behaves w.r.t. the instantaneous Cartesian position $\boldsymbol{\xi}_t$. From a mobile robotics planning approach, the problem to be solved is to yield the automatic search of the unknown Cartesian location $\boldsymbol{\xi}_0$, which is a constant position. The general robot motion equation is given by (7.17), its derivative w.r.t. t yields the linear velocity vector $\mathbf{v}^Q$ exerted by the integration w.r.t. time of the gradient $-\nabla Q$. Where u^o is a constant value that setup velocity amplitude.

$$\mathbf{v}^Q(t) = u^o \int_t \left(\frac{\partial Q_t}{\partial x} \mathbf{i} + \frac{\partial Q_t}{\partial y} \mathbf{j} \right) dt \qquad (7.17)$$

Let us define $x = x_{t_1} - x_0$, and $y = y_{t_1} - y_0$, and substituting the functional form of $Q(\hat{E}, r_t)$ as a function of the actual measurement, and the non linear fitted model $r_t(\hat{E})$,

$$\mathbf{v}^Q(t) = -u^o \int_t \left(\hat{E} \, r_t \, x \, \mathbf{i} + \hat{E} \, r_t \, y \, \mathbf{j} \right) dt \qquad (7.18)$$

algebraically re-arrange and group terms,

$$\mathbf{v}^Q = -u^o \, \hat{E}_t \int_t (r_t x \, \mathbf{i} + r_t y \, \mathbf{j}) \, dt \qquad (7.19)$$

before integrating w.r.t. time, we substitute the Cartesian displacements by averaged component accelerations $r_t = a_t t^2$, so that,

$$\mathbf{v}^Q = -u^o \hat{E}_t \int_t \left(a_t \, t^2 \ddot{x} \, t^2 \, \mathbf{i} + a_t \, t^2 \, \ddot{y} \, t^2 \, \mathbf{j} \right) \mathrm{dt} \tag{7.20}$$

Hence, integrating the equation in time,

$$\mathbf{v}^Q = -u^o \hat{E}_t \left(a_t \ddot{x} \frac{t^5}{5} \mathbf{i} + a_t \ddot{y} \frac{t^5}{5} \mathbf{j} \right) + c \tag{7.21}$$

backing to original distance variables, and algebraically arranging,

$$\mathbf{v}^Q = -u^o \, \hat{E}_t \left(r_t x \frac{t}{5} \mathbf{i} + r_t y \frac{t}{5} \mathbf{j} \right) + c \tag{7.22}$$

and finally, our velocity vector is given by a directional derivative equation,

$$\mathbf{v}^Q = \frac{u^o}{5} \, t \, \hat{E}_t \, r_t \, ((x_{t_1} \pm \varepsilon_1) \, \mathbf{i} + (y_{t_1} \pm \varepsilon_2) \, \mathbf{j}) + c \tag{7.23}$$

Where, ε_1 and ε_2 are discussed in next section. The potential behaviour of E_t w.r.t. Cartesian
positions is the function of attractive potential field to find the goal destination.

7.3   System of non linear equations for searching tasks

Finding a solution to reach the goal or equilibrium point from a vector field perspective, $\boldsymbol{\xi}_0$
is postulated as a search problem (see figure 7.1). In principle, this postulate states that it is
possible to infer the Cartesian values $\boldsymbol{\xi}_0$ automatically by iteratively feed-backing the $Q(\hat{E}, r_t)$
measurements. The distance r_t is calculated by measuring $\hat{E}$, which is the key-issue to pro-
gressively lead the robot towards $\boldsymbol{\xi}_0$.

Postulate 7.3.1. The distance $r_t = \|\boldsymbol{\xi}(t) - \boldsymbol{\xi}_0\|$ is a relationship of time, and a measurement
that is modelled by the equation of the measurement model,

$$\delta_t = M \, u(t) + z_0 \tag{7.24}$$

Where, M is an arbitrary measurement factor, z_0 is a known vertical height of the goal position, and $f(Q, t)$ is any non linear potential function associated to the measurement,

$$u(t) = f(Q, t)$$

In figure 7.1-a) the robot is depicted at two different Cartesian locations at temporal frame bounded by times t_1 and t_2. In time frame of length $|t_2 - t_1|$, a two-dimension robot's displacement Δx and Δy between $\boldsymbol{\xi}_0$ and the robot actual position occurs. At time t_1 the robot is as far/near of $\boldsymbol{\xi}_0$ as $(\Delta x_1, \Delta y_1)^T$. Subsequently at time t_2 the robot is as far/near of $\boldsymbol{\xi}_0$ as $(\Delta x_2, \Delta y_2)^T$. The robot's absolute distance w.r.t. $\boldsymbol{\xi}_0$ changes from δ_1 to δ_2 at t_1 and t_2 respectively. The robot's wandering-like motion is remarkably non linear, however the robot is continuously attracted to $\boldsymbol{\xi}_0$ according to the motion behaviour of the potential function.

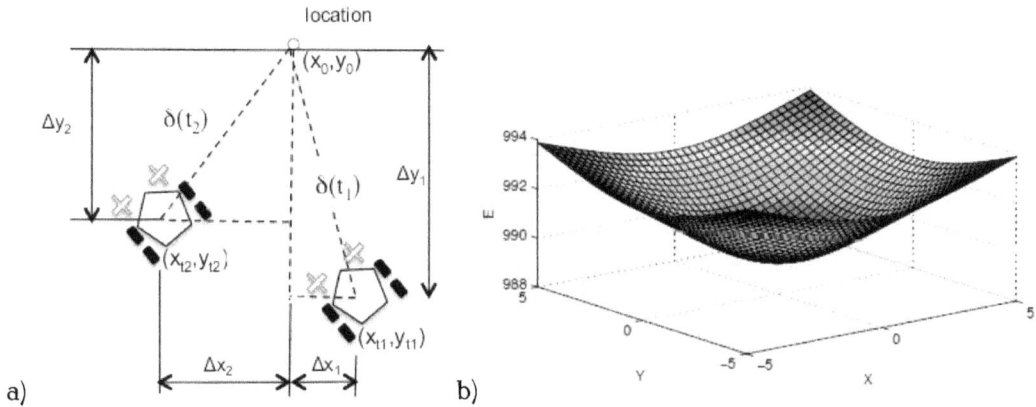

Figure 7.1: a) Search and exploring parameters; b) attractive directional field.

It is desired to find a solution for the vector $\boldsymbol{\xi}_0$. Thus, it is assumed that $\boldsymbol{\xi}(t)$ is iteratively known online just by analysing pairs of sensor measurements with consecutive spatio-temporal differences. Thus, a system of non linear equations with quadratic terms are proposed, which consider pairs of consecutive sensor observations, and pairs of robots' postures combined within the same system of non linear equations. Hence, it gradually approach the robot to the pipe underground fail (parameters are illustrated in figure 7.1),

Postulate 7.3.2. $\|\boldsymbol{\xi}(t) - \boldsymbol{\xi}_0\| \to 0$ as the measures approach to the goal, and by linearising the attractive motion by $\delta(m) = M\, u(t)^2 + z_0^2$. The figure 7.1 is formally described by the next non-linear equations system,

$$\delta_{t_1}^2 = (x_0 - x_{t_1})^2 + (y_0 - y_{t_1})^2 \tag{7.25}$$

and,

$$\delta_{t_2}^2 = (x_0 - x_{t_2})^2 + (y_0 - y_{t_2})^2 \tag{7.26}$$

Proof 7.3.2 Analytic solution of the system of non linear equations: To reduce terms, we define $\varepsilon_1 = x_0 - x_{t_1}$ and $\varepsilon_2 = y_0 - y_{t_1}$, as to have now both equations as,

$$\delta_1^2 = \varepsilon_1^2 + \varepsilon_2^2 \tag{7.27a}$$

$$\delta_2^2 = (\varepsilon_1 - \Delta x)^2 + (\varepsilon_2 - \Delta y)^2 \tag{7.27b}$$

by developing the quadratic terms, we have now both equations as,

$$-\delta_1^2 = -\varepsilon_1^2 - \varepsilon_2^2 \tag{7.28}$$

thus,

$$\delta_2^2 = \varepsilon_1^2 - 2\varepsilon_1 \Delta x + \Delta x^2 + \varepsilon_2^2 - 2\varepsilon_2 \Delta y + \Delta y^2 \tag{7.29}$$

quadratic terms are eliminated, and in order to simplify the equations, it is defined a known term $\Delta_t = \Delta x^2 + \Delta y^2$, now equation (7.29) becomes (7.30)

$$\delta_2^2 - \delta_1^2 = \Delta_t - 2(\varepsilon_1 \Delta x + \varepsilon_2 \Delta y) \tag{7.30}$$

algebraically reordering (7.30),

$$-\left(\frac{\delta_2^2 - \delta_1^2 - \Delta_t}{2}\right) = \varepsilon_1 \Delta x + \varepsilon_2 \Delta y \tag{7.31}$$

simplification is by grouping the known factors,

$$h = - \left(\frac{\delta_2^2 - \delta_1^2 - \Delta_t}{2} \right) \tag{7.32}$$

so then,

$$\varepsilon_1 = \frac{h - \varepsilon_2 \Delta y}{\Delta x} \tag{7.33}$$

Subsequently by substituting ε_1 to solve for ε_2,

$$\delta_1^2 = \left(\frac{h - \varepsilon_2 \Delta y}{\Delta x} \right)^2 + \varepsilon_2^2 \tag{7.34}$$

developing the quadratic term,

$$\delta_1^2 \Delta x^2 = h^2 - 2\varepsilon_2 \Delta y h + \varepsilon_2^2 \Delta y^2 + \varepsilon_2^2 \tag{7.35}$$

$$\varepsilon_2^2 (\Delta y^2 + 1) - \varepsilon_2 (2\Delta y h^2) + (h^2 - \delta_1^2 \Delta x^2) = 0 \tag{7.36}$$

from previous expressions, we now have a general quadratic form equation, and by solving $\delta_1^2 = \varepsilon_1^2 + \varepsilon_2^2$ to have a general solution,

$$\varepsilon_2 = - \frac{\Delta y h \pm \sqrt{(\Delta y h)^2 - (\Delta y^2 + 1)(h - \delta_1^2 \Delta x^2)}}{\Delta y^2 + 1} \tag{7.37}$$

It follows that in next equation, the term $(\Delta y h)^2 > 0$ always because of its quadratic exponent. Only real root are of interest because they represent Cartesian displacements that provide clues on how the mobile robot must navigate.

$$f_1 = \begin{cases} \sqrt{(\Delta y h)^2 - (\Delta y^2 + 1)(h - \delta_1^2 \Delta x^2)}, & (\Delta y h)^2 > (\Delta y^2 + 1)(h - \delta_1^2 \Delta x^2) \\ \sqrt{(\Delta y h)^2 + (\Delta y^2 + 1)(h - \delta_1^2 \Delta x^2)}, & (\Delta y h)^2 < (\Delta y^2 + 1)(h - \delta_1^2 \Delta x^2) \\ 0, & (\Delta y h)^2 = (\Delta y^2 + 1)(h - \delta_1^2 \Delta x^2) \end{cases} \tag{7.38}$$

Hence, by substituting expression f_1 in (7.39), the new equation form is given by,

$$\varepsilon_2 = \pm \frac{-(\Delta y \, h) \pm f_1}{\Delta x^2 + \Delta y^2} \tag{7.39}$$

Thus, a decision engine is formulated for ε_2 to apply it during searching process,

$$\varepsilon_2 = \begin{cases} \min_{f_1}\left(\frac{-\Delta y\,h\pm f1}{\Delta x^2+\Delta y^2}\right), & (-b+f_1>0)\wedge(-b-f_1>0)\wedge(y_{t_2}<0) \\ (-b+f_1>0), & (-b+f_1<0)\wedge(-b-f_1>0)\wedge(y_{t_2}>0) \\ (-b-f_1>0), & (-b+f_1>0)\wedge(-b-f_1<0)\wedge(y_{t_2}>0) \\ \max_{f_1}\left(\frac{-\Delta y\,h\pm f1}{\Delta x^2+\Delta y^2}\right), & (-b+f_1<0)\wedge(-b-f_1<0)\wedge(y_{t_2}>0) \end{cases} \tag{7.40}$$

Thus, for ε_1, if it happens that $\forall\,\delta_1^2\geq 0$, then calculate two function values $g_1(x_t,\varepsilon_2)$ and $g_2(x_t,\varepsilon_2)$,

$$g_1(x_{t_1},\varepsilon_2) = \begin{cases} -\sqrt{\delta_1^2-\varepsilon_2}, & x_{t_1}>0 \\ +\sqrt{\delta_1^2-\varepsilon_2}, & x_{t_1}<0 \\ 0, & x_{t_1}=0 \end{cases} \qquad g_2(x_{t_1},\varepsilon_2) = \begin{cases} -\sqrt{\delta_1^2+\varepsilon_2}, & x_{t_1}>0 \\ +\sqrt{\delta_1^2+\varepsilon_2}, & x_{t_1}<0 \\ 0, & x_{t_1}=0 \end{cases} \tag{7.41}$$

Therefore $\varepsilon_1(g_1,g_2)$ is given in terms of ε_2 and functions g_1 and g_2,

$$\varepsilon_1 = \begin{cases} g_1, & \delta_1^2>\varepsilon_2 \\ g_2, & \delta_1^2<\varepsilon_2 \\ 0, & \delta_1^2=\varepsilon_2 \end{cases} \tag{7.42}$$

The model for prediction of next posture approaching ξ_0 with terms ε_1, an ε_2 is denoted by next equation,

$$y_{t_2} = \begin{cases} y_{t_1}-\varepsilon_2, & y_{t_1}>0 \\ y_{t_1}+\varepsilon_2, & y_{t_1}<0 \\ y_{t_1}, & y_{t_1}=0 \end{cases} \qquad x_{t_2} = \begin{cases} x_{t_1}-\varepsilon_1, & x_{t_1}>0 \\ x_{t_1}+\varepsilon_1, & x_{t_1}<0 \\ x_{t_1}, & x_{t_1}=0 \end{cases} \tag{7.43}$$

Therefore, the theorem 7.3.1 is stated as,

Theorem 7.3.1. The solution for the equilibrium point $\xi_0 = (x_0,y_0,z_0)^\top$ is found by the recursive system, $x_{t+1} = x_t + \varepsilon_1$, and $y_t + \varepsilon_2$; until the boundary distance ϵ is reached by $(x_{t+1}^2 + y_{t+1}^2)^{1/2} \leq \epsilon$.

It follows that behaviours of ε_1 and ε_2 given by equations (7.39) and (7.42) respectively, are depicted in Cartesian space by figure 7.2. Vertical axes represent the distances given by $\varepsilon_{1,2}$, which are calculated at each control loop. ε_1 solves for the x-axis, while ε_2 solves for y-axis. Nevertheless, ε_1 depends on ε_2, according to previous formulation. Depictions in figure are 7.2 left-sided for ε_1, and 7.2 right-sided for ε_2, respectively.

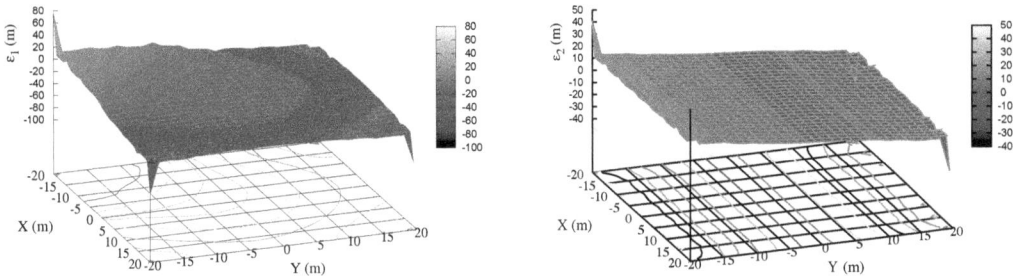

Figure 7.2:   Behaviours of ε_1 (left), and ε_2 (right) are plotted in Cartesian space. Both values are non stationary calculated at every control loop, automatically leading the robot to the goal ξ_0.

7.4   Robot kinematics

With the magnitude of $\varepsilon_{1,2}$ overtime, the robot trajectory is controlled through a sequence of locations, until reaching the end of each $\varepsilon_{1,2}$. Once the distance is reached, new values of $\varepsilon_{1,2}$ are recursively computed, and again a set of line segments must be navigated by the robot. Any geometrical trajectory can be modelled by a sequence of segments of curves. Where, a straight line may be considered as a segment of curve with infinite radius, where such radius coordinates is known as instantaneous centre of rotation. Thus, a model to compute the instantaneous centre of rotation is required.

The sequence of Cartesian locations tracked between the segments namely $\Delta x = |\varepsilon_1 - x_t|$ and $\Delta y = |\varepsilon_2 - y_t|$. The instantaneous robot's speed is defined by $v = \|\mathbf{v}^Q\|$ as well as its direction angle $\theta_t = \arctan(\varepsilon_2 - y_t / \varepsilon_1 - x_t)$,

$$\omega_t = \frac{d\theta_t}{dt} = \frac{(\Delta y - y_t)\dot{x} - (\Delta x - x)\dot{y}}{x_t^2 + y_t^2} \tag{7.44}$$

Thus, projecting at $t + 1$, and combining the velocity $\mathbf{v}^Q$, the controlled trajectory model is,

$$\boldsymbol{\xi}_{t+1} = \boldsymbol{\xi}_t + \frac{v}{\omega} \begin{pmatrix} -\sin(\theta_t) + \sin(\theta + \omega \Delta t) \\ \cos(\theta_t) - \cos(\theta + \omega \Delta t) \\ \frac{\omega^2 \Delta t}{v} \end{pmatrix} \tag{7.45}$$

We refer to a non-holonomic four-wheeled driven robot for all-terrain navigation, figure 7.3. Four asynchronous wheels speed lead the mobile robot to be controlled by a skid-steer modality, providing less kinematic restrictions for robot's manoeuvrability .

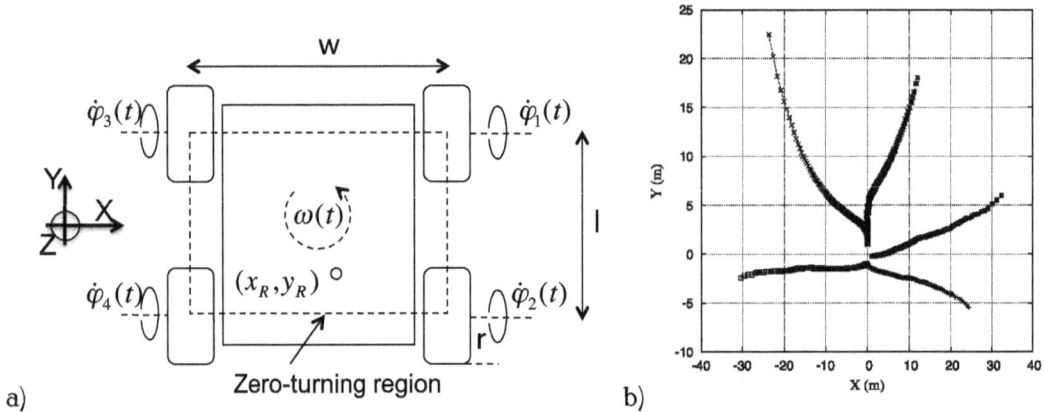

Figure 7.3: a) 4WD mobile robot kinematic parameters with z-turn axis. b) Different trajectories of the robot toward ξ_0 produced by the proposed search approach.

The depiction shows that $\frac{d}{dt}\{\varphi_i(t)\}$ is the ith wheel's rotational velocity. $\omega(t)$ is the robot's yaw rate. The ability of the robot to steer itself through slippage/skid motion effects has its fundamentals on inferring a Z-turn axis at $(x_R, y_R)^T$. The square region scope of figure 7.3 is defined by the wheels' contact point as the boundaries for $(x_R, y_R)^T$. We state that the Z-turn axis can be modelled as a movable axis point, according to the inertial effects suffered by the robot's body by the wheels' velocity configuration. The property of skid-steering that depends on the point (x_R, y_R) gives the robot the ability to change its turning axis, in accordance to the wheels lateral slippages. The authors took advantage of this effects, by calculating a point coordinate called (x_R, y_R). During preliminary motion tests, variations occurred in the motion patterns. A kinematic restriction of this effect establishes that the rover turning Z-axis only moves within a squared area bounded by the wheels' contact point (as depicted in figure 7.3). By modelling the observed effects when parametrizing different angular speeds on the front and the rear wheels, its manoeuvrability equilibrium point is changed. The derivatives to calculate $\dot{x}_R$ and $\dot{y}_R$ (in ms^{-1}) are formulated to model such inertial effects.

$$\frac{\mathrm{d}}{\mathrm{d}t}\{y_R(t)\} = \frac{r\,l}{4v_{max}}\left(\frac{\mathrm{d}^2}{\mathrm{d}t^2}\{\varphi_4(t)\} + \frac{\mathrm{d}^2}{\mathrm{d}t^2}\{\varphi_2(t)\} - \frac{\mathrm{d}^2}{\mathrm{d}t^2}\{\varphi_3(t)\} - \frac{\mathrm{d}^2}{\mathrm{d}t^2}\{\varphi_1(t)\}\right) \qquad (7.46)$$

Where $\dot{y}_R$ is the turning Z-axis displacement along its longitudinal axis; L is the distance between the rear and front side wheels (units m). v_{max} is the maximal allowed robot velocity reached up to a contact point.

$$\frac{\mathrm{d}}{\mathrm{d}t}\{x_R(t)\} = \frac{r\,W}{4v_{max}}\left(\frac{\mathrm{d}^2}{\mathrm{d}t^2}\{\varphi_1(t)\} - \frac{\mathrm{d}^2}{\mathrm{d}t^2}\{\varphi_3(t)\} - \frac{\mathrm{d}^2}{\mathrm{d}t^2}\{\varphi_2(t)\} + \frac{\mathrm{d}^2}{\mathrm{d}t^2}\{\varphi_4(t)\}\right) \qquad (7.47)$$

Similarly, $\dot{x}_R$ is the displacement or shift of the robot reference through its transverse X axis. It follows that a solution for the inverse kinematic parameters depends on four control variables $\ddot{\varphi}_1(t)$, $\ddot{\varphi}_2(t)$, $\ddot{\varphi}_3(t)$ and $\ddot{\varphi}_4(t)$ with physics units rad/s^2. Four linearly independent equations are established to solve the control motion system. The equations of $\dot{v}_t$, $\dot{\omega}_t$, $\dot{x}_R$ and $\dot{y}_R$ meet this requirement, and we just have to rearrange the equations terms to simplify the solution process. If we arrange the linear equation in terms of the wheels parameter (rotational velocities), we found out the following set of equations, also known as the direct kinematics model. As for the 4WD kinematic structure the transition matrix has the following parameters $k_1 = r/4$, $k_2 = \frac{r\,W}{\sqrt{W^2 + l^2}}$, $k_3 = \frac{r\,W}{4v_{max}}$, and $k_4 = \frac{r\,l}{4v_{max}}$. Thus, the forward kinematics equation system model is as follows,

$$\begin{pmatrix} \frac{\mathrm{d}}{\mathrm{d}t}\{v(t)\} \\ \frac{\mathrm{d}}{\mathrm{d}t}\{\omega(t)\} \\ \frac{\mathrm{d}}{\mathrm{d}t}\{x_R(t)\} \\ \frac{\mathrm{d}}{\mathrm{d}t}\{y_R(t)\} \end{pmatrix} = \begin{pmatrix} k_1 & k_1 & k_1 & k_1 \\ k_2 & -k_2 & k_2 & -k_2 \\ k_3 & -k_3 & -k_3 & k_3 \\ -k_4 & -k_4 & k_4 & k_4 \end{pmatrix} \cdot \begin{pmatrix} \frac{\mathrm{d}^2}{\mathrm{d}t^2}\{\varphi_1(t)\} \\ \frac{\mathrm{d}^2}{\mathrm{d}t^2}\{\varphi_2(t)\} \\ \frac{\mathrm{d}^2}{\mathrm{d}t^2}\{\varphi_3(t)\} \\ \frac{\mathrm{d}^2}{\mathrm{d}t^2}\{\varphi_4(t)\} \end{pmatrix} \qquad (7.48)$$

By rearranging the equations system it is analytically solved by an algebraic method. Now, the solution represents the inverse kinematics control vector namely $\dot{\Omega}_t = r(\ddot{\varphi}_1(t), \ddot{\varphi}_2(t), \ddot{\varphi}_3(t), \ddot{\varphi}_4(t))^T$

$$\frac{d}{dt}\{\Omega(t)\} = \begin{pmatrix} k_1^{-1} & k_2^{-1} & k_3^{-1} & -k_4^{-1} \\ k_1^{-1} & -k_2^{-1} & -k_3^{-1} & -k_4^{-1} \\ k_1^{-1} & k_2^{-1} & -k_3^{-1} & k_4^{-1} \\ k_1^{-1} & -k_2^{-1} & k_3^{-1} & k_4^{-1} \end{pmatrix} \cdot \begin{pmatrix} \frac{d}{dt}\{v(t)\} \\ \frac{d}{dt}\{\omega(t)\} \\ \frac{d}{dt}\{p_x(t)\} \\ \frac{d}{dt}\{p_y(t)\} \end{pmatrix} \tag{7.49}$$

The variable $\frac{d}{dt}\{\Omega(t)\}$ is the vector inverse kinematics solution, which contains the wheels linear velocities, used to control the in-wheels actuator. The first term $(d\boldsymbol{\xi}_{t+1}/dt)\tau^{-1}$ leads the robot to local goals. It is the internal robot's search motivation, which refers to the robot's motion exerted through a sequence of small increments between the robot's actual posture until reaching the end of $\varepsilon_{1,2}$ magnitude. The term $(d\boldsymbol{\xi}_{t+1}/dt)\tau^{-1}$ yields a motion behaviour constrained by the non-holonomy of the robot's four-wheel four-drive (4W4D) kinematics. As $\mathbf{a}^e$ is given in terms of acceleration (units of $m \cdot s^{-2}$), the first term yields a set of small consecutive displacements that are tied to a relaxation time τ, which is the robot's time frame required to develop velocity changes. The second term $d\mathbf{v}^Q/dt$ gradually yields the robot to a global goal. It exerts large motion displacements, and it refers to accelerative directional fields. The second term $d\mathbf{v}^Q/dt$ is implicitly effected by calculations of $\varepsilon_{1,2}$. $d\mathbf{v}^Q/dt$ establishes larger displacements to the goal.

$$\mathbf{a}^e = \frac{1}{\tau}\left(\frac{d}{dt}\boldsymbol{\xi}_{t+1}\right) + \frac{d}{dt}\mathbf{v}^Q \tag{7.50}$$

our final equation for the goal searching is,

$$\mathbf{a}^e = \tau^{-1}\frac{d}{dt}\left(\boldsymbol{\xi}_t + \frac{v}{\omega}\begin{pmatrix} -\sin(\theta_t) + \sin(\theta + \omega\Delta t) \\ \cos(\theta_t) - \cos(\theta + \omega\Delta t) \\ \frac{\omega^2\Delta t}{v} \end{pmatrix}\right) + \frac{d}{dt}\frac{u^o}{5}t\left(\frac{\partial Q}{\partial x}\mathbf{i} + \frac{\partial Q}{\partial x}\mathbf{i}\right) \tag{7.51}$$

Some results from potential equations and analysis discussed along the chapter are illustrated in figure 7.3-B). Where, five different search routes were yielded towards the target (source of energy to be found). The nearer the robot approaches, the larger the number of measurements $\hat{E}$ the robot gets (r_t vs $\hat{E}$. The increased number of measurements near the goal is because the values of $\varepsilon_{1,2}$ magnitudes gradually decrease, allowing the robot to obtain more measurements at smaller displacements.

Bibliography

[1] Z. Liu, and Y. Kleiner, *State of the Art Review of Inspection Technologies for Condition Assessment of Water Pipes*, Measurement, Vol. 45(7), pp. 1675-1940, Elsevier, 2012.

[2] R. Bogue, *The role of robotics in non-destructive testing*, Industrial Robot: An International Journal, Vol. 37(5), pp. 421-426, Emerald, 2010.

[3] B.S. Waytt, *Chap.4.21 Practical Application of Cathodic Protection*, Shreir's Corrosion, Vol. 4, Elsevier 2009.

[4] Bard, A. J. & Faulkner, L. R., *Electrochemical methods: fundamental and applications*, 3rd Ed., Wiley & Sons, Inc., 2004.

[5] E. A. Martínez-García and R. Torres-Córdoba, *Exponential Fields Formulation for WMR Navigation*, Journal of Applied Bionics & Biomechanics, Special issue in Personal Care Robotics, Vol. 9, IOS Press, 2012.

[6] E. A. Martínez-García, O. Mar, and R. Torres-Cordoba, *Dead-reckoning inverse and direct kinematic solution of a 4W independent driven rover*, IEEE ANDESCON 2010, Bogotá Colombia, Sep 15-17, 2010.

[7] E. A. Martínez-García and R. Torres-Córdoba, *4WD skid-steer trajectory control of a rover with spring-based suspension analysis*, Intl. Conf. in Intelligent Robotics and Applications, Shanghai, China, Nov 10-12, Part I, LNCS 6424, pp. 453-464, 2010.

Chapter 8

MULTI-ROBOT PATH/TASKS PLANNING

Nilda G. Villanueva Chacón and Edgar A. Martínez García

Laboratorio de Robótica, Institute of Engineering and Technology
Universidad Autónoma de Ciudad Juárez, Mexico.

In this chapter a highly concurrent tasks planner for a distributed multi-robot systems is formulated. Unlike other works ⁻ , the present approach discusses two issues: a) a path-planning model; and b) a robotic-tasks scheduler. A set of kinematic control laws based on directional derivatives allow us to model the robots interaction for dynamic environments. Distributed wheeled mobile robots perform the execution of diverse autonomous tasks concurrently and synchronized just in time. Distributed tasks planning reconfigures and synchronizes the robotic actions throughout exponential functions which dynamically change the priority primitives: sense, plan, and act. The objective is to formulate an automatic planning system using multiple mobile robots to manage the material supply, rubbish recollection for industrial transportation tasks. The task-oriented approach concerns carry-and-fetch, and material collecting, as well as the robots' ability to navigate for battery charging at dock-stations. A diversity of task scenarios such as traffic congestion peaks, orders arrival during the execution of tasks and order modifications have to be considered. When the system is modified, the flow of material during the process differs from the previously used routes.

Mathematical formulation and numerical simulation experiments illustrate the parallel computing performance, and the distributed robots behaviour. Simulation results depict how the robots deal with highly concurrent robotic tasks, and dynamic events by a parallel scheme. A kinematic model for a differential drive robot is formulated. In addition, an acceleration-based model is proposed to provide the messenger robot the ability to navigate and perform transportation tasks. To deal with the computational cost involved in this work, the effect of varying the number of processors executing a job, have been examined. Parallel computing is a capability to manage threads broadcast to different physical processors. The available processing power utilization is maximized to accommodate as many tasks as possible while satisfying the required deadline of each task. Simulations demonstrated the feasibility and efficacy of the proposed task/path planner.

8.1   Robotic tasks scheduler

The tasks and motion actions are synchronized and coordinated by a scheduler designed to be able to synchronize tasks in real-time. The tasks scheduler has the ability to assign multi-threads to different physical processors. All tasks are classified into three types of robotic primitives as traditionally known: planning f_P, sensing f_S, and acting f_A. The real-time system develops online synchronization through shared-memory, execution of multiple threads, and threads priorities are dynamically assigned, depending on which task type f_P, f_S and f_A has more statistical demand.

The function f_S acquires, decides, sorts and stores environmental data to be available for the tasks of planning and acting. The planning task function f_P reads collected sensor data for generating a plan of actions in accordance to the type of event that is occurring at actual time, and f_P prioritizes sensing/acting tasks.

For instance, the presence of multiple dynamic obstacles blocking the actual direction towards the desired goal, online path generation are usually non linear. Each robot re-plan a new local route when new dynamic obstacles block its actual pathway. The Lagrange interpolation polynomial of cubic order arising from numerical Cartesian points (x_i, y_i) that comprise the desired pathway. The interpolation gives a polynomial (see section 1.6.2) with origin at robot's fixed inertial frame by,

$$y(x) = \sum_{j=0}^{k} \left(y_j \prod_{j=0}^{k} \frac{x - x_j}{x_i - x_j} \right) \qquad (8.1)$$

And algebraically expanding with $x0, x_1 \ldots, x_k$, and y_i numerically known,

$$y(x) = \left(\frac{x-x_1}{x_0-x_1}\right)\left(\frac{x-x_2}{x_0-x_2}\right)\left(\frac{x-x_3}{x_0-x_3}\right)y_0 + \left(\frac{x-x_0}{x_1-x_0}\right)\left(\frac{x-x_2}{x_1-x_2}\right)\left(\frac{x-x_3}{x_1-x_3}\right)y_1 + \left(\frac{x-x_0}{x_2-x_0}\right)\left(\frac{x-x_1}{x_2-x_1}\right)\left(\frac{x-x_3}{x_2-x_3}\right)y_2 + \left(\frac{x-x_0}{x_3-x_0}\right)\left(\frac{x-x_1}{x_3-x_1}\right)\left(\frac{x-x_2}{x_3-x_2}\right)y_3 \tag{8.2}$$

Eventually, the developed algebraic polynomial expression becomes

$$y(x) = a_0 + a_1 x + a_2 x^2 + \ldots + a_k x^k \tag{8.3}$$

hence, the segment of distance to navigate is represented by s_i,

$$s_1 = \sqrt{x_1^2 + y_1^2} \tag{8.4}$$

Similarly, in order to reach the goal just-in-time, a new non linear interpolation representing the exact time as a function of the segment of distance s_i is stated by

$$\Gamma = t(s) = \sum_{j=0}^{3}\left(t_j \prod_{j=0}^{3}\frac{s-s_j}{s_i-s_j}\right) \tag{8.5}$$

by substituting the mathematical form of s into Γ,

Postulate 8.1.1. *The segment of distance to be displaced just-in-time*

$$\Gamma = \sum\left(t_1 \prod \frac{s-\sqrt{x_1^2+y_1^2}}{\sqrt{x_0^2+y_0^2}-\sqrt{x_1^2+y_1^2}}\right) \tag{8.6}$$

Re-writing the polynomial equation $\Gamma(s)$ by a third order polynomial,

$$\Gamma = b_0 + b_1 s + b_2 s^2 + b_3 s^3 \tag{8.7}$$

In all situations, time is a factor that is conditioned by the battery timing supply. By using the discharging battery curve of Peukert's Law, it is possible to validate the just in time polynomials previously defined as applicable models.

$$t_P = H \left(\frac{C}{IH} \right)^k \tag{8.8}$$

Where:

    t Time that the battery will last given a particular rate of discharge (hours).

    H The discharge time in hours that the Amp Hour specification is based on.

    C The battery capacity in Amp Hours based on the specified discharge time.

    I Discharging rate (Amp).

    k Peukert number for the battery.

A period of time limit (t_η) that is safe for a robot to reach the charging dock station is defined for each robot based on the distance to the goal Γ. Thus, the robots already charging energy await for the going discharged robots in order to exchange work duties each other.

$$mr = \begin{cases} mr_\beta, & t_P < t_\eta \\ mr_\gamma, & t_P \geq t_\eta \end{cases} \tag{8.9}$$

8.1.1 Tasks scheduling

The scheduler is an algorithm that synchronises the tasks f_P, f_S and f_A, as elements of the set $(\mathcal{U})$.

$$\mathcal{U} = \{f_S, f_A, f_P\} \tag{8.10}$$

In order to automatically select a type of task to be performed, the function $f(x) = \lambda e^{-\lambda x}$ describes the external events behaviour that activates the tasks of $\mathcal{U}$. Therefore, in order to estimate the possible external event occuring at actual time, the inverse solution is defined by

$$F(x) = -\frac{1}{\lambda} \ln \left(\frac{x}{\lambda} \right)$$

It is assigned as a numerical weight, or as a probability value depending on the time that the tasks is occurring.

$$f(x) = \begin{cases} \lambda e^{-\lambda x}, & x \geq 0 \\ 0, & x < 0 \end{cases} \tag{8.11}$$

The cumulative distribution function is modelled by

$$F(x) = \int_0^x \lambda e^{-\lambda x} dx = 1 - e^{-\lambda x} \tag{8.12}$$

The distribution function models occurrences for automatic selection of f_S, f_P by defining a uniform random distribution in interval $[0, 1]$ to produce the number R.

$$1 - e^{-\lambda x} = R \tag{8.13}$$

and

$$R = \begin{cases} \dfrac{1}{b - a}, & a \leq x \leq b \\ 0, & other \end{cases} \tag{8.14}$$

Solving for x,

$$x = -\frac{1}{\lambda} \ln(R) \tag{8.15}$$

The next variable Sem is used for controlling access and indicates if the system is available to execute the acting or sensing task. In this way Sem tracks the status of the resources being assigned, through a status value associated,

Definition 8.1.2. The tasks controlling access has four states

$$Sem = \begin{cases} 0, & wait \\ Sem^P, & \tau_P = g^{-1}(R > \tau_P) \\ Sem^S, & \tau_S = g^{-1}(\tau_A < R < \tau_P) \\ Sem^A, & \tau_A = g^{-1}(\tau_A > R) \end{cases} \tag{8.16}$$

The scheduler P selects a task from a list of waiting tasks, signals the communication center to begin execution of that task, and calls the resource manager to update the dynamic resources list.

$$P = \begin{cases} f_P(\tau_P), & (g^{-1}(R > \tau_P) \Longleftrightarrow B \wedge \neg C \\ f_S(\tau_S), & Sem^S \\ f_A(\tau_A), & (\Delta t_k \leq t_j) \wedge \neg C \end{cases} \tag{8.17}$$

Likewise, next constraint is postulated:

Postulate 8.1.3. *B and C create a condition to satisfy the just-in-time constraint.*

$$B = (\Delta t_{i-1} - t_i) \vee (\Delta t_{j-1} < t_j) \vee (\Delta t_{k-1} - t_i)$$

and

$$C = \neg Sem^P \wedge \neg Sem^A \wedge \neg Sem^S$$

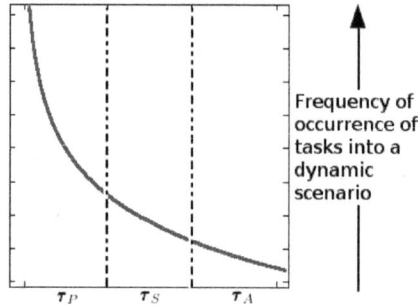

Frequency of
occurrence of
tasks into a
dynamic
scenario

τ_P τ_S τ_A

Figure 8.1:   Scheduler function for the messenger-type robot.

8.1.2 Dispatchers

For dispatching, a mobile robot collects amounts of types of materials from a queue $^-$. The queue is managed by a group of dispatcher robots, where their goal is to maintain available material for the messenger robots. $\mathscr{D}$ defines the set of dispatching robots $(dr$, and let $\mathscr{M}$ be the set of messenger robots (mr), and LR a subset of M with the messenger robots loading material. RR is a subset of M that containing the robots ready for transporting material.

$$\mathscr{D} = \{dr_1, dr_2, ..., dr_n\} \tag{8.18}$$

Where n defines the number of dispatcher robots available.

$$\mathscr{M} = \{mr_1, mr_2, ..., mr_n\} \tag{8.19}$$

and subsets,

$$LR = \{mr_1...mr_j\} \qquad RR = \{mr_{j+1}...mr_k\} \tag{8.20}$$

The following conditions have to be met,

$$RR \subset M \qquad LR \subset M \qquad RR \cap LR = \{\} \tag{8.21}$$

In addition Sem_d is defined as a constraint to access raw material or semi-finished parts,

$$Sem_d = \begin{cases} 0, & loadingRobot \vee dispatching \\ 1, & other \end{cases} \tag{8.22}$$

The $loadingRobot$ function supplies the robot with raw material, and $dispatching$ is the continued activity of robots dispatchers.

$$DispatchingTask = \begin{cases} loadingRobot, & (mat \neq 0) \wedge (loaded < cap) \wedge Sem_d \\ dispatching, & LR \neq \{0\} \wedge (mat = 0) \wedge Sem_d \\ pause, & else \end{cases} \tag{8.23}$$

By using the $DispatchingTask$ function, it possible to create a complete scenario for the process of delivery and loading raw material (Figure 8.2).

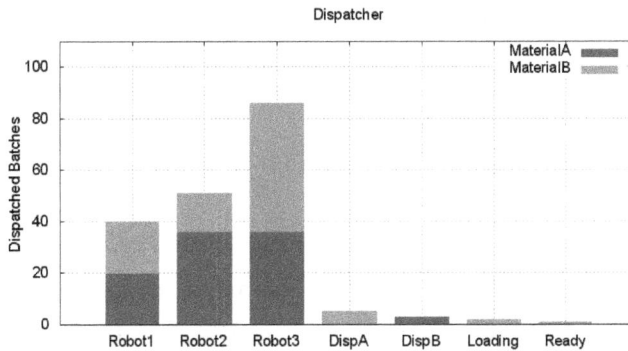

Figure 8.2: Simulation of dispatching activities. Three mobile robots taking different quantities of two types of materials. Simultaneously, two dispatcher-robot put raw materials in holding position.

8.1.3   Parallel Tasks

Clusters computing has emerged as a new paradigm for solving large-scale problems. A cluster of computers is generally defined as a collection of interconnected stand-alone computers working together as a single, integrated computing resource. The most critical software components of the cluster are the allocation and scheduling algorithms. Allocating tasks of a real-time application on a certain processor is the most critical step towards achieving the optimal schedule for the application. Figure 8.3 shows times taken to perform a series of 60 tasks distributed and executed as real-time tasks into different number of processors.

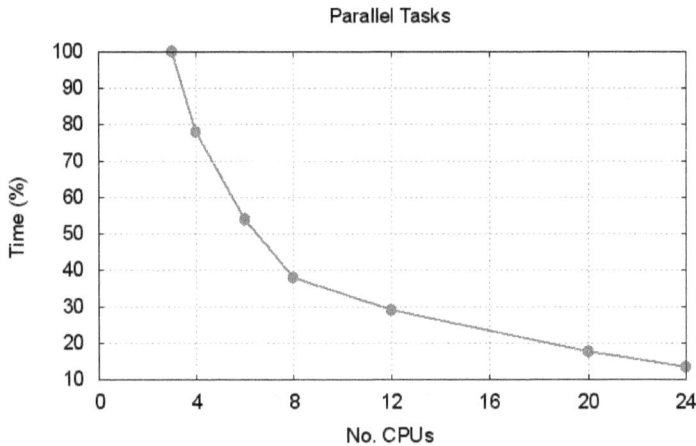

Figure 8.3:   Time to complete 60 real-time tasks using a parallel computing scheme.

8.2   Robot's motion model

In this section we formulate the kinematic models of four-wheel robotic structures with dual differential drives. Thus, inverse and forward kinematics models are provided as functions of the rotational wheels' speed. The robotic platforms are depicted in Figure 8.4. Based on a dead-reckoning approach wheels' speed are directly measured from proprioceptive sensors (rotary encoders). Thus, the robot's displacement Δs is inferred by

$$\Delta s = \frac{\pi r}{f_n}(n_r + n_l) \tag{8.24}$$

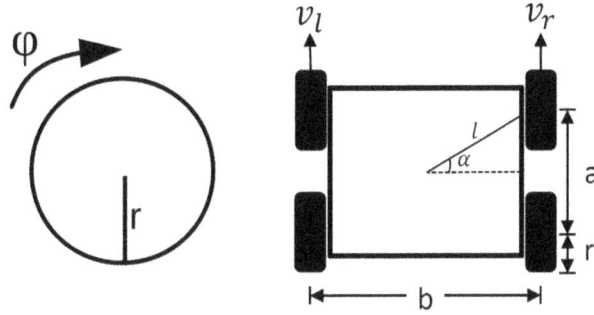

Figure 8.4: Wheel angular position with dual asynchronous velocities (left and right).

Where r is an ideal wheels' radius, f_n is the encoder resolution in pulses per revolution, and $n_{r,l}$ is the number of pulses sensed of the right/left wheel overtime. The tangential velocity $v_{r,l}$, and the angular $\dot{\varphi}_{r,l}$ of the wheels are defined by:

$$v_{r,l} = r \frac{d}{dt} \varphi_{r,l} \tag{8.25}$$

and

$$\dot{\varphi}_{r,l} = \int_{t1}^{t2} \frac{2\pi}{f_n} n \, dt \tag{8.26}$$

Because of the instantaneous robot's velocity approximates an averaged speed along a distance s, velocity and acceleration are described as functions of time,

$$v = \frac{ds}{dt} \tag{8.27}$$

as well as

$$a = \frac{dv}{dt} \tag{8.28}$$

Thus, by equating both expressions through the common term dt, and by integrating the differentials ds and dv.

$$a \int_{s_0}^{s_f} ds = \int_{v_0}^{v_f} v dv \tag{8.29}$$

Hence, the next non linear expression with averaged acceleration is obtained,

$$a(s_f - s_0) = \frac{v_f^2 - v_0^2}{2} \tag{8.30}$$

By considering starting position zero, the acceleration form is simplified,

$$a = \frac{v^2}{2s} = \frac{(v_r + v_l)^2}{8s} \tag{8.31}$$

Since a solely depends on the variations of v for this functional form, its vector form is:

$$\mathbf{v} = \begin{pmatrix} \dot{x} \\ \dot{y} \end{pmatrix} = v \begin{pmatrix} \sin\theta \\ -\cos\theta \end{pmatrix} \tag{8.32}$$

likewise,

$$\theta = \arctan\left(\frac{\dot{y}}{\dot{x}}\right) \tag{8.33}$$

and the magnitude of $\mathbf{v}$ is defined by

$$\|\mathbf{v}\| = \sqrt{\dot{x}^2 + \dot{y}^2} \tag{8.34}$$

Describing the robot's model as depicted in figure 8.4

$$v_r = v_1 + v_2 \tag{8.35}$$

and

$$v_l = v_3 + v_4 \tag{8.36}$$

The wheeled mobile robots forward kinematics is described next,

Proposition 8.2.1. *The wheeled robot's direct and inverse kinematic solutions are provided by*

$$\begin{pmatrix} v \\ \omega \end{pmatrix} = \begin{pmatrix} 1/2 & 1/2 \\ \dfrac{2b}{a^2 + b^2} & -\dfrac{2b}{a^2 + b^2} \end{pmatrix} \cdot \begin{pmatrix} v_r \\ v_l \end{pmatrix} \tag{8.37}$$

and by solving for the inverse solution of equation (8.37), the functional form in terms of v_r and v_l is

$$\begin{pmatrix} v_r \\ v_l \end{pmatrix} = \begin{pmatrix} 1 & \dfrac{a^2 + b^2}{4b} \\ 1 & -\dfrac{a^2 + b^2}{4b} \end{pmatrix} \cdot \begin{pmatrix} v \\ \omega \end{pmatrix} \tag{8.38}$$

8.3 Robots acceleration models

For the messenger-type robots, their navigation control equation is proposed by expression (8.39). The acceleration term a_m speeds up/down in order to reach the zones where they navigate from the warehouse to the recharging dock-station, shipping and station areas.

Proposition 8.3.1. *The robot's general navigation control law is defined by*

$$\mathbf{a}_m = -\nabla_s \mathbf{a}_m(t) = -\nabla_s \left(\frac{v^2}{2s} + [a_{W_t} + \alpha(a^{ref} - a_{W_t})] + a^{cal} e^d + a_{avoid} \right) \qquad (8.39)$$

Where the term $\frac{v^2}{2s}$ guides the robot to the shipping area, the term $a_{Wt} + \alpha(a^{ref} - a_{Wt})$ controls the robot's motion directing it towards the warehouse, and the term a_{avoid} yields a motion behaviour to avoid near static/dynamic obstacles. The tasks are accomplished in time frame constrained by inequality (8.40),

$$\frac{\hat{\mathbf{v}}(t)}{\|\mathbf{a}_m\|} \leq t_{eps1} \qquad (8.40)$$

Where t_{eps1} represents the maximum allowed time in reaching to execute the material transport task successfully on-time. Likewise, $\hat{\mathbf{v}}$ is the robot's translation velocity.

Acceleration to warehouse a_W typically develops as the starting robots' motion, reaching the warehouse zone to collect raw material that is transported to the workstations. Expression (8.41) describes the robot acceleration to warehouse,

$$a_{W_{t+1}} = a_{W_t} + \alpha(a^{ref} - a_{W_t}) \qquad (8.41)$$

Where $0 < \alpha \leq 1$, and a_{Wt+1} is the next desired controlled acceleration. a_{Wt} is the actual measured acceleration, and a^{ref} is a reference acceleration model used to track a desired magnitude together. In addition, an adjustable constant gain factor α is used to attenuate convergence. Likewise, the acceleration to shipping area a_K is provided by equation (8.42) that defines the robot's acceleration required to reach the shipping area.

$$a_K = \frac{v^2}{2s} \tag{8.42}$$

Figure 8.5 depicts the behaviour of the acceleration a_K with respect to distance and speed. The acceleration model acts like an attraction acceleration.

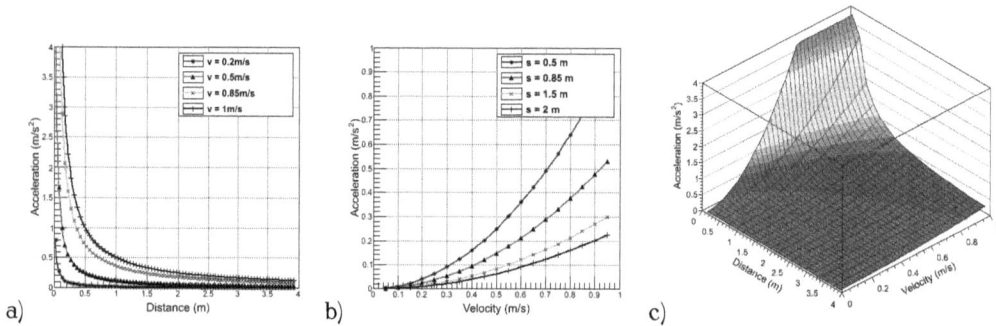

Figure 8.5: a) Distance vs acceleration. b) Acceleration vs velocity (constant distance). c) Acceleration as a function of distance and velocity.

Considering the distance s as the norm or Cartesian distance of the x and y points the next expression is stated,

$$s = \left((x_o - x_r)^2 + (y_o - y_r)^2\right)^{1/2} \tag{8.43}$$

Where x_o, y_o is the x and y position of a reachable point on the global plane, and x_r, y_r is the global position on x and y of the robot. We replace the Cartesian differences $x = (x_o - x_r)$ and $y = (y_o - y_r)$ in order to summarize in next mathematical expressions. Thus, by rewriting,

$$a = \frac{v^2}{2(x^2 + y^2)^{1/2}} \tag{8.44}$$

Considering that the sensor measurement of the distances that might arise from different types of ranging sensors (i.e. LiDAR, ultrasonic sonar),then distance model in Cartesian space is

$$\mathbf{s}_i = \delta_i \begin{pmatrix} \cos \phi_i \\ \sin \phi_i \end{pmatrix} \tag{8.45}$$

Where ϕ_i defines the direction of each sensor's beam w.r.t. the local fixed robot's plane. Therefore, by replacing the distance model in (8.46)

$$a = \sum_j \frac{v^2}{2(x_j^2 + y_j^2)^{1/2}} \tag{8.46}$$

Figure depicts two maps created by using range sensors and odometry, the left-sided map has inconsistencies due to the non linear trajectory of the robot. Please notice that at the right-sided map, it prevails quite consistent due to the robot's simpler trajectory motion.

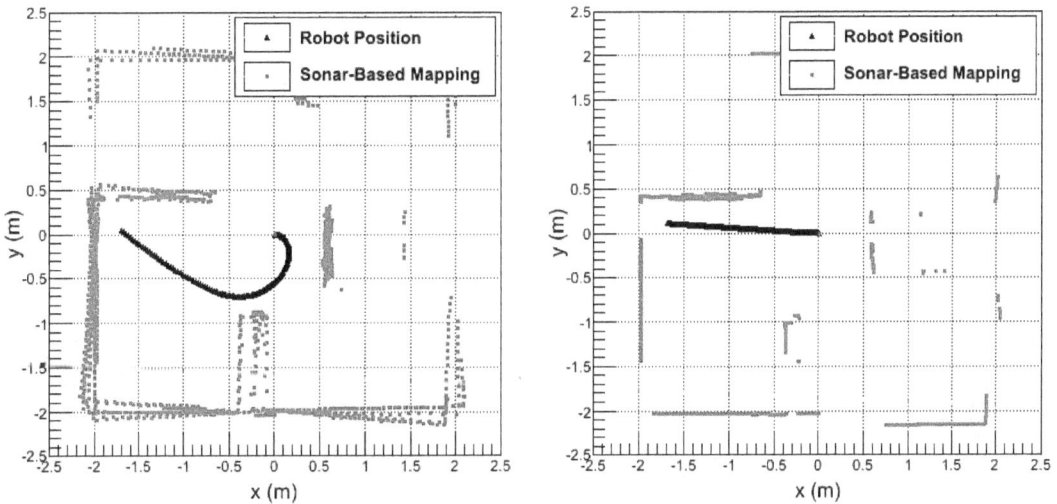

Figure 8.6: Top view of environment model using a ring of sonar on board the robot to build the map. Non linear robot's trajectory (left). Linear robot's trajectory (right).

The scenario is a workspace constrained by walls and corridors. The acceleration towards station a_E is constrained by a route, if no route exists, bottlenecks or stoppages may occur. By defining a general vector field with components of distance to the station $x = |x_r - x_g|$ and $y = |y_r - y_g|$, let $f_E(x, y)$ be an equation of mobility to assure the robot to reach the station. Where x and y are the displacements by Cartesian components required to reach the station zone.

$$f_E(x, y) = \sqrt{x^2 + y^2} - e^{-\sqrt{x^2 + y^2}} \tag{8.47}$$

Hence, we apply the gradient operator with an arbitrary adjustable constant gain κ_E,

$$\mathbf{f}_E = -\nabla_{\|\boldsymbol{\delta}\|}\kappa_E f_E(x,y) \tag{8.48}$$

with $\|\boldsymbol{\delta}\| = \sqrt{x^2 + y^2}$, a deriving w.r.t. x, then the x-component function is denoted by,

$$\frac{\partial f}{\partial x} = -\frac{x}{\sqrt{x^2 + y^2}}\left(1 - e^{-\sqrt{x^2+y^2}}\right) \tag{8.49}$$

Likewise, for the y-component the function is given by,

$$\frac{\partial f}{\partial y} = -\frac{y}{\sqrt{x^2 + y^2}}\left(1 - e^{-\sqrt{x^2+y^2}}\right) \tag{8.50}$$

and

$$v = \| - \nabla f_E(x,y)\| \tag{8.51}$$

Velocity magnitude to the station is provided by (8.51), and the angle equation provided by (8.52),

$$\theta = \theta_r - \arctan\left(\frac{\partial f/\partial y}{\partial f/\partial x}\right) \tag{8.52}$$

Finally multiple mobile robots navigation toward the station are controlled by the equation (8.53) of next proposition.

Proposition 8.3.2. *The robot's navigation vector function to reach the station:*

$$\mathbf{f}_E(x,y) = \frac{-\left(1 - e^{-\sqrt{x^2+y^2}}\right)}{\sqrt{x^2 + y^2}}\begin{pmatrix} x \\ y \end{pmatrix} \tag{8.53}$$

Bibliography

[1] Guizzo, E., *Three engineers, hundreds of robots, one warehouse*, IEEE Spectrum, vol. 45(7), pp. 26–34, 2008, doi:dx.doi.org/10.1109/MSPEC.2008.4547508.

[2] Helleboogh, A., Holvoet, T., Berbers, Y., *Testing AGVâĂŹs in dynamic warehouse environments*, In Environments for multi-agent systems II, pp. 270–290, Academic Press, 2006.

[3] Shengfang, L., Xingzhe, H., *Research on the AGV based robot system used in substation inspection*, In Proceedings of Power System Technology (pp. 1-4). Academic Press, 2006, doi:10.1109/ICPST.2006.321495.

[4] Zhaowei, M., Guojun, J., Rui, Q , Fan, W., *The automated guided vehicle problem in logistics operations*, In Proceedings of Service Systems and Service Management (pp. 1-6). Academic Press, 2008, doi:10.1109/ICSSSM.2008.4598470.

[5] Dang, Q.; Nielsen, I.; Steger-Jensen, K.; Madsen, O., *Scheduling a single mobile robot for part-feeding tasks of production lines*, Journal of Intelligent Manufacturing, pp. 1-17, 2013.

[6] Gerkey, B.,; Mataric, M., *A formal analysis and taxonomy of task allocation in multi-robot systems*, International Journal of Robotics Research, pp. 939-954, 2004.

[7] Hasgl, S.; Saricicek, I.; Ozkan, M.; Parlaktuna, O., *Project-oriented task scheduling for mobile robot team*, J. of Intelligent Manuf., 20(2), pp. 151–158, 2009, doi: dx.doi.org/10.1007/s10845-008-0228-8.

[8] Liu, L.; Shell, D., *Large-scale multi-robot task allocation via dynamic partitioning and distribution*, Autonomous Robots, 33(3), pp.291–307, 2012, doi:dx.doi.org/10.1007/s10514-012-9303-2.

[9] Martinez-Garcia, E., Torres-Cordoba, R., *Exponential fields formulation for WMR navigation*, J. of Applied Bionics & Biomechanics, 9, pp.375–397, 2012.

Part III

Dynamics and Trajectory Control

Chapter 9

NON-LINEAR REFERENCE MODEL
TRAJECTORY CONTROL

Omar Ramírez and Edgar A. Martínez García

Laboratorio de Robótica, Institute of Engineering and Technology
Universidad Autónoma de Ciudad Juárez, Mexico.

Service robotics is strongly tided to highly accurate navigational tasks where pathway tracking is a practise commonly carried out through control algorithms. This study proposes theoretical model references of non-linear pathways presented as kth-degree polynomials. This study establishes proportional controls using variable reference models at the level of second order derivatives, in order for the robot's motion to be adapted on-line. Initially, a Cartesian trajectory model is inversely transformed into wheels' angular acceleration component equations, which function as the ideal reference models. Although, our proposal may be applied to any type of robot's kinematic structure, we are presenting an example for dual asynchronous differential active wheels. Obtained results raised from successful experimental practise, and numerical simulations as well. In order to complete a navigation task, a robot must be capable to follow a desired pathway. The pathway complexity may vary depending on the environmental geometry. In recent years, numerous research have been realised upon the path-following problem [¯]. From control schemes combining conventional integral terms and fuzzy logic for the adjustment of proportional gains, up to control schemes using path following algorithms with back-stepping schemes . As a difference from cited approaches, in the present study we introduced an adaptive non-linear path following control using second order derivatives as a reference model changing overtime, as a scheme for the wheel's acceleration control.

9.1 Polynomial pathway models

Let us assume any non-linear function $s(t)$ defined as the trajectory that a wheeled robot will follow. Firstly stating that a non-linear function represented by a polynomial form fits a set of Cartesian points (i.e. pathway). Such pathway points are fitted by accomplishing a polynomial interpolation approach. Thus, considering that the set of points are given by

$$\{(x_0, y_0, t_0), (x_1, y_1, t_1), \ldots, (x_n, y_n, t_n)\} \tag{9.1}$$

every single coordinate of the pathway is reached by the robot to follow it in a certain time t_n. The Lagrange-based interpolation for polynomials (see section 1.6.2) is developed in order to obtain a functional form of the distances travelled in terms of the two Cartesian components $x(t)$ and $y(t)$, which are given by

Postulate 9.1.1 (pathway components). *Path generation is provided by the Cartesian components as functions of time.*

$$x(t) = \sum_{i=0}^{n} \left(\left(\prod_{j=0, i \neq j}^{n} \frac{t - t_j}{t_i - t_j} \right) x(t_i) \right) \tag{9.2}$$

and

$$y(t) = \sum_{i=0}^{n} \left(\left(\prod_{j=0, i \neq j}^{n} \frac{t - t_j}{t_i - t_j} \right) y(t_i) \right) \tag{9.3}$$

Higher than second order derivatives (accelerations), our equations would be formalised as Jerks. However, our interest is solely on the second order derivative equations, because the proportional model references are treated as linear systems. Hence, the path generation approach consists of a maximal of four Cartesian points, which comprise third degree polynomials.

Such that

$$x(t) = a_0 + a_1 t + a_2 t^2 + a_3 t^3 \tag{9.4}$$

and

$$y(t) = b_0 + b_1 t + b_2 t^2 + b_3 t^3 \tag{9.5}$$

The speed components along the trajectory positions are described by their first derivatives as the travelling speeds, modelled by the general expressions

$$\dot{x} = \frac{d}{dt} x(t) = a_1 + 2a_2 t + 3a_3 t^2 \tag{9.6}$$

and

$$\dot{y} = x \frac{d}{dt} y(t) = b_1 + 2b_2 t + 3b_3 t^2 \tag{9.7}$$

By expressing previous expressions in terms of the cylindrical form of motion (v, θ) for each Cartesian point,

$$v^2 = \dot{x}^2 + \dot{y}^2 \tag{9.8}$$

hence,

$$\theta = \arctan \left(\frac{\dot{y}}{\dot{x}} \right) \tag{9.9}$$

These speed models are easily transformed (roto-translated) into any arbitrary desired reference model. Thus, the acceleration linear polynomial is useful to represents the functional form of the robot's kinematic of motion, and subsequently to obtain an inverse kinematic analytical solution.

9.2 Robot's kinematics

The robot's kinematics is developed approaching a dead-reckoning modality, where the input vector is defined by $\mathbf{u} = (v, \omega)^\top$, and their components are described as functions of the wheels' speed $\dot{\varphi}_i$.

Figure 9.1:   Differential speed control wheeled robot *Franky*.

9.2.1   Forward kinematics

Let us consider a differential wheeled robot with ideal wheels' radius r (figure 9.1). Each wheel's contact region keeps a metric separation by fixed a baseline, namely d. The kinematic model for this structure is described by,

$$\begin{pmatrix} v \\ \dot{\theta} \end{pmatrix} = \begin{pmatrix} r/2 & r/2 \\ 2r/d & -2r/d \end{pmatrix} \cdot \begin{pmatrix} \dot{\varphi}_r \\ \dot{\varphi}_l \end{pmatrix} \tag{9.10}$$

The equation (9.10) describes the robot's linear speed v, and the angular speed $\dot{\theta}$ as mathematical functions of the wheels rotational speed, right and left respectively ($\dot{\varphi}_r, \dot{\varphi}_l$). The robot's translation velocity is decomposed on its speed components $\dot{x}$ and $\dot{y}$, the angle θ as in shown by equation (9.11),

$$\begin{pmatrix} \dot{x} \\ \dot{y} \end{pmatrix} = v \cdot \begin{pmatrix} \cos \theta \\ \sin \theta \end{pmatrix} \tag{9.11}$$

By substituting the functional form of v, we may express previous equation with the speed components, and evaluated in terms of the angular speed of the wheels (matrix form).

$$\begin{pmatrix} \dot{x} \\ \dot{y} \end{pmatrix} = \begin{pmatrix} \cos\theta \\ \sin\theta \end{pmatrix} \cdot \begin{pmatrix} r/2 & r/2 \end{pmatrix} \cdot \begin{pmatrix} \dot{\varphi}_r \\ \dot{\varphi}_l \end{pmatrix} \tag{9.12}$$

The complete direct kinematic model is described by (9.13) as a function of the input vector of wheels velocities. This model has its fundamentals on the robot's geometric configuration in terms of first derivative of the actuators' motion.

$$\begin{pmatrix} \dot{x} \\ \dot{y} \\ \dot{\theta} \end{pmatrix} = \begin{pmatrix} \cos\theta & 0 \\ \sin\theta & 0 \\ 0 & 1 \end{pmatrix} \cdot \begin{pmatrix} r/2 & r/2 \\ 2r/d & -2r/d \end{pmatrix} \cdot \begin{pmatrix} \dot{\varphi}_r \\ \dot{\varphi}_l \end{pmatrix} \tag{9.13}$$

9.2.2 Backward kinematics

The inverse kinematic model is obtained by solving previous equation for the wheels speed vector by equation (9.13), so that we have the inverse kinematic model given by,

$$\begin{pmatrix} \dot{\varphi}_r \\ \dot{\varphi}_l \end{pmatrix} = \begin{pmatrix} 1/r & d/4r \\ 1/r & -d/4r \end{pmatrix} \cdot \begin{pmatrix} \cos\theta & \sin\theta & 0 \\ 0 & 0 & 1 \end{pmatrix} \cdot \begin{pmatrix} \dot{x} \\ \dot{y} \\ \dot{\theta} \end{pmatrix} \tag{9.14}$$

noi This model represents the basis to obtain the second order (acceleration) reference model given a path functional form (arising from Taylor's theorem, sec. 1.3). By assuming that the wheel speed reference model has been computed with constant period of time τ. It is required a precise as possible observation of the real wheels speed. Thus, by using the finite central differences method (see section 1.7) as a wheels sensing model, an inferred numerical version of the angular velocities is enhanced to approach a measure of the wheels acceleration. Described by

$$\ddot{\varphi}_w = \frac{\dot{\varphi}_{wt+1} - \dot{\varphi}_{wt-1}}{2\tau} \tag{9.15}$$

where $w = \{l, r\}$ represents any of left/right wheel. The wheel acceleration reference model is obtained from its direct velocity measurement. Hence, the wheel speed reference model has foundations on the inverse transformation of the first derivative of the polynomial function of the original desired pathway.

9.3   Fitting the actuators model

A common problem found when controlling rotary actuators is the non-linear response, which is yielded by different engineering factors such as the speed-driver electronics. Usually, such devices are manufactured with optoelectronic components that inherently add non-linear power responses. Besides, its output differences are given from commercial product to product. Nevertheless, in our approach a suitable digital set of commands has to fit the desired angular speeds. The theoretical model to control the actuators are not explicitly given; and even if it would exist, it would change with the presence of small loads variation (i.g. frictions, loads). Thus, under such circumstances a manner to model the real actuator's behaviour is to obtain its an empirical model of $\dot{\varphi}_{r,l}$ as a function of the digital control commands δ. Nevertheless, even when having the actuator's empirical model available, the theoretical model is determined by a fitting numerical method (see section 1.6.3). To control the real output speed of the wheels, we fitted a theoretical equation with an experimental motor's model. We highlight the importance on finding a mathematical relationship between digital control commands, with an analytical output speed. Thus, for our home-made robotic platform, the next model fitting the real motors behaviour was calculated.

The theoretical model linking the set of digital control commands to output the rolling speeds is based on the empirical speed measurements. They were inferred by sweeping the range of computer control commands (from -100-0, and 0-100) w.r.t. averaged observed samples. The response of the actuators resulted as depicted by figure 9.2. Because the actuators rotation's sense were similar, the waveforms were approximated by a unique function.

$$\dot{\varphi} = a\,e^{c\delta} + b \qquad (9.16)$$

where δ corresponds to the actuator's digital output control command, $\dot{\varphi}$ corresponds to the output speed, and a, b, c were used to make convergence of the function into the empirical response. Finding the inverse solution for $\delta(\dot{\varphi})$, we obtained the desired output speed digital command (9.17).

$$\delta(\dot{\varphi}) = \frac{1}{c}\ln\left(\frac{\dot{\varphi} - b}{a}\right) \qquad (9.17)$$

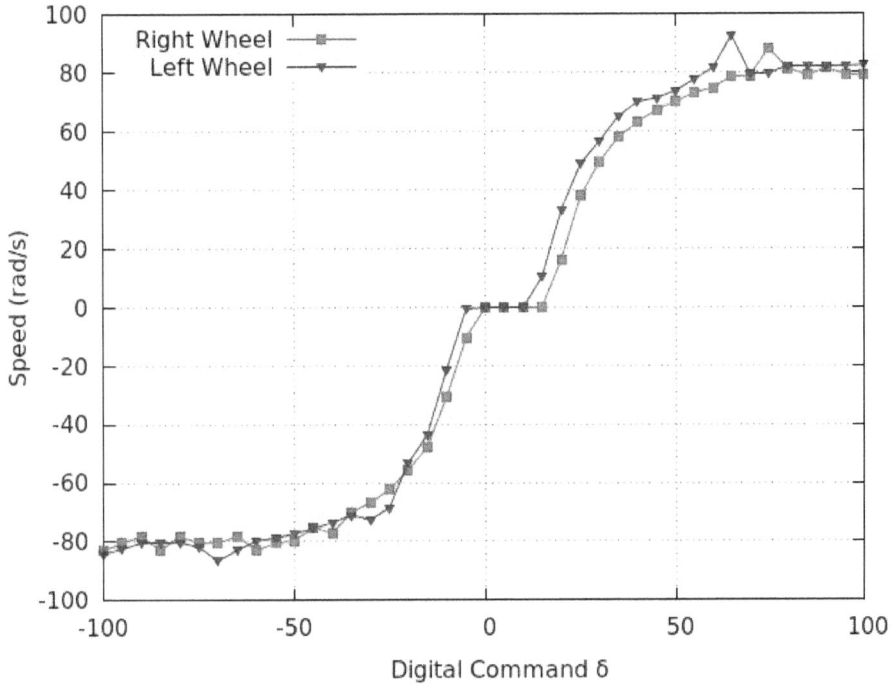

Figure 9.2: Empirical model based of wheels speed measurements.

The theoretical solution depicted in figure 9.3 matches accurately the empirical models shown in figure 9.2, including the inherent perturbations (i.e. frictions).

9.4 Trajectory tracking

The acceleration control model is introduced by equation (9.18) as $\ddot{\varphi}_{t+1}$ that is recursively calculated as a function itself $\ddot{\varphi}_{wt-1}$, with proportional gain β. The instantaneous error is calculated by the difference between a reference model (9.15), and the actual observation $\dot{\hat{\varphi}}$.

$$\ddot{\varphi}_{t+1} = \ddot{\varphi}_{t-1} + \beta \left(\frac{\dot{\varphi}_{t+1} - \dot{\varphi}_{t-1}}{2\tau} - \dot{\hat{\varphi}} \right) \tag{9.18}$$

Hence, the theoretical model $\dot{\varphi}$ is produced as a function of the independent digital variable δ, and the speed control scheme (9.19), which is iteratively used to satisfy the controlled acceleration (9.18) by means of the theoretical model (9.17).

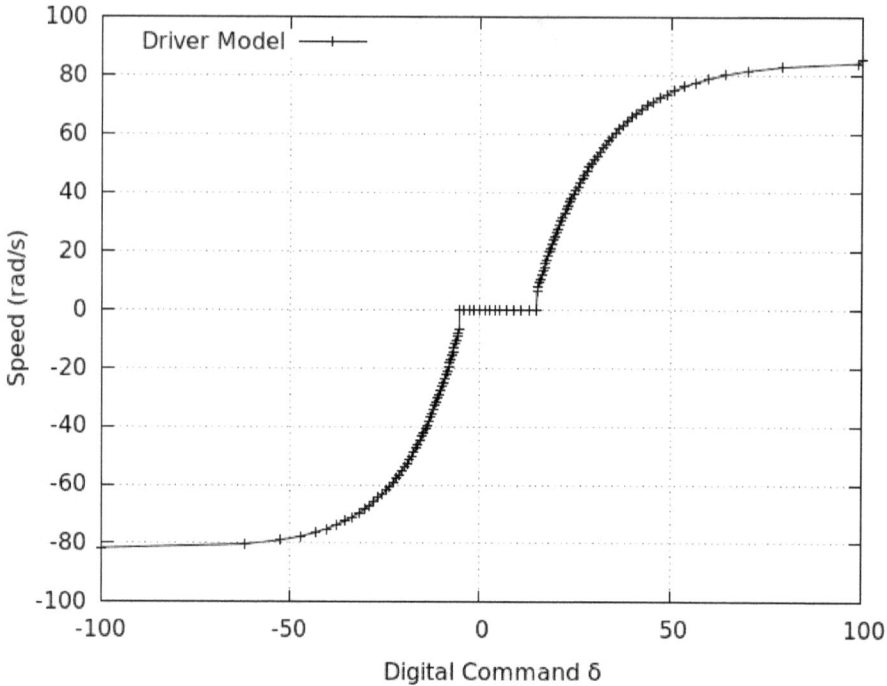

Figure 9.3:   Simulation of wheels speed theoretical model.

$$\dot{\varphi}_{t+1} = \dot{\varphi}_{t-1} + \alpha \left(\frac{d}{dt} s(t) - \int_0^t \ddot{\varphi}_{t+1} dt \right) \tag{9.19}$$

where $\dot{\varphi}_{t+1}$ is the wheel's controlled angular velocity, recursively calculated as function of past $\dot{\varphi}_{t-1}$, and proportional gain is $0 < \alpha \leq 1$. The desired path model is represented by $\frac{d}{dt} s(t)$, and the real velocity observation is obtained by integrating $\ddot{\varphi}_{t+1}$ in the time domain.

Numerous experiments were realized to obtain real-time actuators' response with foundations upon (9.18) - (9.19), and including different non-linear reference models. The first sets of experiments involved ideal paths that when inversely transformed into the wheels' reference model, they represented constant numeric values (figure 9.4-left). Likewise, a second set of experiments involved pathways modelled by second degree polynomials, which when inversely transformed, they represented linear speed models (figure 9.4-right). Moreover, the controlled speed vector is compounded of the left and right angular speeds, as defined by

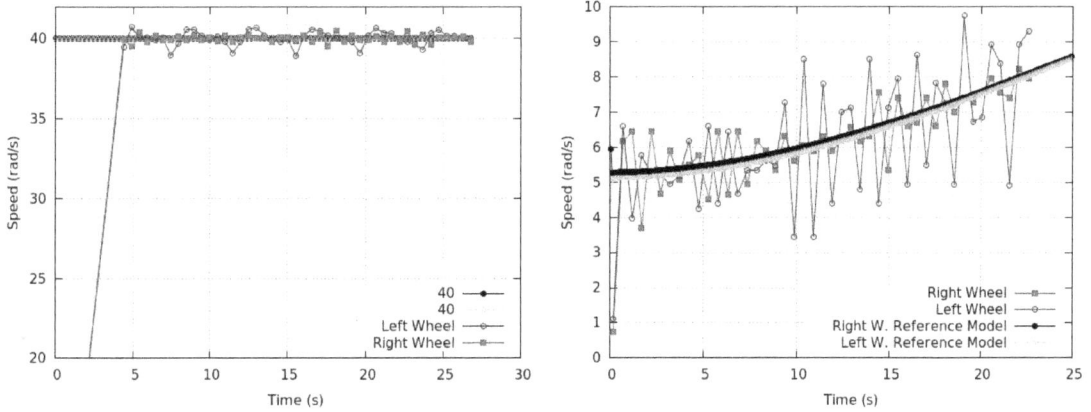

Figure 9.4: Wheels actuator's angular velocity response over time w.r.t. the wheel's reference model.

Proposition 9.4.1 (inverse trajectory tracking vector). *The robot wheels controlled velocity vector is a function of the reference model Cartesian components.*

$$
\begin{pmatrix} \dot\varphi_r \\ \dot\varphi_l \end{pmatrix}_t = \begin{pmatrix} \dot\varphi_r \\ \dot\varphi_l \end{pmatrix}_{t-1} + \frac{\alpha}{r} \begin{pmatrix} (\dot x \cos\theta + \dot y \sin\theta) + \dfrac{d}{4}\dot\theta - \int_0^t \ddot\varphi_{rt+1}\mathrm{d}t \\ (\dot x \cos\theta + \dot y \sin\theta) - \dfrac{d}{4}\dot\theta - \int_0^t \ddot\varphi_{lt+1}\mathrm{d}t \end{pmatrix}
\tag{9.20}
$$

Previous mathematical proposition states that the inverse solution, or wheels angular velocity is obtained from knowing the reference model, which is obtained from actual segment of pathway. The pathway at hand usually (for this work) is generated as third degree polynomials. Therefore, by substituting the controlled speed of equation (9.19) into equation (9.17), the complete control expression (9.21) given as an inverse function of the digital control command is provided.

$$
\delta_{t+1} = \frac{1}{c}\ln\left(\frac{\dot\varphi_{t-1} + \alpha\left(\dfrac{d}{dt}s(t) - \int_0^t \ddot\varphi_{t+1}\mathrm{d}t\right) - b}{a} \right)
\tag{9.21}
$$

The complete control vector involves the left and right wheels' angular speed, which is obtained through an inverse decomposition that obtains the following reference model,

$$\begin{pmatrix} \delta_r \\ \delta_l \end{pmatrix}_{t+1} = \frac{1}{c} \ln \left(\left(\frac{\left(\dot{\varphi}_{rt-1} + \alpha \left(\frac{1}{r}(\dot{x}\cos\theta + \dot{y}\sin\theta) + \frac{d}{4r}\dot{\theta} - \int_0^t \ddot{\varphi}_{rt+1}dt \right) - b \right)}{a} \right. \right.$$
$$\left. \left. \frac{\left(\dot{\varphi}_{lt-1} + \alpha \left(\frac{1}{r}(\dot{x}\cos\theta + \dot{y}\sin\theta) - \frac{d}{4r}\dot{\theta} - \int_0^t \ddot{\varphi}_{lt+1}dt \right) - b \right)}{a} \right) \right) \qquad (9.22)$$

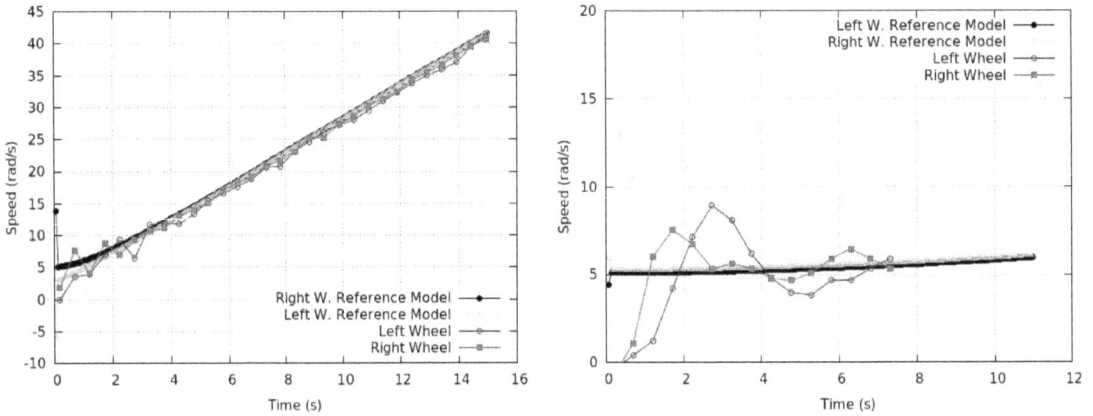

Figure 9.5: Wheels' actuator angular velocity response over time w.r.t. each wheel's reference model, high slope (left), and low slope (right).

Experiments involved theoretical and empirical paths of second degree polynomials, with high slopes reference models. The robot's weight, and its mechanisms friction perturbed the robot trajectory tracking. However the proposed control algorithm acted fast and reliable. The second order adaptive reference model control was tested experimentally with a dual velocity differential robot, and through numerical simulations. Third degree polynomials represented pathways with the minimal degree required. Their second order derivative produce linear functions as acceleration reference models. Second degree polynomials yield linear but constant value reference model with the best performance. Higher than third degree polynomials with low magnitude slopes perform acceptable. The proposed approach is a general solution for any type of wheeled robotic structure of dual or greater asynchronous speeds.

Bibliography

[1] Hsieh M. F., Zguner Ö, *A path following control algorithm for urban driving*, In Proc. of the IEEE Intl. Conf. on Vehicular Electronics and Safety, pp.227–231, 2008.

[2] Tso S.K., fung Y.H., *Intelligent fuzzy switching of control strategies in path control for autonomous vehicles*, In Proc. of the IEEE Intl. Conf. on Robotics and Automation, vol. 1, pp.281–286, 1995.

[3] Soetanto D., Lapierre L., Pascoal A., *Adaptive, non-singular path-following control of dynamic wheeled robots*, 42nd IEEE Decision and Control vol. 2, pp.1765–1770, 2003.

[4] Indiveri G., Nütcher A., Lingemann K., *High speed differential drive mobile robot path following control with bounded wheel speed commands*, IEEE ICRA, pp. 2202–2207, 2007.

[5] Lapierre L., Zapata R., Lepinay P., *Simulatneous path following and obstacle avoidance control of a unicycle-type robot*, IEEE ICRA, pp.2617–2622, 2007, doi: dx.doi.org/10.1109/robot.2007.363860.

[6] Xiang A., Lapierre L., Jouvencel B., Parodi O., *Coordinated path following control of multiple wheeled mobile robots through decentralized speed adaptation*, IEEE/RSJ IROS, pp. 4547–4552, 2009.

[7] Ghommam J., Mehrjerdi H., Saad M., Mnif F., *Adaptive Coordinated Path Following Control of Non-holonomic Mobile Robots with Quantised Communication*, IET Control Theory & Applications, vol.5, pp. 1990–2004, 2011.

[8] Peymani E., Fossen T. I., *A Lagrangian framework to incorporate positional and velocity constraints to achieve path-following control*, in Proc. of the 50th IEEE Conference on Decision and Control and European Control Conference (CDC-ECC), pp. 3940–3945, 2011.

[9] Qi L., Ma B., Li W., *Nonsingular geometric path following control of a wheeled mobile robot*, in Proc. of the Chinese Control and Decision Conference (CCDC), pp. 3068–3072, 2010.

Chapter 10

ALL-ACTIVE 4-WHEEL KINEMATICS

Erik Lerín García and Edgar A. Martínez García

Laboratorio de Robótica, Institute of Engineering and Technology
Universidad Autónoma de Ciudad Juárez, Mexico.

Wheeled mobile robots (WMR) are rolling devices capable of performing locomotive tasks on surfaces solely through the actuation of wheels in contact with the surfaces of displacement. Some link assemblies contain passive suspension, while others contain active suspension [?]. In this chapter a general kinematic control law for multi-configuration of four-wheel active drive/steer robots is discussed. This work models four-wheel drive and steer (4WDS) robotic systems where all wheels drive and steer simultaneously. The control variables are wheel yaw, wheel roll, and suspension pitch by active/passive damper systems. The latter implies that a wheel's contact point translates its position over time collinear with the robot's lateral sides. We present a suspension mechanical system featuring three DOF per wheel. We define wheel's yaw β, wheel's roll angle φ, and an uncommon characteristic regarding mobility based on the wheel contact point location controlled by the suspension angle γ. A possible manner for navigating ground surfaces and avoiding obstacles is to use holonomic or kinematic redundant vehicles[?]. The WMR's degree of holonomy is determined by the value of mobility and manoeuvrability (how quickly the direction of travelling can be changed). This value is measured by the wheels' restriction, and WMRs are classified as systems with holonomic and non-holonomic properties according to their degrees of mobility m and steerability. The holonomic WMRs are able to move in all degrees of freedom available in the workspace, thus this kind of robot does not have mechanical constraints that limits mobility. Accordingly, the non-holonomic robots are not able to move in all directions because they have kinematic constraints in their locomotion structure (one example of this mechanical structure are the like-car robots). In all-terrain WMRs, use of odometry to obtain distance and direction displacement usually is not enough; to improve distance and direction estimations, the addition of inertial measurement units, GPS and optical speed sensors are needed.

10.1   4W kinematic structures

Full active rolling systems explicitly need a function that models for each control variable ′ . With combined passive drive and active steer, the degree of controllability is poorer than full-active driving because wheels lose the ability to move forward/backward at individual contact point speeds. With all-active drive/steer DOF, wheeled systems behave in a nearly fully holonomic manner. Full-active systems have advantages over combined partial-passive systems when feasibility and reliability in manoeuvrability ′ are demanded to navigate complex terrain surfaces. Because all-active drive-steer systems allow multiple kinematic configurations, they provide diverse advantages for self-adaptation to different geological surface features. In all-active 4WD/4WS there are a number of possible locomotive configurations that the kinematics may yield. Some locomotive configurations are illustrated in figure 10.1.

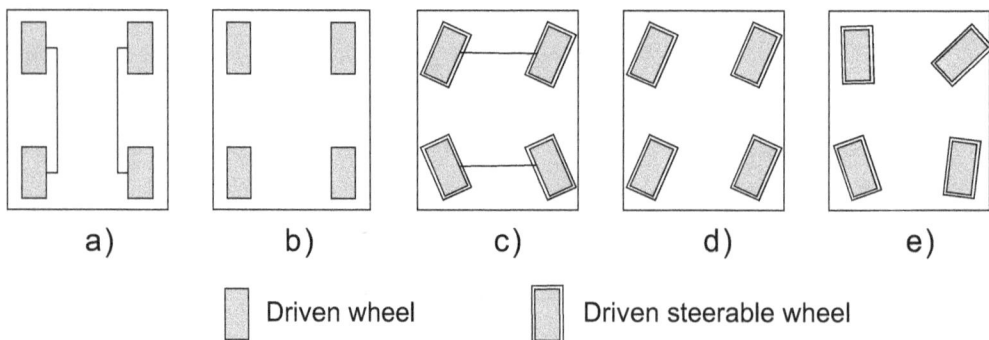

Figure 10.1: 4W locomotion modes. a) dual differential drive; b) asynchronous fix-driven wheels; c) differential drive/steer; d) 4W synchronous drive/steer; e) 4W asynchronous drive/steer.

Figure 10.1 matches the 4W categories described in table 10.1. It shows the different configuration modes of 4W systems and shows how driven wheels and steer wheels are configured.

Table 10.1: 4WD4S main configuration modalities (NH: Nonholonomic, H: Holonomic).

4WD configuration modes					
	Holonomy	Speeds	Description	Steer	Figure 10.1
1	NH	$\dot{\varphi}_r, \dot{\varphi}_l$	2 differential drives.	-	a)
2	NH	$\dot{\varphi}_1, .., \dot{\varphi}_4$	4 differential drives.	-	b)
3	H	$\dot{\varphi}_f, \dot{\varphi}_r$	Rear-wheels, front-wheels steer/drive.	β_f, β_r	c)
4	H	$\dot{\varphi}$	1 speed, 1 steer.	β	d)
5	H	$\dot{\varphi}_1, .., \dot{\varphi}_4$	4 drives, 4 steers.	$\beta_1, \ldots, \beta_4$	e)

According to the categories laid out in table 10.1 we have different locomotive configuration modalities with respect to wheel rolling speeds $\dot{\varphi}$, and steer β. This chapter introduces a mechanical design of a wheeled mobile robot, in the category of a 4WD4S displayed in figure 10.2. The kinematic design possesses 12 DOF, being 3-DOF by wheel, which includes a spring-mass-damper angle (γ), a wheel rolling angle (φ), and a wheel steering angle (β). Depending of the locomotion configuration mode adopted, when the number of this variables in use are larger than three (x, y, θ), the resulting mathematical system is considered with kinematic redundancy. An advantage of this complex structure, is that its wide degree of freedom capability allows different holonomic configuration modes to be featured.

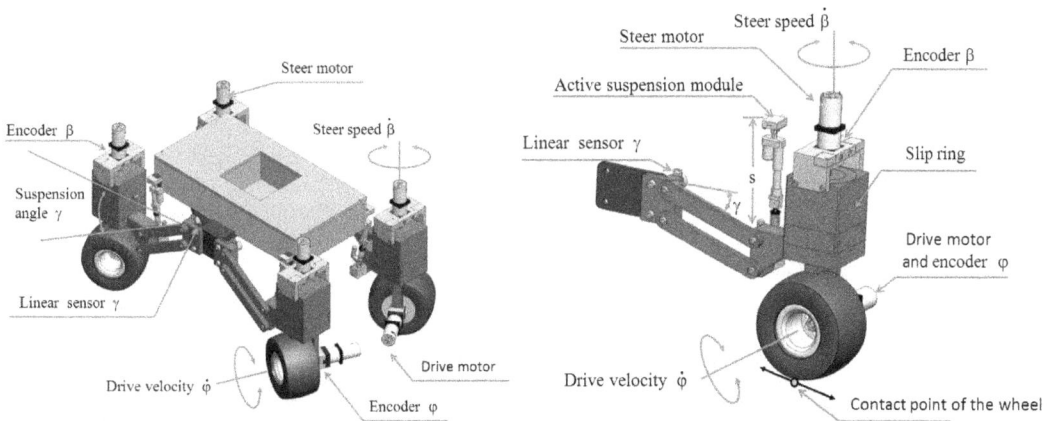

Figure 10.2: Design of the 4W robot mechanical prototype, and one wheel suspension assembly.

Figure 10.2 depicts several suspension elements proposed for the wheel-suspension assembly. The platform is instrumented with two DC motors with encoders, one for the steer (β), another to drive (φ). One slip ring to transmit electrical signals between the robot, the motors and the encoders. Articulated arms with a sensor angle measurement (γ). The lateral mobility restrictions of the robot's wheels (orthogonal forces) are also called the non slip kinematic condition. Figure 10.3 depicts a top view of the 4W structure kinematics. The wheel's instantaneous angular velocities are denoted by $\dot{\varphi}_1, \ldots, \dot{\varphi}_4$. Likewise, the wheels' steering angle are defined by $\beta_1, .., \beta_4$, which represent its value within the robot's attached coordinate frame. The wheels' contact points locations are given in cylindrical form by $\alpha_1, .., \alpha_4$, with their respective distances $l_i, \forall i = \{1, \cdots, 4\}$ w.r.t. robot's geometric centre. Thus, the orthogonal kinematic components constraints s for fixed and centred wheels is described by,

$$\left[\cos(\alpha + \beta) \sin(\alpha + \beta) l \sin\beta \right] \mathbf{R}(\theta) \dot{\xi} = 0 \tag{10.1}$$

where $\dot{\xi}$ is the robot's posture, and $\mathbf{R}(\theta)$ is the rotation Euler matrix. This is the kinematic restriction given by a single wheel yielded to the robot's entire motion behaviour.

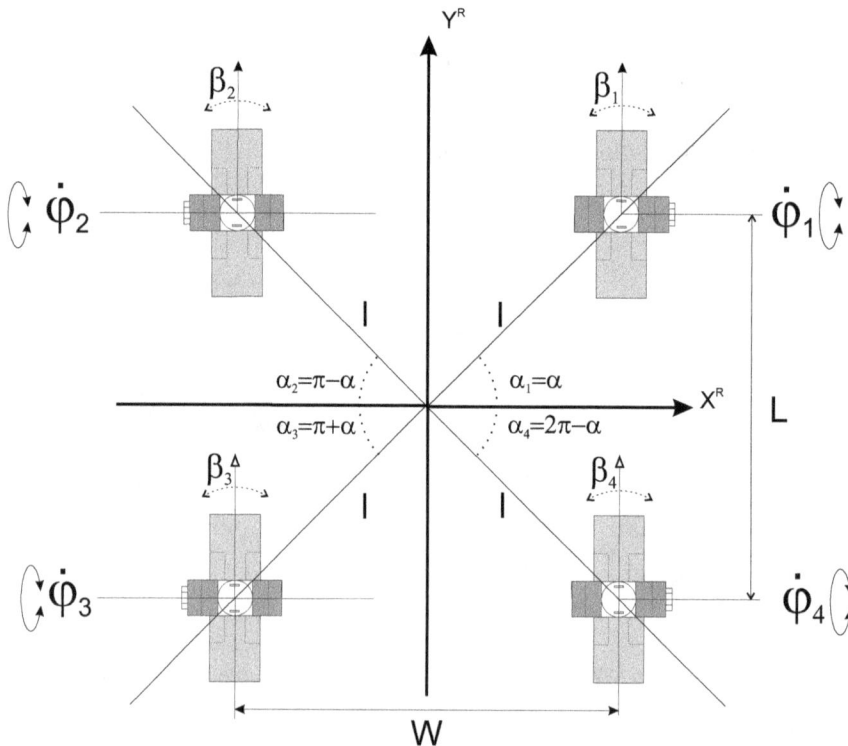

Figure 10.3: Top view of a 4W drive-steer robot's kinematic configuration.

We started our reasoning following the reported method of non-slip kinematic conditions, and the notation established. Thus, working from equation (10.1), the kinematic restrictions matrix is defined by K_1

$$
K_1 = \begin{pmatrix}
\cos(\alpha + \beta_1) & \sin(\alpha + \beta_1) & l\sin(\beta_1) \\
-\cos(\alpha - \beta_2) & \sin(\alpha - \beta_2) & l\sin(\beta_2) \\
-\cos(\alpha + \beta_3) & -\sin(\alpha + \beta_3) & l\sin(\beta_3) \\
\cos(\alpha - \beta_4) & -\sin(\alpha - \beta_4) & l\sin(\beta_4)
\end{pmatrix}
\tag{10.2}
$$

The fixed-wheels matrix K_1 is obtained from the orthogonal kinematic restrictions of all involved fixed conventional wheels in a 4WDS structure. In addition, the kinematic conditions for the centred steerable wheels are given in matrix K_2. The resulting mobility capability using the non slip conditions yields a mobility degree of $\delta_m = 0$, and a steering degree of $\delta_s = 3$

$$
\delta_m = 3 - rank\{K_1\}
$$

and

$$
\delta_s = rank\{K_2\}
$$

The numeric value 3 arises from the number of degrees of freedom in the robot's plane of mobility x,y, θ. In our particular case (i.e. 4WDS structure), this factor uniquely includes centred steerable wheels. Thus, both matrices are $K_1 = K_2$. This is because the rank(K_1)=3. Moreover, hereafter $u(t)$ is the control input vector, so the general formulation for the posture kinematic model is $\dot{z} = B(z)u(t)$, obtained from the product of transposing the orthogonal rotation matrix $R(\theta)^T$ and the vector solutions (the null space vectors) of K_2, described by $\Sigma(K_2)$ as

$$
\dot{z} = R(\theta)^T \Sigma(K_2)u
\tag{10.3}
$$

Since the rank(K_2)=3 it is not possible to obtain the posture kinematic model because the null-space vector for K_2 has dimension zero.

$$
\Sigma(K_2) = 0
\tag{10.4}
$$

Therefore, a real mathematical solution for this algebraic problem is tackled from a different kinematic approach in the next sections.

10.2 Damper kinematic constraints

We model pure rolling conditions affected by the suspensions damper effects. Figure 10.4 depicts the suspension's mechanical parameters. x_{w_i} is the wheel position along the robot's longitudinal axis this value refers to the wheels' contact regions, also defined by equation (10.5). x_{w_i} varies as the suspension's angle γ_i changes overtime. Such wheel variations in translation affects the robot's global controllability and manoeuvrability. The wheels' coordinates along the robot's fixed longitudinal axis are x_i, as depicted by figure 10.4. Coordinate values along the robot's transversal axis are y_i, and for each wheel such values prevails as a constant, as denoted by equation (10.6).

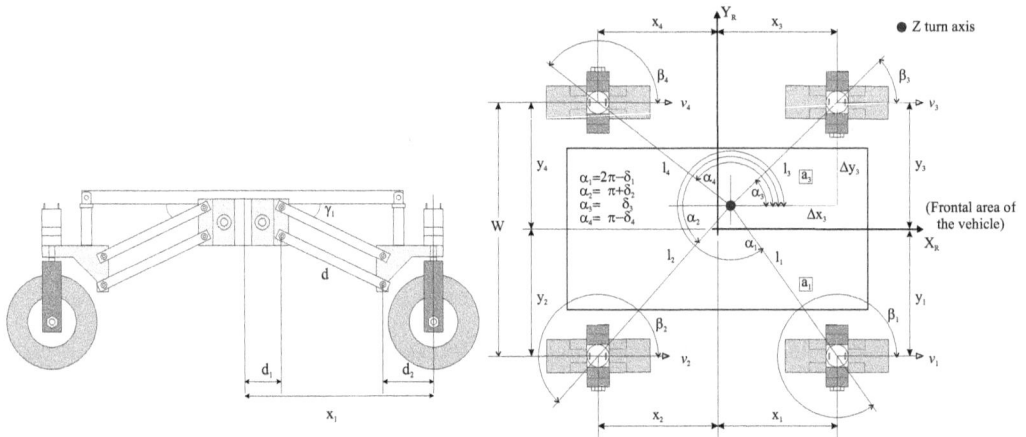

Figure 10.4: Left: robot suspension system (side view). Right: kinematic parameters (top view).

Definition 10.2.1. The robot's wheel position is $(x_{w_i}, y_{w_i}) \, \forall i = \{1, \dots, 4\}$.

$$x_{w_i} = d_1 + d_2 + d \cos(\gamma_i) \tag{10.5}$$

with parameters $d, d_1, d_2,$ and

$$y_{w_i} = \begin{cases} -\frac{W}{2}, & \pi < \alpha_i < 2\pi \\ \frac{W}{2}, & 0 < \alpha_i < \pi \end{cases} \tag{10.6}$$

In addition, the model for γ_i is given by the active suspension system. Let us define the following parameters, Δ_s is the suspension offset that sets the device's fixed height (given in m). m is the spring mass (in kg). κ_r and κ_v are the restitution and viscous coefficients, respectively. g is the gravity acceleration constant (m/s^2). Finally, $\dot{y}_s$ and $\ddot{y}_s$ are the instantaneous velocity (m/s) and acceleration (m/s^2) of the spring-mass elongation. Wheel contact points prevail with no change when steer angles are $|\beta_i| \geq \frac{\pi}{4}$ and no damper effects exist. $\kappa_r = 1$, and $\kappa_v = 0$. Gravity force exerts no affects over wheel contact points.

$$\gamma_i = \arcsin\left(\frac{\Delta s}{d}\right) \tag{10.7}$$

Proposition 10.2.2. *With no damper effects, it is assumed the z-turn axis is placed on the robot's geometric centre. Hence, this location is taken as a common reference through l_i.*

$$l_i \sin(\alpha_i) = x_{w_i} \tag{10.8}$$

Thus, three linear equations that project x^{w_i} are stated. Since the model is already known from a wheel plane perspective, then the expression (10.5) is substituted. A first equation approach is proposed:

$$l_i \sin(\alpha_i) - d_1 - d_2 - d\cos(\gamma_t^i) = 0; \tag{10.9}$$

A second equation approach is defined,

$$y_{w_i} \tan(\alpha_i) - d_1 - d_2 - d\cos(\gamma_i) = 0 \tag{10.10}$$

And a third mathematical approach,

$$\frac{y_{w_i}}{\cos(\alpha_i)} - d_1 - d_2 - d\cos(\gamma_i) = 0 \tag{10.11}$$

The robot's global motion behaviour is given critically by the instantaneous value of γ_i. An electric adjustable resistance (potentiometer) is used as a linear measurement device to obtain direct measurements $\hat{\gamma}_i$. Nevertheless, a set of functional forms for γ_i are proposed, in accordance with the actual terrain and manoeuvrability situation. Thus, two more propositions for γ_i behaviour are stated, suited to different situations. The following propositions assume magnitudes of l_i as variables and converge to the z-turn location (x_z, y_z).

Proposition 10.2.3. *Damper effects are restricted to $v_t = v_{t-1}$, $\forall t$, steer angles $\beta_i = 0$, $\forall i$, and (x_z, y_z) is located at the robot's centroid as a consequence of $\ddot{y} = 0$.*

$$\gamma_i(\dot{y}) = \left(\frac{mg - \kappa_v \dot{y}_s + \Delta s}{\kappa_r d} \right) \tag{10.12}$$

Proposition 10.2.4. *Damper effects have no restrictions, and $v_t \neq v_{t-1}$ $\forall t$, $a_t \neq 0$, $\beta_i \neq 0$, and (x_z, y_z) varies its location.*

$$\gamma_i(\ddot{y}_s, \dot{y}_s) = \arcsin \left(\frac{mg - m\ddot{y}_s - \kappa_v \dot{y}_s + \Delta_s}{\kappa_r d} \right) \tag{10.13}$$

10.3   Instantaneous z-turn axis model

In this section, a mathematical model is proposed to infer the z-turn axis location. The z-turn is a virtual axis that implicitly governs the robot's body yaw speed w.r.t. its rotation point. There is not any existing sensor device to measure (x_z, y_z). However, we introduce an approach to infer this on-line by deploying 2-axis accelerometer devices on board (figure 10.5). This is a contribution approach where the z-axis location is used to control the robot's manoeuvrability, forcing it to reach a desired posture regardless of slip/skid effects. We state in this manuscript that the z-turn region of translation is scoped by the wheels' location (x_{w_i}, y_{w_i}). Inertial accelerometer devices at fixed locations are deployed to infer (x_z, y_z), by means of their instantaneous acceleration measurements overtime. Velocities in local inertial systems are deduced by numerical integration w.r.t. time. Let us define a velocity vector $v = (\dot{x}, \dot{y})^\top$ for each robot's inertial device on board. In accordance with figure 10.4, let us represent the two accelerometer devices a_1 and a_3 as they match the wheel number.

The general model of average acceleration is

$$\mathbf{a}dt = d\boldsymbol{v} \tag{10.14}$$

Thus, integration w.r.t. time in interval $\Delta t = t_2 - t_1$, and highlighting that acceleration $\hat{\mathbf{a}}$ is the sensor measurement,

$$\int_{v_1}^{v_2} d\boldsymbol{v} = \hat{\mathbf{a}} \int_{t_1}^{t_2} dt \tag{10.15}$$

Developing and algebraically arranging previous equation for two dimensions:

$$\boldsymbol{v} = \begin{pmatrix} \dot{x}_t \\ \dot{y}_t \end{pmatrix} = \begin{pmatrix} \dot{x}_{t-1} \\ \dot{y}_{t-1} \end{pmatrix} + \begin{pmatrix} \hat{\dot{x}}_t \\ \hat{\dot{y}}_t \end{pmatrix} \Delta t \tag{10.16}$$

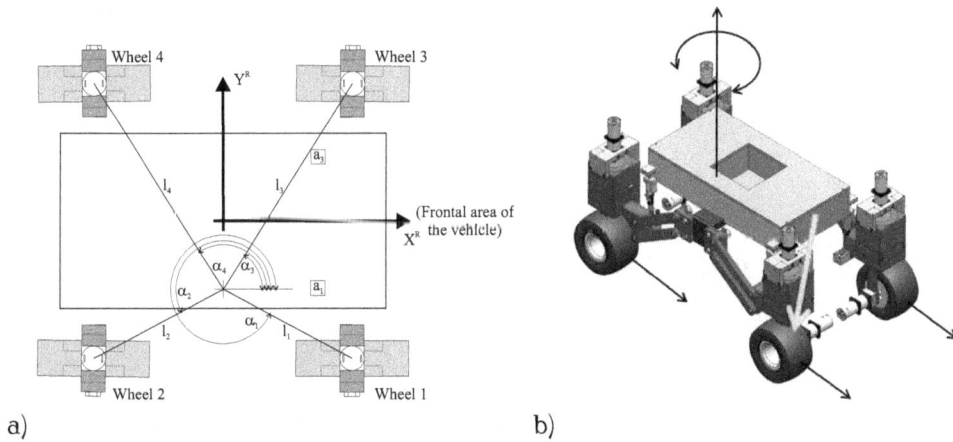

Figure 10.5: a) The z-axis location is displaced from the robot's centroid; b) robot's direction (green arrow) effected by the z-turn and wheels' behaviour.

The instantaneous acceleration $\hat{\mathbf{a}}$ is the sensor's measurement with two components $\hat{\mathbf{a}} = (\hat{\dot{x}}, \hat{\dot{y}})^{\top}$. As depicted by figure 10.5, two accelerometers a_1 and a_3 were fixed to robot's body. a_i the reference of a given inertial device. To find an exact solution for two dimensions, we set two equations as a minimum requirement.

Proposition 10.3.1. *The trigonometric ratio between accelerometer, z-turn axis, and measured speeds defines the next relationship.*

$$\frac{y_z - y^{a_j}}{x_z - x^{a_j}} = \int_t \left(\frac{\ddot{y}^{a_j}}{\ddot{x}^{a_j}} \right) dt \tag{10.17}$$

This proposition satisfies the stipulation that the jth accelerometer location (x^{a_j}, y^{a_y}) w.r.t. (x_z, y_z) has the same geometric ratio as the accelerometer readings $\int \ddot{y}^{a_j}/\ddot{x}^{a_j} dt$, if and only if it is aligned with the robot's fixed frame. Thus, from this proposition, the next theorem is presented, which states that an equation system allows description of a behavioural z-turn axis location.

Theorem 10.3.2. *One linear equation describe a single accelerometer, and at least two speed measurements are required as a necessary and sufficient condition to infer (x_z, y_z).*

$$y_z - x_z \int_t \left(\frac{\ddot{y}_t^{a_1}}{\ddot{x}_t^{a_1}} \right) dt = y^{a_1} - x^{a_1} \int_t \left(\frac{\ddot{y}_t^{a_1}}{\ddot{x}_t^{a_1}} \right) dt \tag{10.18}$$

and

$$y_z - x_z \int_t \left(\frac{\ddot{y}_t^{a_3}}{\ddot{x}_t^{a_3}} \right) dt = y^{a_3} - x^{a_3} \int_t \left(\frac{\ddot{y}_t^{a_3}}{\ddot{x}_t^{a_3}} \right) dt \tag{10.19}$$

This particular set of equations is algebraically rewritten and integrated w.r.t. time, according to equation (10.16). Thus, for device a_1,

$$y_z - \left(\frac{\dot{y}_{t-1}^{a_1} + \hat{\dot{y}}_t^{a_1} \Delta t}{\dot{x}_{t-1}^{a_1} + \hat{\dot{x}}_t^{a_1} \Delta t} \right) x_z = y^{a_1} - \left(\frac{\dot{y}_{t-1}^{a_1} + \hat{\dot{y}}_t^{a_1} \Delta t}{\dot{x}_{t-1}^{a_1} + \hat{\dot{x}}_t^{a_1} \Delta t} \right) x^{a_1} \tag{10.20}$$

Likewise the equation for device a_3 is

$$y_z - \left(\frac{\dot{y}_{t-1}^{a_3} + \hat{\dot{y}}_t^{a_3} \Delta t}{\dot{x}_{t-1}^{a_3} + \hat{\dot{x}}_t^{a_3} \Delta t} \right) x_z = y^{a_3} - \left(\frac{\dot{y}_{t-1}^{a_3} + \hat{\dot{y}}_t^{a_3} \Delta t}{\dot{x}_{t-1}^{a_3} + \hat{\dot{x}}_t^{a_3} \Delta t} \right) x^{a_3} \tag{10.21}$$

Hereafter, by substituting some terms for the sake of easy when treating the equations system algebraically, let us define $\zeta_i = (\dot{y}_{t-1}^{a_i}/\dot{x}_{t-1}^{a_i})$, $\eta_i = y^{a_i}$, and $\varrho_i = x^{a_i}$.

The system of linear equations is therefore arranged in the matrix form, and it is solved for an analytical solution:

$$\begin{pmatrix} 1-\zeta_1 \\ 1-\zeta_3 \end{pmatrix} \cdot \begin{pmatrix} x_z \\ y_z \end{pmatrix} = \begin{pmatrix} \eta_1 - \varrho_1\zeta_1 \\ \eta_3 - \varrho_3\zeta_3 \end{pmatrix} \tag{10.22}$$

The solution for planar coordinates of the z-axis is expressed through the next corollary,

Corollary 10.3.3. *Given a squared matrix, the Cramer theorem yields the z-turn model solution stated by*

$$x_z = \frac{-\zeta_3(\eta_1 - \varrho_1\zeta_1) + \zeta_1(\eta_3 - \varrho_3 - \zeta_3)}{-\zeta_3 + \zeta_1} \tag{10.23}$$

and

$$y_z = \frac{\eta_3 - \varrho_3\zeta_3 - \eta_1 + \varrho_1\zeta_1}{-\zeta_3 + \zeta_1} \tag{10.24}$$

In addition to theorem (10.3.2) and corollary (10.3.3), this idea is complemented by the kinematic effects yielded by the suspension angle behaviour γ_i through the factor l_i.

$$l_i = \sqrt{(\Delta x_i)^2 + (\Delta y_i)^2} \tag{10.25}$$

defined by the terms,

$$\Delta x_i = x_{w_i} - x_z \tag{10.26}$$

as well as the magnitude,

$$\Delta y_i = |y_{w_i} - y_z| \tag{10.27}$$

The angles α_i denote the angular relationship between l_i w.r.t. the x-axis (counter clockwise) as depicted by figure 10.4. Each α_i angle value varies according to l_i values, which are effected by the suspension oscillations. For $-y$, a negative arcsin sign is obtained with $\alpha_1 = 2\pi + \arcsin(\Delta y_1/l_1)$, and $\alpha_2 = \pi - \arcsin(\Delta y_2/l_2)$. For $-y$, a positive arcsin sign is obtained with $\alpha_3 = \arcsin(\Delta y_3/l_3)$, and $\alpha_4 = \pi - \arcsin(\Delta y_4/l_4)$. In addition, we may obtain a simplified expression for α by replacing the terms,

$$\delta_i = \arcsin\left(\frac{\Delta y_i}{l_i}\right) \tag{10.28}$$

Hence, $\alpha_1 = 2\pi + \delta_1$, $\alpha_2 = \pi - \delta_2$, $\alpha_3 = \delta_3$, $\alpha_4 = \pi - \delta_4$ Likewise, the formulation of β_i to control the wheels' steering is $\beta_1 = -\alpha_1 + \phi_1, \beta_2 = -\alpha_2 + \phi_2, \beta_3 = -\alpha_3 + \phi_3, \beta_4 = -\alpha_4 + \phi_4$ where the variables ϕ_i are the direct angle measurement given by the encoders to quantify steering angles. For instance, when the wheels are set fixed with no steer, then $\phi_i = 0$. Without loss of generality, by applying theorem 6.2 and corollary 2, figure 10.6 depicts the robot's pose with its instantaneous z-turn axis relocation during a trajectory.

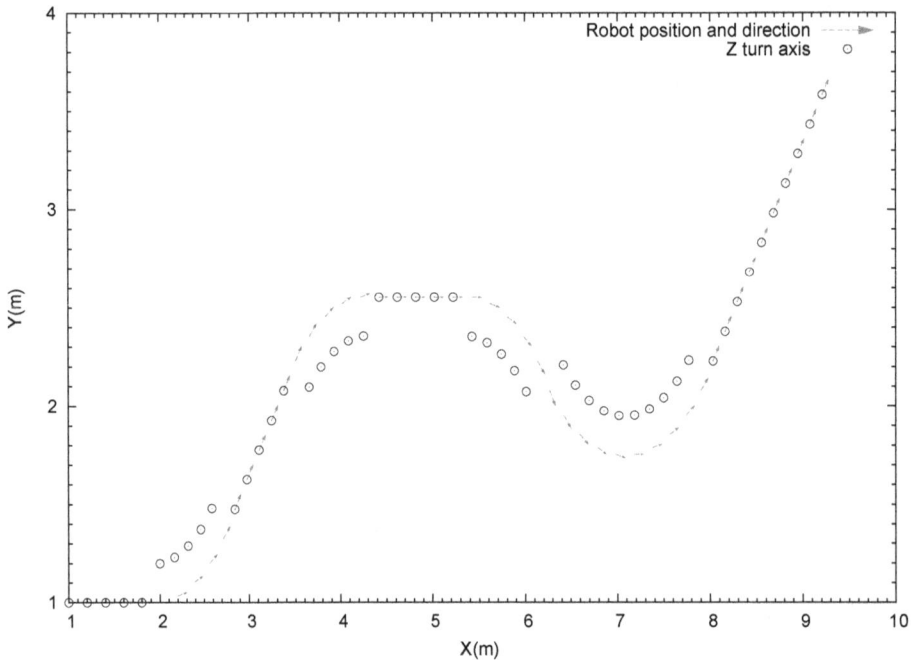

Figure 10.6: Z-turn axis displacement (circles), with respect robot's pose (vectors).

10.4   Motion stability analysis

When the inertial effects exceed the friction force between the four wheels contact points $\mathbf{p}_i = (x_i, y_i)^\top$ and the ground surface, the wheels may lose physical contact with the terrain surface and manoeuvrability efficiency and stability are disturbed. The wheels' contact point separation from the surface occurs due to exceeding magnitudes of z-axis moments of inertia I_z w.r.t. its farthest instantaneous wheel contact point, namely $\mathbf{p}_c$ and obtained by,

$$d_f = \max_{1 \le i \le 4} \| \mathbf{p}_z - \mathbf{p}_i \|$$

Hence, the angle α_i is maximized with d_f and is given by α_f

$$\mathbf{p}_c = d_f \begin{pmatrix} \cos(\alpha_f) \\ \sin(\alpha_f) \end{pmatrix} \qquad (10.29)$$

In figure 10.7 depiction of the turning point $\mathbf{p}_z = (x_z, y_z)^\top$ is given (also represented by circles in figure 10.6), in which an instantaneous moment of inertia I_z is yielded. Considering only the strongest angular moment magnitude $M = I_z \alpha$, among the four wheels' contact points unlike other vehicle's mass point is meaningful.

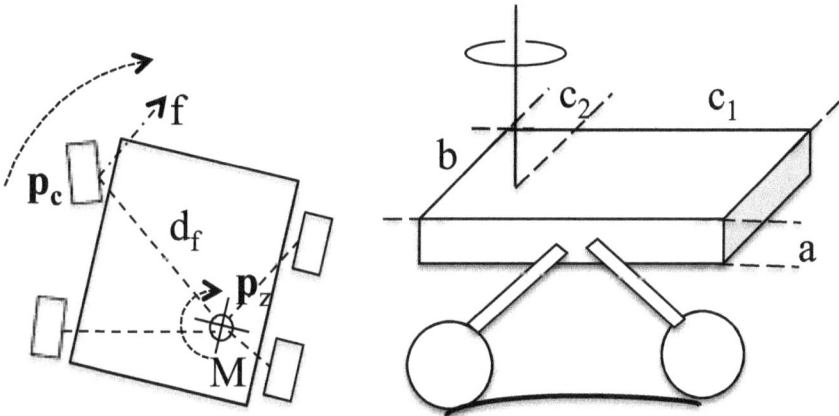

Figure 10.7: Dynamic variables and parameters related to the moment of inertia w.r.t. $\mathbf{p}_z$.

This is assumed to be the case when wheels strike no obstacles on the terrain surface that would forcibly separate the wheels from the ground. Nevertheless, our concerns is with the magnitudes of inertial effects that may yield instability. A vehicle's swift actions might continuously yield strong inertial moment effects. As a consequence, instability rises as $|\gamma_i|$ increases in magnitude. Thus the case that better suits the situation was previously given in proposition 10.2.4, for the angle γ_i as a function of $\dot{y}$ and $\ddot{y}$.

$$\gamma_i(\ddot{y}_s, \dot{y}_s) = \arcsin\left(\frac{mg - m\ddot{y}_s - \kappa_v \dot{y}_s + \Delta_s}{\kappa_r d} \right)$$

Variations of I_z and $|\gamma_i(\dot{y}, \ddot{y})|$ contribute to separating the robot's centre of gravity from the ground, and damper effects occur more frequently impacting the contact points' friction with the ground's surface. Thus, wheels vector point location $\mathbf{p}_i$ is given,

$$\mathbf{p}_i = \left(\frac{d \cos(\gamma_i(\dot{y}, \ddot{y},))}{2}, \frac{b}{2} \right)^\top \tag{10.30}$$

hence, according to figure 10.7 the vehicles geometric parameters are defined as

$$c_{1,3} = \| \mathbf{p}_z - \mathbf{p}_{1,3} \| \cos(\alpha_{1,3}) \tag{10.31}$$

and

$$c_{2,4} = \| \mathbf{p}_z - \mathbf{p}_{2,4} \| \cos(\alpha_{2,4}) \tag{10.32}$$

Likewise, considering that the vehicle's height $a(\gamma_i)$ changes overtime in terms of γ_i,

$$a(\gamma_i) = r + d \cos(\gamma_i) + \Delta_s \tag{10.33}$$

The robot's inertial moment is a mass property of a rigid non-uniform body, in which its inertial moment is located around its z-turn axis $\mathbf{p}_z$. Thus, the parallel theorem is applied for an arbitrary axis.

$$I = \int_{-c_2}^{c_1} \left(\frac{b^2}{12} \right) \frac{m}{c} dc \tag{10.34}$$

Then, after integrating the general expression,

$$I = \frac{m}{12}(b^2 + c^2) \tag{10.35}$$

Hence, substituting the variable limits, $c = c_1 + c_2$ (from figure 10.7), the z-turn axis inertial moment is given by,

$$I = \frac{m}{12} \left(b^2 + (\| \mathbf{p}_z - \mathbf{p}_1 \| \cos(\alpha_1) + \| \mathbf{p}_z - \mathbf{p}_2 \| \cos(\alpha_2))^2 \right) \tag{10.36}$$

Since the angular moment is equivalent to the torsional moment $M = I_z d\omega/dt$, we refer to the tangential effects on $\mathbf{p}_c$,

Proposition 10.4.1. *For the farthest wheel contact point* $\mathbf{p}_c$ *from the instantaneous spinning point* $\mathbf{p}_z$, *the tangential force* f_T *has the equivalence* $I_z d\omega/dt \equiv f_T d_f$.

$$I_z \frac{d\omega_c}{dt} = m \cdot a_T \|\mathbf{p}_z - \mathbf{p}_c\| \tag{10.37}$$

The tangential acceleration a_T *of the wheel point w.r.t. z-turn point* $\|\mathbf{p}_z - \mathbf{p}_c\|$ *is,*

$$a_T = \frac{I_z \dot{\omega}_c}{m\|\mathbf{p}_z - \mathbf{p}_c\|} \tag{10.38}$$

Furthermore, a second proposition terms of the energies associated with the tangential force f_T is defined,

Proposition 10.4.2. *The instantaneous torsional moment* M_z *yielded w.r.t.* $\mathbf{p}_z$ *as a function of the kinetic and potential energies,*

$$M_z = f_T \|\mathbf{p}_z - \mathbf{p}_c\| \tag{10.39}$$

The torsional moment expressed in terms of the Euler-Lagrange form,

$$M_z = \frac{d}{dt}\left(\frac{\partial \mathcal{L}}{\partial \dot{\mathbf{q}}}\right) - \frac{\partial \mathcal{L}}{\partial \mathbf{q}} \tag{10.40}$$

The generalized coordinate vector $\mathbf{q} = (v_c, \omega_c)^\top$ *represents actual wheel's contact point* $\mathbf{p}_c$ *whith highest inertial value.* v_c *and omega$_c$ are tangential and angular velocities respectively.*

$$\mathcal{L} = \frac{1}{2}mv_c^2 + \frac{1}{2}I_z\omega_c^2$$

Therefore, the tangential force f_T *of* $\mathbf{p}_c$ *if defined by,*

$$f_T = \frac{\left\| \frac{d}{dt}\left(\frac{\partial \mathcal{L}}{\partial \dot{\mathbf{q}}}\right) - \frac{\partial \mathcal{L}}{\partial \mathbf{q}} \right\|}{\|\mathbf{p}_z - \mathbf{p}_c\|} \tag{10.41}$$

Therefore, from propositions (10.4.1) and (10.4.2), the inertial equilibrium conditions arises, and its defined in lemma (10.4.3),

Lemma 10.4.3. *The system has inertial stabillty when either of following equilibrium conditions occur:*

a) *If $a_c < \varepsilon_a$, $\mathbf{p}_c$ remains in contact with the ground's surface. Where ε_a is the limit acceleration magnitude, in terms of moment of inertia (Prop.10.4.1).*

$$\left| \frac{I_z \dot{\omega}}{m \| \mathbf{p}_z - \mathbf{p}_c \|} \right| < \varepsilon_a \tag{10.42}$$

b) *If $f_T < \varepsilon_f$, $\mathbf{p}_c$ remains in contact with the ground's surface. Where ε_f is the limit force magnitude, in terms of energies (Prop.10.4.2).*

$$\frac{\| M_z \|}{\| \mathbf{p}_z - \mathbf{p}_c \|} < \varepsilon_f \tag{10.43}$$

10.5   Wheels kinematic control law

This section concerns the robot's kinematic analysis on the wheels' degrees of freedom (DOF) not yet discussed in $\dot{\varphi}_i$ and $\dot{\beta}_i$, and how they contribute in effecting the robot's posture. Figure 10.4 depicts a description of the wheels steer angle $\beta_i \, \forall i = \{1, \ldots, 4\}$.

Lemma 10.5.1. *The robot's body motion $(\dot{x}_i, \dot{y}_i, \dot{\theta}_i)$ is partially contributed to by each ith with wheel tangential velocity $r\dot{\varphi}_i$, of nominal radius r, and of angular speed $\dot{\varphi}_i$. A single-wheel contribution to the robot's motion is described by equation (10.44),*

$$\begin{pmatrix} \dot{x}_i \\ \dot{y}_i \\ \dot{\theta}_i \end{pmatrix} = \frac{r}{4} \begin{pmatrix} \cos(\alpha_i + \beta_i) \\ \sin(\alpha_i + \beta_i) \\ \frac{\sin(\beta_i)}{l_i} \end{pmatrix} \varphi_i \tag{10.44}$$

However, the whole robot's translational and rotational velocities $\dot{\xi} = (\dot{x}, \dot{y}, \dot{\theta})^\top$ are given as an average of all wheels fixed to the system. The four wheel restrictions along the centred wheel plane are described by expressions (10.45)-(10.47).

$$\dot{x} = \frac{r}{4} \sum_{i=1}^{4} \cos(\alpha_i + \beta_i)\dot{\varphi}_i \tag{10.45}$$

$$\dot{y} = \frac{r}{4} \sum_{i=1}^{4} \sin(\alpha_i + \beta_i)\dot{\varphi}_i \tag{10.46}$$

In addition, see the model for $\dot{\theta}$ in terms of l_i, the latter discussed in previous section,

$$\dot{\theta} = \frac{r}{4} \sum_{i=1}^{4} \frac{\sin(\beta_i)}{l_i}\dot{\varphi}_i \tag{10.47}$$

Thus, in stating the wheels kinematic constraints in vector form for a 4W4D system, the wheel's plane restriction is governed by the vectors κ_i.

This is described in the following equations (10.48)-(10.51).

$$\kappa_1 = \left(\cos(\alpha_1 + \beta_1), \sin(\alpha_1 + \beta_1), \frac{\sin(\beta_1)}{l_1}\right)^\top \tag{10.48}$$

$$\kappa_2 = \left(\cos(\alpha_2 + \beta_2), \sin(\alpha_2 + \beta_2), \frac{\sin(\beta_2)}{l_2}\right)^\top \tag{10.49}$$

$$\kappa_3 = \left(\cos(\alpha_3 + \beta_3), \sin(\alpha_{3(t)} + \beta_3), \frac{\sin(\beta_3)}{l_3}\right)^\top \tag{10.50}$$

$$\kappa_4 = \left(\cos(\alpha_4 + \beta_4), \sin(\alpha_4 + \beta_4), \frac{\sin(\beta_4)}{l_4}\right)^\top \tag{10.51}$$

Such that, the four wheels rolling condition vectors comprise the transition non-squared matrix $\mathbf{K}$ containing all-wheel restrictions, given by,

$$\mathbf{K} = (\kappa_1 \quad \kappa_2 \quad \kappa_3 \quad \kappa_4) \tag{10.52}$$

Similarly, we define a vector of rotational velocities $\dot{\boldsymbol{\Phi}}$ to simplify the four-wheel system,

$$\dot{\boldsymbol{\Phi}} = r\,(\dot{\varphi}_1 \quad \dot{\varphi}_2 \quad \dot{\varphi}_3 \quad \dot{\varphi}_4)^\top \tag{10.53}$$

The functional form for each wheel speed $\dot{\varphi}_i$ is described by the third degree polynomial,

$$\dot{\varphi}_i(d_i) = a_0 + a_1 d_i + a_2 d_i^2 + a_3 d_i^3 \tag{10.54}$$

Obtained from the response curve of the actuator device (DC motor), with calibrated coefficient terms $a_0,\ldots,a_3$. d_i, the ith digital control variable at computer level is associated to its rotational speed $\dot{\varphi}_i(d_i)$. Thus, without loss of generality, we state the next corollary:

Corollary 10.5.2. *The control vector of a 4W active system is described by*

$$\mathbf{u} = \frac{\mathbf{K}\cdot\dot{\boldsymbol{\Phi}}}{4} \tag{10.55}$$

Furthermore, the $\mathbf{K}$ matrix is not a square and, except for the fixed suspension and synchronous and differential steering mode the matrix inverse does not have a trivial solution, and for this reason the inverse matrix $\mathbf{K}_A^{-1}$ is obtained numerically. The inverse kinematic solution for the generalized system is given through a general inverse form of the non-squared transition matrix $\mathbf{K}$, namely pseudo-inverse for linearly independent columns, where,

$$\dot{\boldsymbol{\Phi}} = 4\mathbf{K}^\top\cdot(\mathbf{K}\cdot\mathbf{K}^\top)^{-1}\cdot\mathbf{u} \tag{10.56}$$

where the Moore - Penrose pseudo-inverse exists and is unique $\mathbf{K}^+\cdot\mathbf{K} = \mathbf{I}$. Such that, $\mathbf{K}^+ = \mathbf{K}^\top\cdot(\mathbf{K}\cdot\mathbf{K}^\top)^{-1}$.

Moreover, regarding the robot's forward kinematics solution, it is in principle given within a local inertial system. Nevertheless, its enhanced description of a global system may be described by,

Theorem 10.5.3. *The posture $\dot{\xi}$ of a 4W system with active $\dot{\varphi}$, $\dot{\beta}$, and $\dot{\gamma}$ variables is controllable through the state equation (10.57),*

$$\dot{\xi}_{t+1} = \mathbf{R}(\theta) \cdot \dot{\xi}_t + \mathbf{B}(\theta)\mathbf{u}_t + \mathbf{t} \tag{10.57}$$

where $\mathbf{t} = (t_x, t_y, 0)^\top$ is a translation vector, and $\mathbf{R}$ the squared orthogonal Euler rotation matrix, $\mathbf{R}(\theta)^\top = \mathbf{R}(\theta)^{-1}$.

$$\mathbf{R}(\theta) = \begin{pmatrix} \cos(\theta) & -\sin(\theta) & 0 \\ \sin(\theta) & \cos(\theta) & 0 \\ 0 & 0 & 1 \end{pmatrix} \tag{10.58}$$

Hence, the squared transition matrix $\mathbf{B}$ is defined by,

$$\mathbf{B}(\theta) = \begin{pmatrix} \cos(\theta) & 0 & 0 \\ \sin(\theta) & 0 & 0 \\ 0 & 0 & 1 \end{pmatrix} \tag{10.59}$$

For instance, the recursive form to calculate the robot's position at time $t + 1$, expanding the control vector $\mathbf{u}$ is stated. By assuming $\mathbf{t} = (0, 0, 0)^\top$, which means the common coordinate system origin is given at the robot's initial posture, we have

$$\dot{\xi}_{t+1} = \mathbf{R} \cdot \dot{\xi}_t + \frac{1}{4}\mathbf{B}(\theta) \cdot \mathbf{K}(\alpha, \beta) \cdot \dot{\mathbf{\Phi}} \tag{10.60}$$

Thus, by algebraically expanding the expression (10.60), the general equation (10.61) for a four-wheel robot rises to allow multi-configuration of a diversity of four-drive/four-steer kinematic modalities:

$$\dot{\xi}^I_{t+1} = \mathbf{R} \cdot \begin{pmatrix} \dot{x}_t \\ \dot{y}_t \\ \dot{\theta}_t \end{pmatrix} + \frac{r}{4}\mathbf{B}(\theta) \cdot \begin{pmatrix} \cos(\alpha_1 + \beta_1) & \cos(\alpha_2 + \beta_2) & \cos(\alpha_3 + \beta_3) & \cos(\alpha_4 + \beta_4) \\ \sin(\alpha_1 + \beta_1) & \sin(\alpha_2 + \beta_2) & \sin(\alpha_3 + \beta_3) & \sin(\alpha_4 + \beta_4) \\ \frac{\sin(\beta_1)}{l_1} & \frac{\sin(\beta_2)}{l_2} & \frac{\sin(\beta_3)}{l_3} & \frac{\sin(\beta_4)}{l_4} \end{pmatrix} \cdot \begin{pmatrix} \dot{\varphi}_1 \\ \dot{\varphi}_2 \\ \dot{\varphi}_3 \\ \dot{\varphi}_4 \end{pmatrix} \tag{10.61}$$

Finally, the next statement (10.5.4) is an inverse general solution, which is consequent of the previous theorem.

Corollary 10.5.4. *The wheel velocity $\Delta\dot{\varphi}$ required to displace the robot from ξ_t to ξ_{t+1} in a global inertial system is modelled by* (10.62),

$$\dot{\Phi} = 4\mathbf{K}^{\top} \cdot (\mathbf{K} \cdot \mathbf{K}^{\top})^{-1} \cdot \mathbf{B}^{-1} \cdot (\dot{\xi}_{t+1} - \dot{\xi}_t - \mathbf{t}) \tag{10.62}$$

This last equation (10.62) provides the wheels speed magnitudes required between the postures $\dot{\xi}_{t+1}$ and $\dot{\xi}_t$. However, the major contribution of this mathematical solution is that we now can manipulate the variables α_i, and β_i to associate their configuration with different 4-drive/4-steer kinematic modalities as will be discussed in next section.

10.6   Kinematic multi-configuration

In this section, we deduce the kinematic control law transition matrix $\mathbf{K}$ for some structures to describe their locomotive modalities. To provide a better understanding, figure 10.1 depicts some related kinematic variables and parameters. Configuration with fixed suspensions and wheels. Lateral velocities are $v_1 = v_2 = v_r$, and $v_3 = v_4 = v_l$. For fixed suspensions, as well as the z-turn axis aligned in the robot's centre, the magnitudes of $\Delta x_i = \Delta x$, $\Delta y_i = \Delta y$ and the values of $l_i = l$ are set as constants.

The values are set as $\alpha_1 = 2\pi - \alpha$, $\alpha_2 = \pi + \alpha$, $\alpha_3 = \alpha$, and $\alpha_4 = \pi - \alpha$. In order to align the wheels' orientation with the robots local reference axis X, the value of the β angles are equal to the negative value of the respective α_n angles (with this assumption, it is not necessary to consider clockwise rotation of the left-side wheels). Thus, $\beta_1 = -\alpha_1$, $\beta_2 = -\alpha_2$, $\beta_3 = -\alpha_3$, and $\beta_4 = -\alpha_4$. By replacing the values β, α angles, the kinematic control matrix is obtained accordingly. Additionally, if we assign the value of $\alpha = \frac{\pi}{2}$, the resultant equation yields a kinematic model for a 2W differential-drive robot. This locomotion and steering configuration is depicted in Figure 10.1-a), and is included in table 10.1 in the non-holonomic group 1:NH. The values α_n, β_n angles and l used for differential mode, are restricted to fixed suspension and steering angles. Thus, $\alpha_1 = 1/4\pi$, $\alpha_2 = 7/4\pi$, $\alpha_3 = 3/4\pi$, and $\alpha_4 = 5/4\pi$, and $\beta_i = -\alpha_i$.

For this configuration mode the vector of velocities for the right (wheels 1 and 2) and left (wheels 3 and 4) is described by the wheels' speed vector $\mathbf{\Phi} = r(\dot{\varphi}_r, \dot{\varphi}_l)^\top$. Through equation (10.52), a simplified restriction matrix (10.63) is obtained.

$$\mathbf{K} = \begin{pmatrix} 2 & 2 \\ 0 & 0 \\ \frac{\sqrt{2}}{l} & -\frac{\sqrt{2}}{l} \end{pmatrix} \qquad (10.63)$$

Furthermore, the matrix $\mathbf{K}$ is conversely described with its inverse solution $\mathbf{K}^{-1}$ provided by the next expression (10.64),

$$\mathbf{K}^{-1} = \frac{1}{4} \begin{pmatrix} 10 & l\sqrt{2} \\ 10 & -l\sqrt{2} \end{pmatrix} \qquad (10.64)$$

For simulation purposes, the wheels' tangential speed ranged from 0 to 0.2m/s in alternate directions on sides of the 4WD4S. The trajectory obtained in the global plane is depicted in figure 10.8 for two types of robotic structures: a two-differential speeds, and a four-differential speeds.

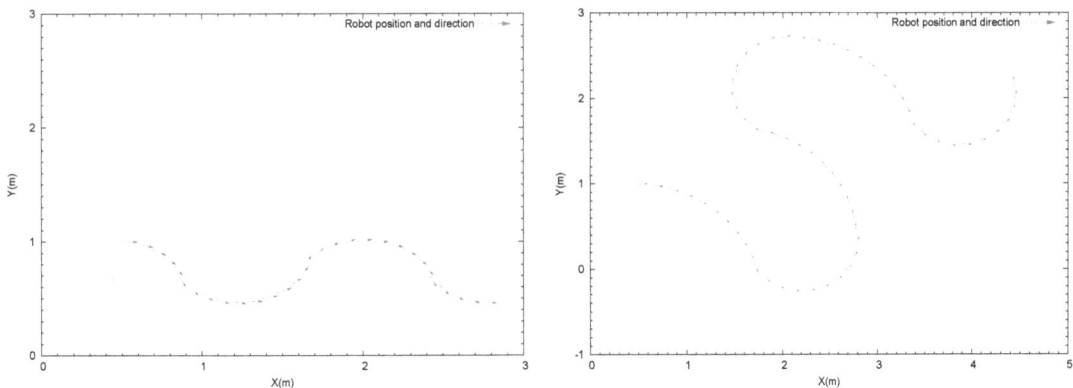

Figure 10.8: Two-differential drive locomotion configuration trajectory; and 4W-differential drive (four asynchronous speeds) trajectory.

The change of kinematic properties from 2WD to 4WD, improves the robot manoeuvrability by reducing kinematic mobility constraints, as depicted by previous figure, accordingly.

Another numerical analysis case is depicted by figure 10.1-b), which concerns case 2 of table 10.1, 4W four drive, and fixed wheels with no steering control. A simplified restriction vectors matrix (10.65) is algebraically obtained:

$$
\mathbf{K} = \begin{pmatrix} 1 & 1 & 1 & 1 \\ 0 & 0 & 0 & 0 \\ \frac{1}{2}\frac{\sqrt{2}}{l} & \frac{1}{2}\frac{\sqrt{2}}{l} & -\frac{1}{2}\frac{\sqrt{2}}{l} & -\frac{1}{2}\frac{\sqrt{2}}{l} \end{pmatrix}
\tag{10.65}
$$

Likewise, the inverse matrix $\mathbf{K}$ is described by (10.66). An issue with the differences between this configuration mode and the previous one already depicted by figure 10.1-a) is the difference in the wheel's velocities vector. For this case, all-wheel systems are explicitly described in $\dot{\mathbf{\Phi}} = r(\dot{\varphi}_1, \dot{\varphi}_2, \dot{\varphi}_3, \dot{\varphi}_4)^\top$.

$$
\mathbf{K}^{-1} = \frac{1}{4} \begin{pmatrix} 10\ \sqrt{2}l \\ 10\ \sqrt{2}l \\ 10-\sqrt{2}l \\ 10-\sqrt{2}l \end{pmatrix}
\tag{10.66}
$$

For the sake of simplicity in obtaining the matrix solution, we again considered the case with fixed suspension, with no variations of α_i and β_i angles 10.8-right.

The values for l_i, α_i, and β_i in each wheel are the same over time. The angle β_t was the only angle showing synchronization of all wheels. Again for the sake of algebraic simplicity in the solution form, the suspension is considered fixed. Thus, α_i and l_i have the following constant values $\alpha_1 = 1/4\pi$, $\alpha_2 = 7/4\pi$, $\alpha_3 = 3/4\pi$, $\alpha_4 = 5/4\pi$, and $\beta_i = -\alpha_i + \phi_i$. The transition matrix of four wheels kinematic restriction for an asynchronous robotic structure is given in a simplified form by equation (10.67).

$$
\mathbf{K} = \begin{pmatrix} \cos(\beta_t) & \cos(\beta_t) & \cos(\beta_t) & \cos(\beta_t) \\ \sin(\beta_t) & \sin(\beta_t) & \sin(\beta_t) & \sin(\beta_t) \\ \frac{\sin(\frac{\pi}{4}+\beta_t)}{l} & \frac{\cos(\frac{\pi}{4}+\beta_t)}{l} & -\frac{\cos(\frac{\pi}{4}+\beta_t)}{l} & -\frac{\sin(\frac{\pi}{4}+\beta_t)}{l} \end{pmatrix}
\tag{10.67}
$$

Likewise, the inverse matrix $\mathbf{K}$ is described by (10.68).

$$\mathbf{K}^{-1} = \frac{1}{4} \begin{pmatrix} \cos(\beta_t)\sin(\beta_t) & 2\sin(\frac{\pi}{4} + \beta_t)l \\ \cos(\beta_t)\sin(\beta_t) & 2\cos(\frac{\pi}{4} + \beta_t)l \\ \cos(\beta_t)\sin(\beta_t) & -2\cos(\frac{\pi}{4} + \beta_t)l \\ \cos(\beta_t)\sin(\beta_t) & -2\sin(\frac{\pi}{4} + \beta_t)l \end{pmatrix} \tag{10.68}$$

The figure 10.9 a numerical simulation of trajectory obtained by the synchronous steer/drive locomotion configuration, starts at a steering angle of $-90°$, with increments of $15°/s$, and all-wheel speed of 0.2m/s. The robot's initial location is $x_0 = 0.5$m, $y_0 = 1.0$m, $\theta_0 = 0$ degrees.

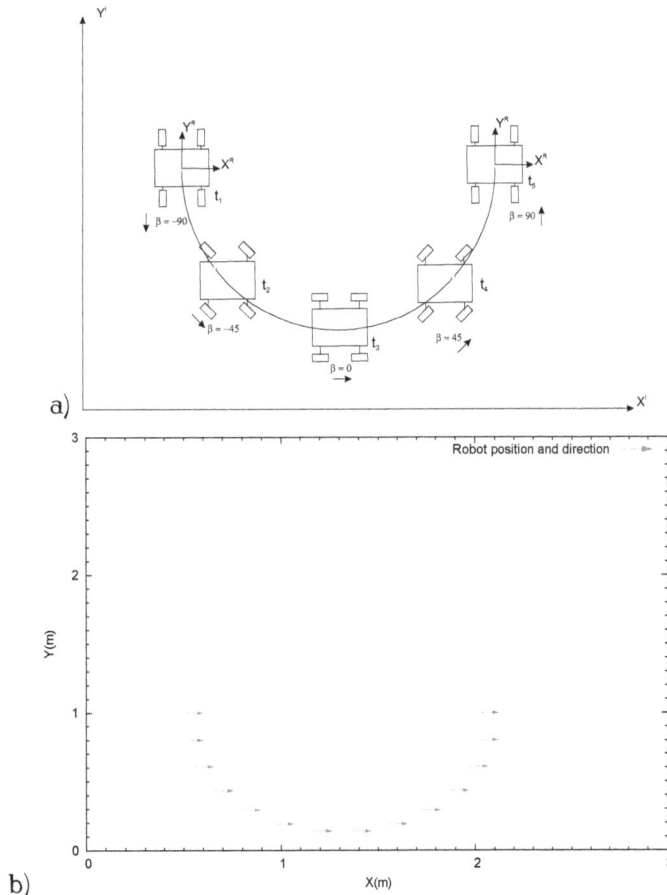

Figure 10.9: Robot's synchronous mode (above); trajectory simulation (below).

Bibliography

[1] Arslan, S., and Temeltas, H., *Robust motion control of a four wheel drive skid-steered mobile robot*, In 7th Intl. Conf. on Electrical and Electronics Engineering (2011), vol. 2, pp. 415–419.

[2] Campion, G., Bastin, G., D'Andrea N.B., *Structural properties and classification of kinematic and dynamic models of wheeled mobile robots*, IEEE Trans. on Robotics and Automation, vol.12(1), pp. 47–62, 1996, doi: http://dx.doi.org/10.1109/70.481750.

[3] Cordes, F., Dettmann, A., Kirchner, F., *Locomotion modes for a hybrid wheeled-leg planetary rover*, IEEE Intl. Conf. on Robotics and Biomimetics, pp. 2586–2592, 2011 doi: 10.1109/ROBIO.2011.6181694.

[4] Duan Song He-nan Chen Dan-yong Li, Y., *Visibility-based fault-tolerant lateral and longitudinal control of 4w-steering vehicles*, IEEE Transactions on intelligent transportation systems 12(4), 2011.

[5] Filipescu, R. S. A. F. V. M. S., *Sliding-mode trajectory-tracking control for a four-wheel-steering vehicle*, In 2010 8th IEEE Intl. Conf. on Control and Automation, pp. 382–387.

[6] Freitas, G., Gleizer, G., Lizarralde, F., Hsu, L., Reis, N. R. S., *Kinematic reconfigurability control for an environmental mobile robot operating in the amazon rain forest*, J. Field Robot, vol.27(2), 197–216, 2010, doi: 10.1002/rob.20334.

[7] Grepl, R., Vejlupek, J., Lambersky, V., Jasansky, M., Vadlejch, F., Coupek, P., *Development of 4ws/4wd experimental vehicle: platform for research and education in mechatronics*, IEEE Intl. Conf. on Mechatronics, pp. 893–898, 2011, doi: 10.1109/ICMECH.2011.5971241.

[8] Iagnemma, K., Rzepniewski, A., Dubowsky, S., Pirjanian, P., Hunts-berger, T., Schenker, P., *Mobile robot kinematic reconfigurability for rough-terrain*, 2000.

[9] Iagnemma, K., Rzepniewski, A., Dubowsky, S., Schenker, P., *Control of robotic vehicles with actively articulated suspensions in rough terrain*, Autonomous Robots, vol.14, pp.5–16, 2003, doi: 10.1023/A:1020962718637.

[10] Kasahua, M., Mori, Y., *Trajectory tracking control of the four-wheel vehicle according to speed change*, In SICE Annual Conf. 201, pp. 3449–3452, 2010.

Chapter 11

CONTROL OF A SELF-CONFIGURABLE QUADRUPED

Manuel Vega Heredia and Edgar A. Martínez García

Laboratorio de Robótica, Institute of Engineering and Technology
Universidad Autónoma de Ciudad Juárez, Mexico.

This chapter describes the mechanical design, an Euler-Lagrange analysis, and the motion model of a self-reconfigurable quadruped robot. The quadruped robot changes its limbs locomotive configuration through different kinematic configurations. The proposed limb mechanism poses 5 independent rotational control variables, and are discussed through a kinematic, and an energy-based analysis. The design of a leg-wheel that is self-reconfigurable allows the robot to change its locomotive settings according to the type of terrain. Thus, the basic approach of this chapter is on the dynamic modelling of the limb required for control of locomotive functions. In recent years there have been a series of rules based on space technologies, among the most recognizable is the Athlete robot of NASA , which is a hybrid robot with limbs and wheels, it uses both forms of locomotion found simultaneously. Likewise, a tele-operated hybrid reconfigurable robot with limbs was developed where the wheel becomes a 4 degrees of freedom mechanism. In previous reported work artificial locomotion control with human-robot interaction has been developed . Applications of industrial inspection deploying hexapod robots with laser range and vision as been reported . The kinematic control play a fundamental roll in redundantly actuated robots ' , as well as the slip measurement control of leg/wheel mobile robots . Realization of biped leg-wheeled robots , and leg-wheel hybrid quadruped   has been reported.

11.1   Limb's mechanism description

In this section the design of a self-reconfigurable leg-wheel mechanism is disclosed. The limb system has the ability to drive the links through the joints to adopt positions as depicted in figure 11.1-right. Figure 11.1-left depicts the limb's kinematic diagram, which consists of 5 rotational driven joints, providing the capability to configure as wheel, leg, hook, foot, and aquatic fin.

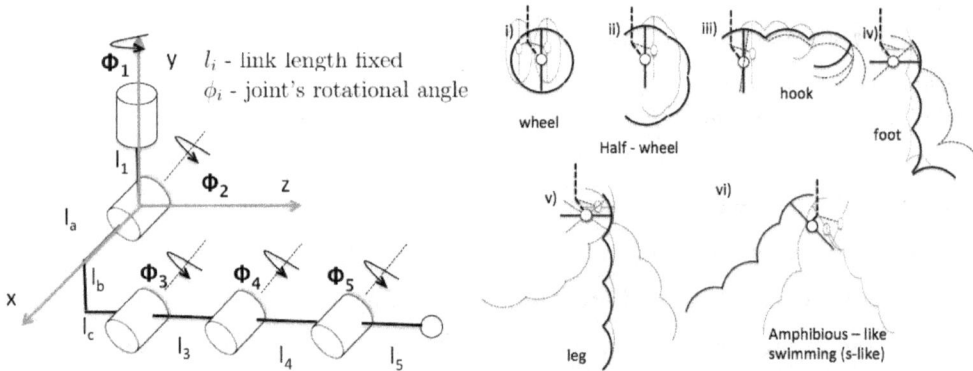

Figure 11.1:   Kinematic diagram of the proposed limb (left.) Locomotive limb's configurations (right).

Some of the rotational servomotors allow 2πrad of motion, while other servomotors only work within a range of motion of πrad, accordingly as required. According to figure 11.2, the position of the wheel configuration depends on the ensemble secured by a magnetic docking device located at the end-point of each rigid link forming a kinematic chain, which may be closed with the rear section of the first link. Likewise, the mechanical design of the limb includes a magnetic device that allows a safe posture capable to support loads, connecting the core with the middle of the third link.

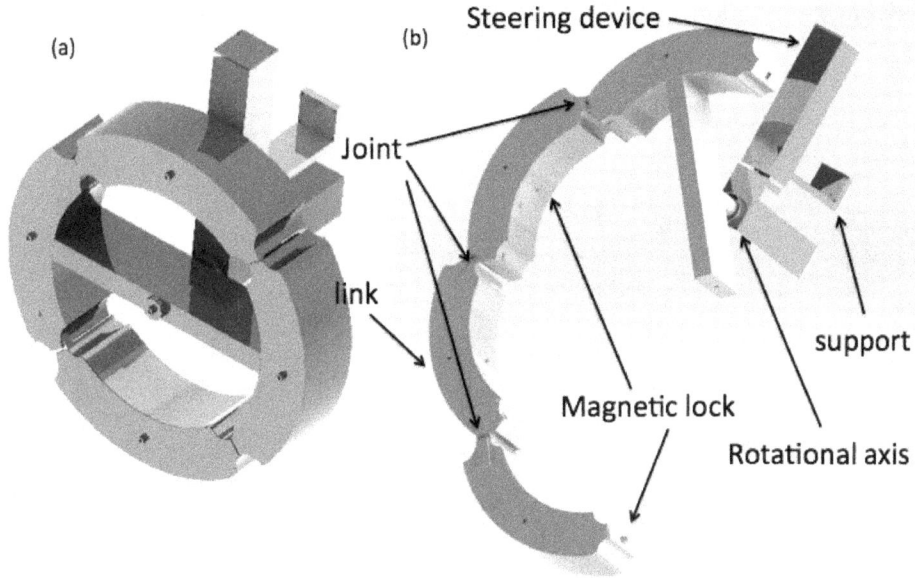

Figure 11.2:   Self-reconfigurable limb's mechanism design.

11.2   Limb kinematic analysis

The mobility of a mechanism is defined by the number of independent variables known as driven joints. Each structure component is considered a rigid link (ideally non deformable). The analysis is developed with foundations on a generalized coordinate system for the entire limb, shown in figure 11.1-left. The general kinematic vector equation with inclusion of all active joints represent the limb's contact point with the surface $\mathbf{p} \in \mathbb{R}^3$, such that $\mathbf{p} = (x, y, z)^\top$,

Axiom 11.2.1 (Limb's position model). *The algebraic limbs model with all active joints is*

$$\mathbf{p} = \begin{pmatrix} \{l_a + l_b \cos(\theta_2) + l_c \sin(\theta_2) + l_3 \cos(\theta_2 + \theta_3) + l_4 \cos(\theta_2 + \theta_3 + \theta_4) \\ \qquad\qquad + l_5 \cos(\theta_2 + \theta_3 + \theta_4 + \theta_5)\} \cos(\theta_1) \\ l_b \sin(\theta_2) + l_c + l_3 \sin(\theta_2 + \theta_3) + l_4 \sin(\theta_2 + \theta_3 + \theta_4) + l_5 \sin(\theta_2 + \theta_3 + \theta_4 + \theta_5) \\ \{l_a + l_b \sin(\theta_2) + l_c \sin(\theta_2) + l_3 \cos(\theta_2 + \theta_3) + l_4 \cos(\theta_2 + \theta_3 + \theta_4) \\ \qquad\qquad + l_5 \cos(\theta_2 + \theta_3 + \theta_4 + \theta_5)\} \sin(\theta_1) \end{pmatrix} \qquad (11.1)$$

By deriving the position equations w.r.t. time, the following equations are obtained:

$$\dot{\mathbf{p}} = \frac{d}{dt}\mathbf{p} \tag{11.2}$$

and

$$\dot{x} = \frac{d}{dt}x; \qquad \dot{y} = \frac{d}{dt}y; \qquad \dot{z} = \frac{d}{dt}z \tag{11.3}$$

Therefore,

$$
\begin{aligned}
\dot{x} = {}& (l_a + l_b \cos(\theta_2) + l_c \sin(\theta_2) + l_3 \cos(\theta_2 + \theta_3) + l_4 \cos(\theta_2 + \theta_3 + \theta_4) \\
& + l_5 \cos(\theta_2 + \theta_3 + \theta_4 + \theta_5)) \sin(\theta_1)(\dot{\theta}_1) + \Big(-l_b \sin(\theta_2)(\dot{\theta}_2) + l_c \cos(\theta_2) \\
& - l_3 \sin(\theta_2 + \theta_3)(\dot{\theta}_2 + \dot{\theta}_3) - l_4 \sin(\theta_2 + \theta_3 + \theta_4)(\dot{\theta}_2 + \dot{\theta}_3 + \dot{\theta}_4) \\
& - l_5 \sin(\theta_2 + \theta_3 + \theta_4 + \theta_5)(\dot{\theta}_2 + \dot{\theta}_3 + \dot{\theta}_4 + \dot{\theta}_5)\Big) \cos(\theta_1)
\end{aligned}
\tag{11.4}
$$

and

$$
\begin{aligned}
\dot{y} = {}& l_b \cos(\theta_2)(\dot{\theta}_2) + l_c + l_3 \cos(\theta_2 + \theta_3)(\dot{\theta}_2 + \dot{\theta}_3) + l_4 \cos(\theta_2 + \theta_3 + \theta_4) \\
& (\dot{\theta}_2 + \dot{\theta}_3 + \dot{\theta}_4) + l_5 \cos(\theta_2 + \theta_3 + \theta_4 + \theta_5)(\dot{\theta}_2 + \dot{\theta}_3 + \dot{\theta}_4 + \dot{\theta}_5)
\end{aligned}
\tag{11.5}
$$

as well as

$$
\begin{aligned}
\dot{z} = {}& (l_a + l_b \sin(\theta_2) + l_c \sin(\theta_2) + l_3 \cos(\theta_2 + \theta_3) + l_4 \cos(\theta_2 + \theta_3 + \theta_4) \\
& + l_5 \cos(\theta_2 + \theta_3 + \theta_4 + \theta_5)) \cos(\theta_1)(\dot{\theta}_1) + \Big(l_b \cos(\theta_2)(\dot{\theta}_2) + l_c \cos(\theta_2)(\dot{\theta}_2) \\
& + l_3 \sin(\theta_2 + \theta_3)(\dot{\theta}_2 + \dot{\theta}_3) + l_4 \sin(\theta_2 + \theta_3 + \theta_4)(\dot{\theta}_2 + \dot{\theta}_3 + \dot{\theta}_4) \\
& + l_5 \sin(\theta_2 + \theta_3 + \theta_4 + \theta_5)(\dot{\theta}_2 + \dot{\theta}_3 + \dot{\theta}_4 + \dot{\theta}_5)\Big) \sin(\theta_1)
\end{aligned}
\tag{11.6}
$$

Figure 11.1-right shows the different kinematic configurations, where the limb's kinematics is reconfigured by cancelling motion in some joints, but still uses other joints that are necessary for a given locomotive configuration. Thus, next sections will define the kinematics of those locomotive configurations.

11.2.1 Limb's posture configured as wheel

For the wheel configuration, only steering angle and driving joint are enabled to be controlled for each limb configured as a wheel. The rest of the joints will prevail with constant positions over time, while the kinematic configuration is used. The next vector equation is postulated and describes how its kinematic models a wheel.

Postulate 11.2.2 (Kinematic model for wheel configuration). *The kinematic vector model for wheel configuration is postulated by*

$$\mathbf{p_i} = \begin{pmatrix} \cos\theta_1 (la + l_b \sin(\theta_2) + l_c \cos\theta_2) \\ \\ l_b \cos\theta_2 + l_c \sin\theta_2 \\ \\ \sin\theta_1 (la + l_b \cos(\theta_2) + l_c \cos(\theta_2)) \end{pmatrix} \tag{11.7}$$

By defining some expressions, the following algebraic process is simplified, $l_{1a} = (la+lb)\sin(\theta_2)$, $l_{1b} = (la+lb)\cos(\theta_2)$, $l_{2a} = lc\cos(\theta_2)$, and $l_{2b} = lc\sin(\theta_2)$. Hence, the wheel's velocity vector,

$$\mathbf{v_t} = \begin{pmatrix} \dot{x} \\ \dot{y} \\ \dot{z} \end{pmatrix} = \begin{pmatrix} -\sin(\theta_1)\dot{\theta}_1(l_{1a} + l_{2a}) + \cos(\theta_1)(l_{1b}\dot{(\theta_2)} - l_{2b}\dot{\theta}_2) \\ \\ -\sin(\theta_2)\dot{\theta}_2 lb + l_{2a}\dot{\theta}_2 \\ \\ \cos(\theta_1)\dot{\theta}_1(l_{1b} + l_{2a}) + \sin(\theta_1)(-l_{1a}\dot{\theta}_2 - l_{2b}\dot{\theta}_2) \end{pmatrix} \tag{11.8}$$

Equation (11.1) describes the general kinematics of the entire limb. However, by replacing the angular constant values combined along the joints, we can set special postures to exert variants of locomotion over walking cycles. Thus, hereafter such special positions are established.

11.2.2   Limb's posture configured as half-wheel

Similarly, as previous section, for the half-wheel configuration, some angular values are required to be defined to set the joints posture accordingly. Thus,

$$\theta_a = -15 + \theta_2; \qquad \theta_b = -30 + \theta_2; \qquad \theta_c = -45 + \theta_2$$

And by substituting previous expressions, the vector position is stated,

Postulate 11.2.3 (Kinematic model for half-wheel configuration). *The kinematic vector model for half-wheel configuration is postulated by*

$$\mathbf{p}_{ii} = \begin{pmatrix} \cos\theta_1 \left(la + l_b \sin(\theta_2) + l_c \cos\theta_2 + l3\cos(\theta_a) + l_4\cos(\theta_b) \right. \\ \left. + l_5\cos(\theta_c) \right) \\ \\ (l_b \cos\theta_2 + l_c \sin\theta_2 + l3\sin(\theta_a) + l_4\sin(\theta_b) + l_5\sin(\theta_c)) \\ \\ \sin\theta_1 \left(la + (l_b \cos(\theta_2) + l_c \cos(\theta_2) + l3\cos(\theta_a) + l_4\cos(\theta_b) + l_5\cos(\theta_c) \right) \end{pmatrix} \tag{11.9}$$

Therefore, its derivation w.r.t. time produces the limb's velocity vector,

$$\mathbf{v}_{ii} = \begin{pmatrix} \dot{x} \\ \dot{y} \\ \dot{z} \end{pmatrix} = \begin{pmatrix} -\sin(\theta_1)\dot{\theta}_1(la + \sin(\theta_2)lb + lc\cos(\theta_2) + l3\cos(\theta_a) \\ + l4\cos(\theta_b) + l5\cos(\theta_c)) + \cos(\theta_1)\cos(\theta_2)\dot{\theta}_2 lb \\ -lc\sin(\theta_2)\dot{\theta}_2 - l3\sin(\theta_a)\dot{\theta}_2 - l4\sin(\theta_b)\dot{\theta}_2 - l5\sin(\theta_c)\dot{\theta}_2) \\ \\ -\sin(\theta_2)\dot{\theta}_2 lb + lc\cos(\theta_2)\dot{\theta}_2 + l3\cos(\theta_a)\dot{\theta}_2 \\ + l4\cos(\theta_b)\dot{\theta}_2 + l5\cos(\theta_c)\dot{\theta}_2 \\ \\ \cos(\theta_1)\dot{\theta}_1(la + lb\cos(\theta_2) + lc\cos(\theta_2) + l3\cos(\theta_a) \\ + l4\cos(\theta_b) + l5\cos(\theta_c)) + \sin(\theta_1) - \sin(\theta_2)\dot{\theta}_2 lb \\ -lc\sin(\theta_2)\dot{\theta}_2 - l3\sin(\theta_a)\dot{\theta}_2 - l4\sin(\theta_b)\dot{\theta}_2 - l5\sin(\theta_c)\dot{\theta}_2 \end{pmatrix} \tag{11.10}$$

11.2.3 Limb's posture configured as walking hook

For the walking hook configuration, a different set of joints need different numerical constant angular values, which are required to be defined as to set the posture accordingly. The links $l_{1a} = l_a + l_b \sin(\theta_2)$, $l_{1b} = l_a + l_b \cos(\theta_2)$, $l_2 = l_c + l_3 + l_4$. In addition, the joints' angle $\theta_{25} = \theta_2 + \theta_5$, and $\dot{\theta}_{25} = \dot{\theta}_2 + \dot{\theta}_5$ And by substituting previous expressions in the position vector, the following postulation defining positions of a walking hook in Cartesian space is stated.

Postulate 11.2.4 (Kinematic model for walking hook configuration). *The kinematic vector model for walking hook configuration is postulated by*

$$\mathbf{p}_{iii} = \begin{pmatrix} \cos\theta_1 (l_{1a} + (l_2)\cos(\theta_2) + l_5\cos(\theta_{25})) \\[2em] (l_b \cos\theta_2 (l_2)\sin(\theta_2) + l_5\sin(\theta_{25})) \\[2em] \sin\theta_1 (l_{1b}(l_2)\cos(\theta_2) + l_5\cos(\theta_{25}))) \end{pmatrix} \tag{11.11}$$

Likewise, its first order derivative vector equation is given next,

$$\mathbf{v}_{iii} = \begin{pmatrix} \dot{x} \\ \dot{y} \\ \dot{z} \end{pmatrix} = \begin{pmatrix} -\sin(\theta_1)\dot{\theta}_1(l_{1a} + (l_2)\cos(\theta_2) + l_5\cos(\theta_{25})) + \cos(\theta_1)(\cos(\theta_2)\dot{\theta}_2 \\ l_b - (l_2)\sin(\theta_2)\dot{\theta}_2 - l_5\sin(\theta_{25})(\dot{\theta}_{25})) \\[2em] -l_b\sin(\theta_2)^2\dot{\theta}_2 l_2 + l_b\cos(\theta_2)^2 \\ (l_2)\dot{\theta}_2 + l_5\cos(\theta_{25})(\dot{\theta}_{25}) \\[2em] \cos(\theta_1)\dot{\theta}_1 l_{1b} l_2 \cos(\theta_2) + l_5\cos(\theta_{25}) \\ + \sin(\theta_1)(-l_b\cos(\theta_2)l_2\sin(\theta_2)\dot{\theta}_2 - l_5\sin(\theta_{25})\dot{\theta}_{25}) \end{pmatrix} \tag{11.12}$$

11.2.4   Limb's posture configured as leg

For the leg configuration, a different set of joints need different numerical constant angular values, which are required to be defined as to set the posture accordingly. The links $l_{1a} = l_a + l_b \sin(\theta_2)$, $l_{1b} = l_a + l_b \cos(\theta_2)$ $l_2 = l_c + l_3$, and $l_{45} = l_4 + l_5$. The joints' angle $\theta_{24} = \theta_2 + \theta_4$, and $\dot{\theta}_{24} = \dot{\theta}_2 + \dot{\theta}_4$. Thus, by substituting previous expressions in the position vector equation, the next is obtained:

Postulate 11.2.5 (Kinematic model for leg configuration). *The kinematic vector model for leg configuration is postulated by*

$$
\mathbf{p}_v =
\begin{pmatrix}
\cos\theta_1 (l_{1a} + (l_2)\cos(\theta_2) + (l_{45})\cos(\theta_{24})) \\
\\
(l_b \cos\theta_2 (l_2)\sin(\theta_2) + (l_{45})\sin(\theta_{24}))) \\
\\
\sin\theta_1 (l_{1b}(l_2)\cos(\theta_2) + (l_{45})\cos(\theta_{24})))
\end{pmatrix}
\tag{11.13}
$$

Furthermore, the first order derivative w.r.t. time is stated by

$$
\mathbf{v}_v =
\begin{pmatrix}
\dot{x} \\
\dot{y} \\
\dot{z}
\end{pmatrix}
=
\begin{pmatrix}
-\sin(\theta_1)\dot{\theta}_1(l_{1a} + (l_2)\cos(\theta_2) + (l_{45})\cos(\theta_{24})) + \cos(\theta_1)(\cos(\theta_2)\dot{\theta}_2 \\
lb - (l_2)\sin(\theta_2)\dot{\theta}_2 - (l_{45})\sin(\theta_{24})\dot{\theta}_{24}) \\
\\
\\
-lb\sin(\theta_2)^2\dot{\theta}_2(l_2) \\
+lb\cos(\theta_2)^2(l_2)\dot{\theta}_2 + (l_{45})\cos(\theta_{24})\dot{\theta}_{24} \\
\\
\\
\cos(\theta_1)\dot{\theta}_1(l_{1b}(l_2)\cos(\theta_2) + (l_{45})\cos(\theta_{24})) \\
+ \sin(\theta_1)(-lb(\cos(\theta_2))(l_2)\sin(\theta_2)\dot{\theta}_2 - (l_{45})\sin(\theta_{24})\dot{\theta}_{24})
\end{pmatrix}
\tag{11.14}
$$

11.2.5 Limb's posture configured as foot

For the leg using foot configuration, a different set of joints need different numerical constant angular values, which are required to be defined as to set the posture accordingly. Thus,

$$l_{1a} = l_a + l_b \sin(\theta_2); \quad l_{1b} = l_a + l_b \cos(\theta_2); \quad l_2 = l_c + l_3$$

and

$$\theta_{24} = \theta_2 + \theta_4; \quad \theta_{245} = \theta_2 + \theta_4 + \theta_5; \quad \dot{\theta}_{24} = \dot{\theta}_2 + \dot{\theta}_4; \quad \dot{\theta}_{245} = \dot{\theta}_2 + \dot{\theta}_4 + \dot{\theta}_5$$

By substituting previous expressions, the position vector is stated by

Postulate 11.2.6 (Kinematic model for foot configuration). *The kinematic vector model for foot configuration is postulated by*

$$\mathbf{p}_{iv} = \begin{pmatrix} \cos\theta_1(l_{1a} + (l_2)\cos(\theta_2) + (l_4)\cos(\theta_{24}) + (l_5)\cos(\theta_{245})) \\ (l_b\cos\theta_2(l_2)\sin(\theta_2) + (l_4)\sin(\theta_{24}) + (l_5)\cos(\theta_{245})) \\ \sin\theta_1(l_{1b}(l_2)\cos(\theta_2) + (l_4)\cos(\theta_{24}) + (l_5)\cos(\theta_{245})) \end{pmatrix} \tag{11.15}$$

likewise, the velocity vector is defined next,

$$\mathbf{v}_{iv} = \begin{pmatrix} \dot{x} \\ \dot{y} \\ \dot{z} \end{pmatrix} = \begin{pmatrix} -\sin(\theta_1)\theta_1(l_{1_a} + l_2\cos(\theta_2) + l_4\cos(\theta_{24}) + l_5\cos(\theta_{245})) \\ +\cos(\theta_1)(\cos(\theta_2)\dot{\theta}_2 l_b - (l_2)\sin(\theta_2)\dot{\theta}_2) - l_4\sin(\theta_{24})\dot{\theta}_{24} \\ -l_5\sin(\theta_{245})\dot{\theta}_{245}) \\ \\ l_b\cos(\theta_2)l_2\cos(\theta_2)\dot{\theta}_2 + l_4\cos(\theta_{24})\dot{\theta}_{24} \\ -l_5\sin(\theta_{245})\dot{\theta}_{245} \\ \\ \cos(\theta_1)\dot{\theta}_1 l_{1b}l_2\cos(\theta_2) + l_4\cos(\theta_{24}) + l_5\cos(\theta_{245}) \\ +\sin(\theta_1)(-lb\cos(\theta_2)l_2\sin(\theta_2)\dot{\theta}_2) - l_4\sin(\theta_{24})\dot{\theta}_{24} \\ -l_5\sin(\theta_{245})\dot{\theta}_{245} \end{pmatrix} \tag{11.16}$$

11.2.6 Limb's posture configured as aquatic fin

For the aquatic fin configuration, there are more active needed joints. Similarly, a different set of joints are set with different numerical constant values to define the posture accordingly. With links $l_{1a} = l_a + l_b \sin(\theta_2)$, $l_{1b} = l_a + l_b \cos(\theta_2)$, and $l_2 = l_c + l_3$. And joints' angle definition $\theta_{24} = \theta_2 + \theta_4$, $\theta_{245} = \theta_2 + \theta_4 + \theta_5$, $\dot{\theta}_{24} = \dot{\theta}_2 + \dot{\theta}_4$, $\dot{\theta}_{245} = \dot{\theta}_2 + \dot{\theta}_4 + \dot{\theta}_5$. Therefore, by substituting previous expression in the next position vector,

Postulate 11.2.7 (Kinematic model for aquatic fin configuration). *The kinematic vector model for aquatic fin configuration is postulated by*

$$\mathbf{p}_{vi} = \begin{pmatrix} \cos\theta_1 (l_{1a} + (l_2)\cos(\theta_2) + (l_4)\cos(\theta_{24}) + (l_5)\cos(\theta_{245})) \\ (l_b \cos\theta_2 (l_2)\sin(\theta_2) + (l_4)\sin(\theta_{24}) + (l_5)\cos(\theta_{245})) \\ \sin\theta_1 (l_{1b}(l_2)\cos(\theta_2) + (l_4)\cos(\theta_{24}) + (l_5)\cos(\theta_{245}))) \end{pmatrix} \tag{11.17}$$

From previous postulate, by deriving w.r.t. time the velocity vector is defined,

$$\mathbf{v}_{vi} = \begin{pmatrix} \dot{x} \\ \dot{y} \\ \dot{z} \end{pmatrix} = \begin{pmatrix} -\sin(\theta_1)\dot{\theta}_1(l_{1b} + l_2 \sin(\theta_2) + l_4 \cos(\theta_{24}) + l_5 \cos(\theta_{245})) \\ + \cos(\theta_1)(\cos(\theta_2)\dot{\theta}_2 l_b - l_2 \sin(\theta_2)\dot{\theta}_2 \\ -l_4 \sin(\theta_{24})\dot{\theta}_{24} - l_5 \sin(\theta_{245})\dot{\theta}_{245}) \\ \\ l_b \cos(\theta_2)l_2 \cos(\theta_2)\dot{\theta}_2 + l_4 \cos(\theta_{24})\dot{\theta}_{24} \\ -l_5 \sin(\theta_{245})\dot{\theta}_{245} \\ \\ \cos(\theta_1)\dot{\theta}_1(l_{1b}l_2 \cos(\theta_2) + l_4 \cos(\theta_{24}) + l_5 \cos(\theta_{245})) \\ + \sin(\theta_1)(-l_b \cos(\theta_2)l_2 \sin(\theta_2)\dot{\theta}_2 - l_4 \sin(\theta_{24})\dot{\theta}_{24} \\ -l_5 \sin(\theta_{245})\dot{\theta}_{245}) \end{pmatrix} \tag{11.18}$$

11.2.7   Numerical simulations

Thus, the numerical simulations of vector positions for each kinematic configuration that were modelled previously are depicted now in figure 11.3. Each curve was plotted assuming a step of a natural walking gait. Thus, the plots represent the limb's contact point track in the way a given locomotive configuration would move. For instance, the wheel configuration motion (green color) only rotates, and each plotted point is the track of the limb's contact point as shown in figure 11.3. Likewise, the half-wheel configuration rotates touching the ground only during a half circumference (π rad) of the wheel, thus it is depicted by the red curve.

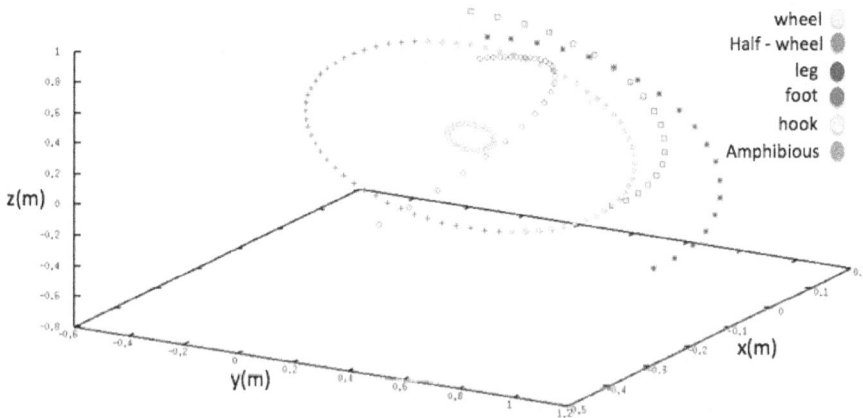

Figure 11.3:   Limb trajectory of different kinematic configurations.

Considering the same simulated Cartesian positions depicted in figure 11.3, the gait behaviours are simulated in terms of the velocity space. The Cartesian components XYZ are illustrated separately in figure 11.4 for the same limb's kinematic configurations.

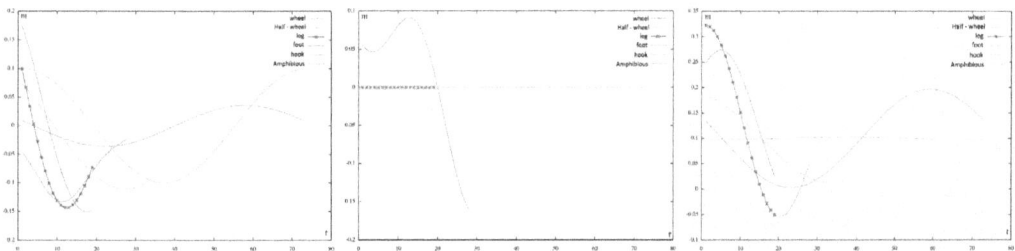

Figure 11.4:   Distance vs. Time of the Cartesian components, X (left), Y (centre), and Z (right).

11.3 Energy-based analysis

In this section an analysis of the evolution during a cycle of displacement of a leg-wheel system is provided. Likewise, the instantaneous kinetic and potential energy behaviour during a motion tasks is modelled. Thus, let us define the kinetic energy for a limb's link assuming it poses a cylindrical shape,

$$K = \frac{1}{2}mv^2 + \frac{1}{2}mr^2\omega^2 \tag{11.19}$$

where m is the mass of the link in kg. v is the translation velocity in m/s. and ω is the link angular velocity in rad/s. Likewise, a link potential energy is modelled by

$$P = mgh \tag{11.20}$$

where g is the gravity acceleration constant in m/s^2. h is the height of the mass of the link in m. Hence, the total energy of the mechanical system is stated the Langrange operator $\mathcal{L}$, which consider the total kinetic energies and the total potential energies of the limb.

$$\mathcal{L} = \sum_i^5 k_i - \sum_j^5 p_j \tag{11.21}$$

Hereafter, equivalences are indistinctly used $\dot{q}_1 \equiv v$ and $\dot{q}_2 \equiv \omega$. Thus, by expanding the Lagrange expression,

$$\mathcal{L} = \sum_{i=1}^5 \frac{1}{2}m_i v_i^2 + \frac{1}{2}m_i r^2 \omega^2 - \sum_{i=1}^5 \frac{1}{2}m_i g \left(\frac{l_i}{2}\right) \sin q_i \tag{11.22}$$

and

$$\mathcal{L} = \sum_{i=1}^5 \frac{1}{2}m_i(x^2 + y^2 + z^2) + \frac{1}{2}m_i(\frac{l_i}{2})^2 \dot{\theta}_i^2 - \sum_{i=1}^5 \frac{1}{2}m_i g \left(\frac{l_i}{2}\right) \sin q_i \tag{11.23}$$

11.3.1 Limb's posture configured as wheel

Lagrangian calculation for the wheel configuration on each driven joint used. Thus, from velocity equation 11.8 and figure 11.3,

$$\mathcal{L}_i = \frac{1}{2} m_i v_i^2 + \frac{1}{2} m_i \left(\frac{l_i}{2}\right)^2 \dot{\theta}_i^2 - \frac{1}{2} m_i g \left(\frac{l_i}{2}\right) \sin q_i \tag{11.24}$$

and analysing each link separately,

$$\mathcal{L}_{i_1} = \frac{m}{2} l_1^2 \dot{\theta}_1^2 - \frac{m}{2} g l_b \tag{11.25}$$

Thus, making some substitutions in order to simplify the next equations

$$f_1 = -l_b \cos(\theta_2) \dot{\theta}_2 + l_c \cos(\theta_2) \dot{\theta}_2; \qquad f_2 = l_a + (\cos(\theta_2)) l_b + l_c (\sin(\theta_2))$$

$$f_3 = l_b (\cos(\theta_2)) \dot{\theta}_2 + l_c (\cos(\theta_2)) \dot{\theta}_2; \qquad f_4 = -l_b (\sin(\theta_2)) \dot{\theta}_2 + l_c (\cos(\theta_2)) \dot{\theta}_2$$

$$f_5 = l_b \sin(\theta_2) + l_c \sin(\theta_2)$$

and $m = m_1 + m_2 + m_3 + m_4 + m_5$. Thus, substituting previous expressions in next equation,

$$\mathcal{L}_{i_2} = \frac{m}{2} (\cos(\theta_1)(f_1 - f_2 \sin(\theta_1)\dot{\theta}_1)(\cos(\theta_1)(f_1 - f_2 + (\sin(\theta_1)f_4 + f_2 \sin(\theta_1)\dot{\theta}_1))(\cos(\theta_1))\dot{\theta}_1)$$
$$(\sin(\theta_1)f_4 + f_2 \cos(\theta_1)\dot{\theta}_1) + f_3^2) + \frac{m}{2} (\frac{f_2}{2})^2 (\dot{\theta}_1^2 + \dot{\theta}_2^2) - \frac{m}{2} g f_5 \tag{11.26}$$

11.3.2 Limb's posture configured as half-wheel

The Lagrangian calculation of the half wheel configuration for each DOF is developed. It follows from velocity equation (11.10) and figure 11.3, that

$$\mathcal{L}_{ii} = \frac{1}{2} m_i v_{halfwheel}^2 + \frac{1}{2} m_i \left(\frac{l_i}{2}\right)^2 \dot{\theta}_i^2 - \frac{1}{2} m_i g \left(\frac{l_i}{2}\right) \sin q_i \tag{11.27}$$

likewise,

$$\mathcal{L}_{ii_1} = \frac{m}{2} l_1^2 \dot{\theta}_1^2 - \frac{m}{2} g(\sin(\theta_2)(l_c + l_3 \cos(45) + l_4 \cos(45) + l_5 \cos(45))) \tag{11.28}$$

Likewise, by defining some expressions for subsequent algebraic simplifications,

$$f_1 = -l_b(\sin(\theta_2))\dot{\theta}_2 + l_c(\cos(\theta_2)\dot{\theta}_2); \quad f_2 = l_a + (\cos(\theta_2))l_b + l_c(\sin(\theta_2))$$

As well as $f_3 = l_b(\cos(\theta_2))\dot{\theta}_2 + l_c(\cos(\theta_2))\dot{\theta}_2$, with $m = m_1 + m_2 + m_3 + m_4 + m_5$, and $l_2 = \sqrt{(l_a^2 + l_b^2 + l_c^2)}$. Therefore,

$$\mathcal{L}_{ii_2} = \frac{m}{2}((\cos(\theta_1)(-f_3 - (f_2)\sin(\theta_1)\dot{\theta}_1))(\cos(\theta_1)(-f_3 - (f_2)\sin(\theta_1)\dot{\theta}_1)) + ((\sin(\theta_1))(f_1)$$

$$+ (f_2)(\cos(\theta_1))\dot{\theta}_1)((\sin(\theta_1))(f_1) + (f_2)(\cos(\theta_1))\dot{\theta}_1) + (f_3)(f_3) + \frac{m}{2}(((l_2)$$

$$+ l_3\cos(45) + l_4\cos(45) + l_5\cos(45))/2)\frac{(l_2)}{2}(\dot{\theta}_1^2 + \dot{\theta}_2^2)$$

$$- \frac{1}{2}m2g(l_c + l_3\cos(45) + l_4\cos(45) + l_5\cos(45))$$

11.3.3   Limb's posture configured as walking hook

Likewise, Lagrangian calculation of the hook configuration for each driven joint, the velocity equation (11.12), and figure 11.3 are considered to state the following expression,

$$\mathcal{L}_{iii} = \frac{1}{2}m_i v_{Hook}^2 + \frac{1}{2}m_i \left(\frac{l_i}{2}\right)^2 \dot{\theta}_i^2 - \frac{1}{2}m_i g \left(\frac{l_i}{2}\right) \sin q_i \qquad (11.29)$$

and defining $m = m_1 + m_2 + m_3 + m_4 + m_5$, and $l_2 = \sqrt{(l_a^2 + l_b^2 + l_c^2)}$. Hence, modelling links in motion separately,

$$\mathcal{L}_{iii_1} = \frac{m}{2}l_1^2\dot{\theta}_1^2 - \frac{m}{2}g(\sin(\theta_2)(l_c + l_3\cos(45) + l_4\cos(45) + l_5\cos(45))) \qquad (11.30)$$

and $m = m_2 + m_3 + m_4 + m_5$, with links $l_2 = \sqrt{(l_a^2 + l_b^2 + l_c^2)}$ and $l_{234} = l_3 + l_4 + (l_2)$. In addition, for the following link in motion,

$$\mathcal{L}_{iii_2} = \frac{m}{2}(((-\sin(\theta_1)(\dot{\theta}_1)((l_a + \cos(\theta_2)l_b + (l_{234})\cos(\theta_2) + l_3\cos(\theta_3) + \cos(\theta_1)(-\sin(\theta_2)(\dot{\theta}_2))l_b$$

$$-(l_{234})\sin(\theta_2)\dot{\theta}_2 - l_3\sin(\theta_3)(\dot{\theta}_3))(l_a + \cos(\theta_2)l_b + (l_{234})\cos(\theta_2) + l_3\cos(\theta_3)$$

$$+\cos(\theta_1)(-\sin(\theta_2)(\dot{\theta}_2))l_b - (l_{234})\sin(\theta_2)(\dot{\theta}_2) - l_3\sin(\theta_3)(\dot{\theta}_3)))) + ((((\cos(\theta_1)(\dot{\theta}_1))(l_a + \cos(\theta_2)l_b$$

$$+(l_{234})\cos(\theta_2) + l_3\cos(\theta_3))) + \sin(\theta_1)(-\sin(\theta_2)(\dot{\theta}_2))l_b - (l_{234})\sin(\theta_2)(\dot{\theta}_2) - l_3\sin(\theta_3)(\dot{\theta}_3))(((\cos(\theta_1)(\dot{\theta}_1))$$

$$(l_a + \cos(\theta_2)l_b + (l_{234})\cos(\theta_2) + l_3\cos(\theta_3))) + \sin(\theta_1)(-\sin(\theta_2)(\dot{\theta}_2))l_b - (l_{234})\sin(\theta_2)(\dot{\theta}_2) - l_3\sin(\theta_3)(\dot{\theta}_3)))$$

$$+((l_b\cos(\theta_2)\dot{\theta}_2 + (l_{234})\cos(\theta_2)\dot{\theta}_2 + l_3\cos(\theta_3)\dot{\theta}_3)(l_b\cos(\theta_2)\dot{\theta}_2 + (l_{234})\cos(\theta_2)\dot{\theta}_2 + l_5\cos(\theta_5)\dot{\theta}_5))))$$

$$+\frac{m}{2}((l_{234})/2)(((l_2))/2)(\dot{\theta}_1^2 + \dot{\theta}_2^2) - \frac{m}{2}g((\sin(\theta_2)(l_2 + l_3 + l_4)))$$

and $l_2 = \sqrt{(l_a^2 + l_b^2 + l_c^2)}$, and $l_{234} = l_3 + l_4 + (l_2)$. Thus, for the next link in motion,

$$\mathcal{L}_{iii_3} = \frac{m_5}{2}((-\sin(\theta_1)(\dot{\theta}_1)((l_a + \cos(\theta_2)l_b + (l_{234})\cos(\theta_2) + l_3\cos(\theta_5)$$

$$+\cos(\theta_1)(-\sin(\theta_2)\dot{\theta}_2)l_b - (l_{234})\sin(\theta_2)\dot{\theta}_2 - l_3\sin(\theta_5)\dot{\theta}_5)(l_a + \cos(\theta_2)l_b + (l_{234})\cos(\theta_2) + l_3\cos(\theta_5)$$

$$+\cos(\theta_1)(-\sin(\theta_2)\dot{\theta}_2)l_b - (l_{234})\sin(\theta_2)\dot{\theta}_2 - l_3\sin(\theta_5)\dot{\theta}_5))) + ((((\cos(\theta_1)\dot{\theta}_1)$$

$$(l_a + \cos(\theta_2)l_b + (l_{234})\cos(\theta_2) + l_3\cos(\theta_5))) + \sin(\theta_1)(-\sin(\theta_2)\dot{\theta}_2)l_b - (l_{234})\sin(\theta_2)\dot{\theta}_2$$

$$-l_3\sin(\theta_5)(\dot{\theta}_5))(((\cos(\theta_1)(\dot{\theta}_1))(l_a + \cos(\theta_2)l_b + (l_{234})\cos(\theta_2) + l_3\cos(\theta_5)))$$

$$+\sin(\theta_1)(-\sin(\theta_2)(\dot{\theta}_2))l_b - (l_{234})\sin(\theta_2)(\dot{\theta}_2) - l_3\sin(\theta_5)(\dot{\theta}_5))) + ((l_b\cos(\theta_2)(\dot{\theta}_2)$$

$$+(l_{234})\cos(\theta_2)(\dot{\theta}_2) + l_5\cos(\theta_5)(\dot{\theta}_5))(l_b\cos(\theta_2)(\dot{\theta}_2) + (l_{234})\cos(\theta_2)(\dot{\theta}_2) + l_5\cos(\theta_5)(\dot{\theta}_5))))$$

$$+\frac{m_5}{2}\frac{l_5}{2}^2(\dot{\theta}_1^2 + \dot{\theta}_2^2 + \dot{\theta}_5^2) - \frac{m_5}{2}g(\sin(\theta_2)(\sin(\theta_5)(l_5)))$$

11.3.4 Limb's posture configured as leg

The Lagrangian calculation of the hook configuration for each driven joint starts from the velocity equation (11.14). Please note its trajectory in figure 11.3. Thus, stating the following expression,

$$\mathcal{L}_{v_i} = \frac{1}{2}m_i v_{Leg}^2 + \frac{1}{2}m_i \left(\frac{l_i}{2}\right)^2 \dot{\theta}_i^2 - \frac{1}{2}m_i g \left(\frac{l_i}{2}\right)\sin q_i \qquad (11.31)$$

defining the mass expression, $m = m_1 + m_2 + m_3 + m_4 + m_5$, and the links $l_2 = \sqrt{(l_a^2 + l_b^2 + l_c^2)}$, and $l_{234} = l_3 + l_4 + (l_2)$. Hence,

$$\mathcal{L}_{v_1} = \frac{m}{2} l_1^2 (\dot{\theta}_1^2 - \frac{m}{2} g \sin(\theta_2))(\sqrt{l_a^2 + l_b^2 + l_c^2} + l_3 + l_4 + l_5) \tag{11.32}$$

The Lagrangian L_{v_1} represent the first set of links in the leg configuration that yield rotational motion for providing leg's steering. Please notice that previous equation contains only a description of kinetic energy that is relevant to rotational movement. As for the potential energy description, the height model is for the leg's first joint, which yield steering motion.

Furthermore, the mass expressions are defined to simplify subsequent algebraic processes as $m = m_2 + m_3 + m_4 + m_5$, and the links $l_2 = \sqrt{(l_a^2 + l_b^2 + l_c^2)}$, $l_{23} = l_2 + l_3$, and $4l_{45} = l_4 + l_5$. In this leg configuration, the motion includes some joints, which are required to walk without foot. Only the last link's contact point is assumed to touch the ground with minimal to ankle's bend.

Therefore, by substituting previous expressions of mass, and links' length, then the second Lagrangian equation is formulated for the following set of links and joints,

$$\begin{aligned}
\mathcal{L}_{v_2} = \frac{m}{2}(((&- \sin(\theta_1)(\dot{\theta}_1)((la + \cos(\theta_2)l_b + (l_{23})\cos(\theta_2) + (l_{45})\cos(\theta_4) \\
&+ \cos(\theta_1)(-\sin(\theta_2)(\dot{\theta}_2))l_b - (l_{23})\sin(\theta_2)(\dot{\theta}_2) - l_{45}\sin(\theta_4)\dot{\theta}_4) \\
&(la + \cos(\theta_2)l_b + (l_{23})\cos(\theta_2) + (l_{45})\cos(\theta_4) + \cos(\theta_1)(-\sin(\theta_2)\dot{\theta}_2) \\
&l_b - (l_{23})\sin(\theta_2)(\dot{\theta}_2) - (l_{45})\sin(\theta_4)(\dot{\theta}_4)))) + ((((\cos(\theta_1)(\dot{\theta}_1))(la + \cos(\theta_2)l_b \\
&+ (l_{23})\cos(\theta_2) + (l_{45})\cos(\theta_4))) + \sin(\theta_1)(-\sin(\theta_2)(\dot{\theta}_2))l_b - (l_{23})\sin(\theta_2)(\dot{\theta}_2) \\
&- (l_{45})\sin(\theta_4)(\dot{\theta}_4))(((\cos(\theta_1)(\dot{\theta}_1))(la + \cos(\theta_2)l_b + (l_{23})\cos(\theta_2) + (l_{45})\cos(\theta_4))) \\
&+ \sin(\theta_1)(-\sin(\theta_2)(\dot{\theta}_2))l_b - (l_{23})\sin(\theta_2)(\dot{\theta}_2) - (l_{45})\sin(\theta_4)(\dot{\theta}_4))) + ((l_b\cos(\theta_2)(\dot{\theta}_2) \\
&+ (l_{23})\cos(\theta_2)\dot{\theta}_2 + l_{45}\cos(\theta_4)\dot{\theta}_4)(l_b\cos(\theta_2)\dot{\theta}_2 + (l_{23})\cos(\theta_2)\dot{\theta}_2 + l_{45}\cos(\theta_4)\dot{\theta}_4)))) \\
&+ \frac{m}{2}((l_{23})/2)((l_{23})/2)(\dot{\theta}_1^2 + \dot{\theta}_2^2) - \frac{m}{2}g((\sin(\theta_2)(l_{23} + l_{45})))
\end{aligned} \tag{11.33}$$

For the next link in motion, the mass definition is $m = m_4 + m_5$, the links $l_2 = \sqrt{(l_a^2 + l_b^2 + l_c^2)}$, $l_{23} = l_2 + l_3$, and $l_{45} = l_4 + l_5$. Thus, by substituting such expressions in next equation,

$$\mathcal{L}_{v_3} = \frac{m}{2}((-\sin(\theta_1)(\dot{\theta}_1)((l_a + \cos(\theta_2)l_b + (l_{23})\cos(\theta_2) + l_{45}\cos(\theta_4)$$

$$+ \cos(\theta_1)(-\sin(\theta_2)(\dot{\theta}_2))l_b - (l_{23})\sin(\theta_2)(\dot{\theta}_2) - l_{45}\sin(\theta_4)\dot{\theta}_4)(l_a + \cos(\theta_2)l_b + l_{23}\cos(\theta_2)$$

$$+ (l_{45})\cos(\theta_4) + \cos(\theta_1)(-\sin(\theta_2)(\dot{\theta}_2))l_b - l_{23}\sin(\theta_2)\dot{\theta}_2 - l_{45}\sin(\theta_4)\dot{\theta}_4)))$$

$$+ (((\cos(\theta_1)\dot{\theta}_1(l_a + \cos(\theta_2)l_b + l_{23}\cos(\theta_2) + l_{45}\cos(\theta_4))) + \sin(\theta_1)$$

$$(-\sin(\theta_2)\dot{\theta}_2)l_b - l_{23}\sin(\theta_2)\dot{\theta}_2 - l_{45}\sin(\theta_4)\dot{\theta}_4)((\cos(\theta_1)\dot{\theta}_1(l_a + \cos(\theta_2)l_b + l_{23}\cos(\theta_2) + l_{45}\cos(\theta_4)))$$

$$+ \sin(\theta_1)(-\sin(\theta_2)\dot{\theta}_2)l_b - l_{23}\sin(\theta_2)\dot{\theta}_2 - l_{45}\sin(\theta_4)\dot{\theta}_4)) + ((l_b\cos(\theta_2)\dot{\theta}_2 + l_{23}\cos(\theta_2)\dot{\theta}_2 + l_{45}\cos(\theta_4)\dot{\theta}_4)$$

$$(l_b\cos(\theta_2)\dot{\theta}_2 + l_{23}\cos(\theta_2)\dot{\theta}_2 + l_{45}\cos(\theta_4)\dot{\theta}_4))) + \frac{m}{2}\left(\frac{l_{45}}{2}\right)^2(\dot{\theta}_1^2 + \dot{\theta}_2^2 + \dot{\theta}_4^2) - \frac{m}{2}g\sin(\theta_2)\sin(\theta_4)l_{45}$$

$$(11.34)$$

11.3.5 Limb's posture configured as foot

The Lagrangian model of the foot configuration is produced from the velocity equation (11.16), with trajectory depicted in figure 11.3. Thus, the Lagrangian operator is defied by the next equation,

$$\mathcal{L}_{iv} = \frac{1}{2}m_i v_{Foot}^2 + \frac{1}{2}m_i\left(\frac{l_i}{2}\right)^2\dot{\theta}_i^2 - \frac{1}{2}m_i g\left(\frac{l_i}{2}\right)\sin q_i \qquad (11.35)$$

This foot configuration is actually a leg with a link functioning as foot. The last rigid link works as foot with active angular motion in its joint to resemble an ankle's motion. Previous equation includes potential energy description, as well as translational and rotational kinetic description. The set of rigid links in this linkage configuration evolves with translation and angular motions.

Thus, let us define the following mass expression for further algebraic simplification $m = m_1 + m_2 + m_3 + m_4 + m_5$. In addition, let us define the rigid link's distance $l_2 = \sqrt{(l_a^2 + l_b^2 + l_c^2)}$. Hence by substituting previous definitions in next expression,

$$\mathcal{L}_{iv_1} = \frac{m}{2}l_1^2(\dot{\theta}_1^2) - \frac{m}{2}g((\sin(\theta_2)((l_2) + l_3 + l_4))) \qquad (11.36)$$

And by defining $m = m_2 + m_3 + m_4 + m_5$ and $l_2 = \sqrt{(l_a^2 + l_b^2 + l_c^2)}$, and $l_{23} = l_2 + l_3$, the next equation is stated,

$$\mathcal{L}_{iv_2} = \frac{m}{2}(((-\sin(\theta_1)(\dot{\theta}_1)((l_a + \cos(\theta_2)l_b + (l_{23}\cos(\theta_2) + l_4\cos(\theta_3) + \cos(\theta_1)$$

$$(-\sin(\theta_2)\dot{\theta}_2)l_b - (l_{23}\sin(\theta_2)\dot{\theta}_2 - l_4\sin(\theta_3)(\dot{\theta}_3))(l_a + \cos(\theta_2)l_b + (l_{23}\cos(\theta_2)$$

$$+l_4\cos(\theta_3) + \cos(\theta_1)(-\sin(\theta_2)(\dot{\theta}_2))l_b - (l_{23}\sin(\theta_2)\dot{\theta}_2 - l_4\sin(\theta_3)\dot{\theta}_3)))$$

$$+((((\cos(\theta_1)\dot{\theta}_1)(l_a + \cos(\theta_2)l_b + (l_{23}\cos(\theta_2) + l_4\cos(\theta_3))) + \sin(\theta_1)(-\sin(\theta_2)\dot{\theta}_2)l_b$$

$$-(l_{23}\sin(\theta_2)(\dot{\theta}_2) - l_4\sin(\theta_3)(\dot{\theta}_3))(((\cos(\theta_1)(\dot{\theta}_1))(l_a + \cos(\theta_2)l_b + (l_{23}\cos(\theta_2) + l_4\cos(\theta_3)))$$

$$+ \sin(\theta_1)(-\sin(\theta_2)\dot{\theta}_2)l_b - (l_{23}\sin(\theta_2)\dot{\theta}_2 - l_4\sin(\theta_3)\dot{\theta}_3)) + (l_b\cos(\theta_2)\dot{\theta}_2$$

$$+(l_{23}\cos(\theta_2)\dot{\theta}_2 + l_4\cos(\theta_3)\dot{\theta}_3)(l_b\cos(\theta_2)\dot{\theta}_2 + (l_{23}\cos(\theta_2)\dot{\theta}_2 + l_4\cos(\theta_3)\dot{\theta}_3))))$$

$$+\frac{m}{2}\left(\frac{l_2}{2}\right)(\dot{\theta}_1^2 + \dot{\theta}_2^2) - \frac{m}{2}g((\sin(\theta_2)(l_{23} + 14)))$$

(11.37)

Likewise, $m = m_4 + m_5$, and the links $l_2 = \sqrt{(l_a^2 + l_b^2 + l_c^2)}$ as well as $l_{23} = l_2 + l_3$. Thus,

$$\mathcal{L}_{iv_3} = \frac{m}{2}((-\sin(\theta_1)(\dot{\theta}_1)((l_a + \cos(\theta_2)l_b + (l_{23}\cos(\theta_2) + l_4\cos(\theta_3) + \cos(\theta_1)$$

$$(-\sin(\theta_2)\dot{\theta}_2)l_b - (l_2\sin(\theta_2)\dot{\theta}_2 - l_4\sin(\theta_3)\dot{\theta}_3)(l_a + \cos(\theta_2)l_b$$

$$+(l_{23}\cos(\theta_2) + l_4\cos(\theta_3) + \cos(\theta_1)(-\sin(\theta_2)(\dot{\theta}_2))l_b$$

$$-(l_{23}\sin(\theta_2)(\dot{\theta}_2) - l_4\sin(\theta_3)(\dot{\theta}_3)))) + ((((\cos(\theta_1)(\dot{\theta}_1))(l_a + \cos(\theta_2)l_b + (l_2\cos(\theta_2) + l_4\cos(\theta_3)))$$

$$+ \sin(\theta_1)(-\sin(\theta_2)\dot{\theta}_2)l_b - (l_3 + l_2\sin(\theta_2)\dot{\theta}_2 - l_4\sin(\theta_3)\dot{\theta}_3)(((\cos(\theta_1)\dot{\theta}_1)(l_a + \cos(\theta_2)l_b$$

$$+(l_{23}\cos(\theta_2) + l_4\cos(\theta_3))) + \sin(\theta_1(-\sin(\theta_2)\dot{\theta}_2)l_b - (l_2\sin(\theta_2)\dot{\theta}_2 - l_4\sin(\theta_3)\dot{\theta}_3))$$

$$+((l_b\cos(\theta_2)\dot{\theta}_2 + (l_{23}\cos(\theta_2)\dot{\theta}_2 + l_4\cos(\theta_3)\dot{\theta}_3)(l_b\cos(\theta_2)\dot{\theta}_2 + (l_{23}\cos(\theta_2)\dot{\theta}_2$$

$$+l_4\cos(\theta_3)\dot{\theta}_3))) + \frac{m}{2}\frac{l_4}{2}(\dot{\theta}_1^2 + \dot{\theta}_2^2 + \dot{\theta}_3^2) - \frac{m}{2}g(\sin(\theta_2)(\sin(\theta_3)l_4))$$

(11.38)

11.3.6 Limb's posture configured as aquatic fin

The Lagrangian modelling of the aquatic fin configuration considers starting from the velocity equation (11.18), and depiction of its trajectory in figure 11.3. Thus, the Lagrange equation is stated by

$$\mathcal{L}_{Aquatic_i} = \frac{1}{2}m_i v_{vi}^2 + \frac{1}{2}m_i\left(\frac{l_i}{2}\right)^2\dot{\theta}_i^2 - \frac{1}{2}m_i g\left(\frac{l_i}{2}\right)\sin q_i$$

(11.39)

Rather than an aquatic fin, this configuration resembles a swimmer limb. Although, gaits patterns are not a matter of this manuscript, but a swimmer limb may be capable to perform numerous types of patters for underwater swimming. Thus, this configuration's motion is developed in 3D Cartesian space, where only the initial and final angles of motion may be set to change multiple gaits.

Let us define the following mass expression $m = m_1 + m_2 + m_3 + m_4 + m_5$, and links $l_2 = \sqrt{(l_a^2 + l_b^2 + l_c^2)}$, $l_{23} = l_2 + l_3$, as well as the rigid link expression $l_{45} = l_4 + l_5$. Thus, by substituting in the next expression,

$$\mathcal{L}_{vi_1} = \frac{m}{2}l_1^2\dot{\theta}_1^2 - \frac{m}{2}g(\sin(\theta_2)(l_2 + l_3 + l_4 + l_5)) \tag{11.40}$$

likewise, consider that $m = m_2 + m_3 + m_4 + m_5$, $l_2 = \sqrt{(l_a^2 + l_b^2 + l_c^2)}$, $l_{23} = l_2 + l_3$, and $l_{45} = l_4 + l_5$. Hence,

$$
\begin{aligned}
\mathcal{L}_{vi_2} = \frac{m}{2}(((&-\sin(\theta_1)(\dot{\theta}_1)((l_a + \cos(\theta_2)l_b + (l_{23}))) \cos(\theta_2) + (l_{45}) \cos(\theta_4) \\
&+ \cos(\theta_1)(-\sin(\theta_2)(\dot{\theta}_2))l_b - (l_{23}))) \sin(\theta_2)(\dot{\theta}_2) - (l_{45}) \sin(\theta_4)(\dot{\theta}_4)) \\
&(l_a + \cos(\theta_2)l_b + (l_{23})) \cos(\theta_2) + (l_{45}) \cos(\theta_4) + \cos(\theta_1) \\
&(-\sin(\theta_2)(\dot{\theta}_2))l_b - (l_{23}))) \sin(\theta_2)(\dot{\theta}_2) - (l_{45}) \sin(\theta_4)(\dot{\theta}_4)))+ \\
&((((\cos(\theta_1)(\dot{\theta}_1))(l_a + \cos(\theta_2)l_b + (l_{23}))) \cos(\theta_2) + (l_{45}) \cos(\theta_4))) \\
&+ \sin(\theta_1)(-\sin(\theta_2)(\dot{\theta}_2))l_b - (l_{23}))) \sin(\theta_2)(\dot{\theta}_2) - (l_{45}) \sin(\theta_4)(\dot{\theta}_4)) \\
&(((\cos(\theta_1)(\dot{\theta}_1))(l_a + \cos(\theta_2)l_b + (l_{23}))) \cos(\theta_2) + (l_{45}) \cos(\theta_4))) \\
&+ \sin(\theta_1)(-\sin(\theta_2)(\dot{\theta}_2))l_b - (l_{23}))) \sin(\theta_2)(\dot{\theta}_2) - (l_{45}) \sin(\theta_4)(\dot{\theta}_4))) \\
&+((l_b \cos(\theta_2)(\dot{\theta}_2) + (l_{23}))) \cos(\theta_2)(\dot{\theta}_2) + (l_{45}) \cos(\theta_4)(\dot{\theta}_4))(l_b \cos(\theta_2)(\dot{\theta}_2) \\
&+(l_{23}))) \cos(\theta_2)(\dot{\theta}_2) + (l_{45}) \cos(\theta_4)(\dot{\theta}_4))))) \\
&+\frac{m}{2}((l_{23})))/2)((l_{23})))/2)(\dot{\theta}_1^2 + \dot{\theta}_2^2) - \frac{m}{2}g((\sin(\theta_2)((l_2) + (l_{45}))))
\end{aligned} \tag{11.41}
$$

The mass expression is formulated for the next set of rigid links,

$$m = m_4 + m_5$$

and the definitions of the rigid links are provided by

$$l_2 = \sqrt{(l_a^2 + l_b^2 + l_c^2)}$$

and

$$l_{23} = l_2 + l_3$$

as well as

$$l_{45} = l_4 + l_5$$

Therefore, the last Lagrangian operator is defined by

$$
\begin{aligned}
\mathcal{L}_{vi_3} = \frac{m}{2}((- \sin(\theta_1)(\dot{\theta}_1)((l_a + \cos(\theta_2)l_b + (l_{23}))) \cos(\theta_2) \\
+ (l_{45}) \cos(\theta_4) + \cos(\theta_1)(- \sin(\theta_2)(\dot{\theta}_2))l_b - (l_{23}))) \sin(\theta_2)(\dot{\theta}_2) \\
- (l_{45}) \sin(\theta_4)(\dot{\theta}_4)(l_a + \cos(\theta_2)l_b + (l_{23}))) \cos(\theta_2) + (l_{45}) \cos(\theta_4) \\
+ \cos(\theta_1)(- \sin(\theta_2)(\dot{\theta}_2))l_b - (l_{23}))) \sin(\theta_2)(\dot{\theta}_2) - (l_{45}) \sin(\theta_4)(\dot{\theta}_4)))) \\
+ ((((\cos(\theta_1)(\dot{\theta}_1))(l_a + \cos(\theta_2)l_b + (l_{23}))) \cos(\theta_2) + (l_{45}) \cos(\theta_4))) \\
+ \sin(\theta_1)(- \sin(\theta_2)(\dot{\theta}_2))l_b - (l_{23}))) \sin(\theta_2)(\dot{\theta}_2) - (l_{45}) \sin(\theta_4)(\dot{\theta}_4)) \\
(((\cos(\theta_1)(\dot{\theta}_1))(l_a + \cos(\theta_2)l_b + (l_{23}))) \cos(\theta_2) + (l_{45}) \cos(\theta_4))) \\
+ \sin(\theta_1)(- \sin(\theta_2)(\dot{\theta}_2))l_b - (l_{23}))) \sin(\theta_2)(\dot{\theta}_2) - (l_{45}) \sin(\theta_4)(\dot{\theta}_4))) \\
+ ((l_b \cos(\theta_2)(\dot{\theta}_2) + (l_{23}))) \cos(\theta_2)(\dot{\theta}_2) + (l_{45}) \cos(\theta_4)(\dot{\theta}_4)) \\
(l_b \cos(\theta_2)(\dot{\theta}_2) + (l_{23}))) \cos(\theta_2)(\dot{\theta}_2) + (l_{45}) \cos(\theta_4)(\dot{\theta}_4)))) \\
+ \frac{m}{2} \left(\frac{l_{45}}{2} \right)^2 (\dot{\theta}_1^2 + \dot{\theta}_2^2 + \dot{\theta}_4^2) - \frac{m}{2} g \sin(\theta_2) \sin(\theta_4)l_{45}
\end{aligned}
\tag{11.42}
$$

11.4 Limb's configurations Euler-Lagrange analysis

In this section, an Euler-Lagrange analysis particularly describes each joint individually. Where $\mathcal{L}_1$ represents the Lagrangian equation of the 1st joint used for steering. $\mathcal{L}_2$ is the Lagrangian of the 2nd rotational joint. $\mathcal{L}_3$ is the Lagrangian of the first link. Likewise, $\mathcal{L}_4$ is the Lagrangian of the second rigid link. And $\mathcal{L}_5$ is the Lagrangian of the third rigid link. An analysis of energies may be observed for each driven joint without explicitly describing a particular gait trajectory.

For the first driven joints and links the Lagrangian is defined by

$$\mathcal{L}_1 = \frac{1}{2}m_1 r^1(\dot{\theta}_1^2) - \frac{1}{2}m_1 g \left(\left(\frac{l_2 \sin(\theta_2) + l_3 \sin(\theta_2 + \theta_3) + l_4 \sin(\theta_2 + \theta_3 + \theta_4)}{2} \right) \right. \\ \left. + \left(\frac{l_5 \sin(\theta_2 + \theta_3 + \theta_4 + \theta_5)}{2} \right) \right) \tag{11.43}$$

And subsequently,

$$\mathcal{L}_2 = \frac{1}{2}m_2 v_2^2 + \frac{1}{2}m_2 r^2(\dot{\theta}_1^2 + \dot{\theta}_2^2) - \frac{1}{2}m_2 g \\ \left(\frac{l_2 \sin\theta_2 + l_3 \sin(\theta_2 + \theta_3) + l_4 \sin(\theta_2 + \theta_3 + \theta_4) + l_5 \sin(\theta_2 + \theta_3 + \theta_4 + \theta_5)}{2} \right)$$

and

$$\mathcal{L}_3 = \frac{1}{2}m_2 v_3^2 + \frac{1}{2}m_3 r^2(\dot{\theta}_1^2 + \dot{\theta}_2^2 + \dot{\theta}_3^2) - \frac{1}{2}m_3 g \\ \left(\frac{l_3 \sin(\theta_2 + \theta_3) + l_4 \sin(\theta_2 + \theta_3 + \theta_4) + l_5 \sin(\theta_2 + \theta_3 + \theta_4 + \theta_5)}{2} \right)$$

and

$$\mathcal{L}_4 = \frac{1}{2}m_4 v_4^2 + \frac{1}{2}m_4 r^2(\dot{\theta}_1^2 + \dot{\theta}_2^2 + \dot{\theta}_3^2 + \dot{\theta}_4^2) - \frac{1}{2}m_4 g \\ \left(\frac{l_4 \sin(\theta_2 + \theta_3 + \theta_4) + l_5 \sin(\theta_2 + \theta_3 + \theta_4 + \theta_5)}{2} \right)$$

as well as

$$\mathcal{L}_5 = \frac{1}{2}m_5 v_5^2 + \frac{1}{2}m_5 r^2(\dot{\theta}_1^2 + \dot{\theta}_2^2 + \dot{\theta}_3^2 + \dot{\theta}_4^2 + \dot{\theta}_5^2) - \frac{1}{2}m_5 g \left(\frac{l_5 \sin(\theta_2 + \theta_3 + \theta_4 + \theta_5)}{2} \right) \tag{11.44}$$

Thus, the behaviour of the energies involved in the mechanical system of the limb for each locomotive configuration is described by the general Euler-Lagrange differential equation,

$$\boldsymbol{\tau}_i = \frac{d}{dt}\left(\frac{\partial \mathcal{L}_i}{\partial \dot{\mathbf{q}}_i} \right) - \frac{\partial \mathcal{L}_i}{\partial \mathbf{q}_i} \tag{11.45}$$

for instance, torque calculation of the wheel configuration is provided by

$$\boldsymbol{\tau}_{i_1} = \frac{m}{4}l_1^2\ddot{\theta}_1 - \frac{m}{2}g\sin(\theta_1)\dot{\theta}_1 l_b \tag{11.46}$$

The following expressions are defined, $f_1 = l_a + \cos(\theta_2)l_b + l_c \sin(\theta_2)$, $f_2 = l_c \cos(\theta_2)$, $f_3 = l_c \sin(\theta_2)$, $f_4 = l_b \sin(\theta_2)$, and $f_5 = l_b \cos(\theta_2)$.

And by substituting previous expressions in next equation,

$$\boldsymbol{\tau}_{i_2} = (\frac{m}{2}(-4\cos(\theta_1)(-f_5\dot{\theta}_2 + f_2\dot{\theta}_2 - (f_1)\sin(\theta_1)\dot{\theta}_1)(-\cos(\theta_2)l_b + f_2)\sin(\theta_1)\dot{\theta}_1$$

$$+2\cos(\theta_1)\cos(\theta_1)(f_4\dot{\theta}_2^2 - \cos(\theta_2)\ddot{\theta}_2 l_b - f_3\dot{\theta}_2^2 + f_2\ddot{\theta}_2 - (-\sin(\theta_2)\dot{\theta}_2 l_b + f_2\dot{\theta}_2)\sin(\theta_1)\dot{\theta}_1$$

$$-(f_1)\cos(\theta_1)\dot{\theta}_1\dot{\theta}_1 - (f_1)\sin(\theta_1)d\dot{\theta}_1)(-\cos(\theta_2)l_b + f_2) + 2\cos(\theta_1)\cos(\theta_1)(-f_5)\dot{\theta}_2$$

$$+f_2\dot{\theta}_2 - (f_1)\sin(\theta_1)\dot{\theta}_1)(\sin(\theta_2)\dot{\theta}_2 l_b - f_3\dot{\theta}_2) + (2(\cos(\theta_1)\dot{\theta}_1(-f_4\dot{\theta}_2 + f_2\dot{\theta}_2) + \sin(\theta_1)$$

$$(-f_5\dot{\theta}_2^2 - f_4\ddot{\theta}_2 - f_3\dot{\theta}_2^2 + f_2\ddot{\theta}_2) + (-f_4\dot{\theta}_2 + f_2\dot{\theta}_2)\cos(\theta_1)\dot{\theta}_1 - (f_1)\sin(\theta_1)\dot{\theta}_1\dot{\theta}_1$$

$$+(f_1)\cos(\theta_1)\ddot{\theta}_1))\sin(\theta_1)(-f_4 + f_2) + (2(\sin(\theta_1)(-f_4\dot{\theta}_2 + f_2\dot{\theta}_2) + (f_1)\cos(\theta_1)\dot{\theta}_1))$$

$$\cos(\theta_1)\dot{\theta}_1(-f_4 + f_2) + (2(\sin(\theta_1)(-f_4\dot{\theta}_2 + f_2\dot{\theta}_2) + (f_1)\cos(\theta_1)\dot{\theta}_1))\sin(\theta_1)$$

$$(-f_5\dot{\theta}_2 - f_3\dot{\theta}_2) + (2(-f_4\dot{\theta}_2^2 + f_5\ddot{\theta}_2 - f_3\dot{\theta}_2^2 + f_2\ddot{\theta}_2))(f_5 + f_2) + (2(\cos(\theta_2)\dot{\theta}_2 l_b + f_2\dot{\theta}_2))$$

$$(-f_4\dot{\theta}_2 - f_3\dot{\theta}_2)) + m2/4(l_2)\ddot{\theta}_2 - m2/2(-4\cos(\theta_1)(-\cos(\theta_2)\dot{\theta}_2 l_b + f_2\dot{\theta}_2$$

$$-(f_1)\sin(\theta_1)\dot{\theta}_1)(f_4\dot{\theta}_2 - f_3\dot{\theta}_2 - (-f_4 + f_2)\sin(\theta_1)\dot{\theta}_1)\sin(\theta_1)\dot{\theta}_1 + 2\cos(\theta_1)\cos(\theta_1)(f_4\dot{\theta}_2^2$$

$$-\cos(\theta_2)\ddot{\theta}_2 l_b - f_3\dot{\theta}_2^2 + f_2\ddot{\theta}_2 - (-f_4\dot{\theta}_2 + f_2\dot{\theta}_2)\sin(\theta_1)\dot{\theta}_1 - (f_1)\cos(\theta_1)\dot{\theta}_1\dot{\theta}_1$$

$$-(f_1)\sin(\theta_1)\ddot{\theta}_1)(f_4\dot{\theta}_2 - f_3\dot{\theta}_2 - (-f_4 + f_2)\sin(\theta_1)\dot{\theta}_1) + 2\cos(\theta_1)\cos(\theta_1)(-\cos(\theta_2)\dot{\theta}_2 l_b$$

$$+f_2\dot{\theta}_2 - (f_1)\sin(\theta_1)\dot{\theta}_1)(f_5\dot{\theta}_2^2 + f_4\ddot{\theta}_2 - f_2\dot{\theta}_2^2 - f_3\ddot{\theta}_2 - (-f_5\dot{\theta}_2 - f_3\dot{\theta}_2)\sin(\theta_1)\dot{\theta}_1$$

$$-(-f_4 + f_2)\cos(\theta_1)\dot{\theta}_1\dot{\theta}_1 - (-f_4 + f_2)\sin(\theta_1)\ddot{\theta}_1) + (2(\cos(\theta_1)\dot{\theta}_1(-f_4\dot{\theta}_2 + f_2\dot{\theta}_2)$$

$$+\sin(\theta_1)(-f_5\dot{\theta}_2^2 - f_4\ddot{\theta}_2 - f_3\dot{\theta}_2^2 + f_2\ddot{\theta}_2) + (-\sin(\theta_2)\dot{\theta}_2 l_b + f_2\dot{\theta}_2)\cos(\theta_1)\dot{\theta}_1$$

$$-(f_1)\sin(\theta_1)\dot{\theta}_1\dot{\theta}_1 + (f_1)\cos(\theta_1)\ddot{\theta}_1))(\sin(\theta_1)(-\cos(\theta_2)\dot{\theta}_2 l_b - f_3\dot{\theta}_2) + (-f_4 + f_2)$$

$$\cos(\theta_1)\dot{\theta}_1) + (2(\sin(\theta_1)(-f_4\dot{\theta}_2 + f_2\dot{\theta}_2) + (f_1)\cos(\theta_1)\dot{\theta}_1))(\cos(\theta_1)\dot{\theta}_1(-f_5\dot{\theta}_2 - f_3\dot{\theta}_2)$$

$$+\sin(\theta_1)(f_4\dot{\theta}_2^2 - \cos(\theta_2)\ddot{\theta}_2 l_b - f_2\dot{\theta}_2^2 - f_3\ddot{\theta}_2) + (-f_5\dot{\theta}_2 - f_3\dot{\theta}_2)\cos(\theta_1)\dot{\theta}_1$$

$$-(-f_4 + f_2)\sin(\theta_1)\dot{\theta}_1\dot{\theta}_1 + (-f_4 + f_2)\cos(\theta_1)\ddot{\theta}_1) + (2(-\sin(\theta_2)\dot{\theta}_2^2 l_b + f_5\ddot{\theta}_2 - f_3\dot{\theta}_2^2 + f_2\ddot{\theta}_2))$$

$$(-f_4\dot{\theta}_2 - f_3\dot{\theta}_2) + (2(\cos(\theta_2)\dot{\theta}_2 l_b + f_2\dot{\theta}_2))(-f_5\dot{\theta}_2^2 - f_4\ddot{\theta}_2 - f_2\dot{\theta}_2^2 - f_3\ddot{\theta}_2))$$

$$+\frac{m}{2}g(-\sin(\theta_2)\dot{\theta}_2 l_b - f_3\dot{\theta}_2))$$

Similarly, torque analysis for the half wheel configuration may be described by

$$\boldsymbol{\tau}_{ii_1} = ml_1^2\ddot{\theta}_1 - hmg\sin(\theta_1)\dot{\theta}_1 \tag{11.47}$$

Thus, following definitions are stated,

$$f_1 = l_a + \cos(\theta_2)l_b + l_c \sin(\theta_2); \quad f_2 = l_c \cos(\theta_2); \quad f_3 = l_c \sin(\theta_2)$$

and

$$f_4 = l_b \sin(\theta_2); \quad f_5 = l_b \cos(\theta_2)$$

hence,

$$
\begin{aligned}
\boldsymbol{\tau}_{ii_2} = {}& (2\sin(\theta_1)\sin(\theta_1)\dot{\theta}_1(-\cos(\theta_2)\dot{\theta}_2 + f_2\dot{\theta}_2 - (f_1)\sin(\theta_1)\dot{\theta}_1)(-f_5\dot{\theta}_2 + f_2\dot{\theta}_2 - (f_1)\sin(\theta_1)\dot{\theta}_1) \\
& -4\cos(\theta_1)(-f_5\dot{\theta}_2 + f_2\dot{\theta}_2 - (f_1)\sin(\theta_1)\dot{\theta}_1)\sin(\theta_1)(\sin(\theta_2)\dot{\theta}_2\dot{\theta}_2 l_b - f_5\ddot{\theta}_2 - f_3\dot{\theta}_2\dot{\theta}_2 + f_2\ddot{\theta}_2 \\
& -(-f_4\dot{\theta}_2 + f_2\dot{\theta}_2)\sin(\theta_1)(\dot{\theta}_1) - (f_1)\cos(\theta_1)(\dot{\theta}_1\dot{\theta}_1) - (f_1)\sin(\theta_1)\ddot{\theta}_1) - 2\cos(\theta_1)\cos(\theta_1)(-f_5\dot{\theta}_2 \\
& +f_2\dot{\theta}_2 - (f_1)\sin(\theta_1)\dot{\theta}_1)(-f_5\dot{\theta}_2 + f_2\dot{\theta}_2 - (f_1)\sin(\theta_1)\dot{\theta}_1)(\dot{\theta}_1) - 6(\cos(\theta_1))(-f_5\dot{\theta}_2 + f_2\dot{\theta}_2 \\
& -(f_1)\sin(\theta_1)\dot{\theta}_1)(-f_5\dot{\theta}_2 + f_2\dot{\theta}_2 - (f_1)\sin(\theta_1)\dot{\theta}_1)(f_1)\dot{\theta}_1((-\sin(\theta_1))\dot{\theta}_1)(-f_5\dot{\theta}_2 + f_2(\dot{\theta}_2) \\
& -(f_1)\sin(\theta_1)\dot{\theta}_1)(\sin(\theta_2)(\dot{\theta}_2\dot{\theta}_2)l_b - f_5\ddot{\theta}_2 - f_3\dot{\theta}_2\dot{\theta}_2 + f_2\ddot{\theta}_2 - (-f_4\dot{\theta}_2 + f_2\dot{\theta}_2)\sin(\theta_1)\dot{\theta}_1 \\
& -(f_1)\cos(\theta_1)\dot{\theta}_1\dot{\theta}_1 - (f_1)\sin(\theta_1)\ddot{\theta}_1) - 2(\cos(\theta_1))(-f_5\dot{\theta}_2 + f_2\dot{\theta}_2 - (f_1)\sin(\theta_1)\dot{\theta}_1)(-f_5\dot{\theta}_2 \\
& +f_2\dot{\theta}_2 - (f_1)\sin(\theta_1)\dot{\theta}_1)(-f_5\dot{\theta}_2 + f_2\dot{\theta}_2 - (f_1)\sin(\theta_1)\dot{\theta}_1)(-f_4\dot{\theta}_2 + f_2\dot{\theta}_2)\dot{\theta}_1 - 2(\cos(\theta_1)) \\
& (-f_5\dot{\theta}_2 + f_2\dot{\theta}_2 - (f_1)\sin(\theta_1)\dot{\theta}_1)(-f_5\dot{\theta}_2 + f_2\dot{\theta}_2 - (f_1)\sin(\theta_1)\dot{\theta}_1)(-f_5\dot{\theta}_2 + f_2\dot{\theta}_2 - (f_1)\sin(0_1)\dot{\theta}_1) \\
& (f_1)\ddot{\theta}_1 + (2(\cos(\theta_1)\dot{\theta}_1(-f_4\dot{\theta}_2 + f_2\dot{\theta}_2) + \sin(\theta_1)(-l_b\cos(\theta_2)\dot{\theta}_2\dot{\theta}_2 - f_4\ddot{\theta}_2 - f_3\dot{\theta}_2^2 + f_2\ddot{\theta}_2) \\
& +(-f_4\dot{\theta}_2 + f_2\dot{\theta}_2)\cos(\theta_1)\dot{\theta}_1 - (f_1)\sin(\theta_1)\dot{\theta}_1^2 + (f_1)\cos(\theta_1)\ddot{\theta}_1))(\cos(\theta_1)(-f_4\dot{\theta}_2 + f_2\dot{\theta}_2) \\
& -(f_1)\sin(\theta_1)\dot{\theta}_1) + (2(\sin(\theta_1)(-f_4\dot{\theta}_2 + f_2\dot{\theta}_2) + (f_1)\cos(\theta_1)\dot{\theta}_1))(-\sin(\theta_1)\dot{\theta}_1(-f_4\dot{\theta}_2 + f_2\dot{\theta}_2) \\
& +\cos(\theta_1)(-l_b\cos(\theta_2)\dot{\theta}_2\dot{\theta}_2 - f_4\ddot{\theta}_2 - f_3\dot{\theta}_2\dot{\theta}_2 + f_2\ddot{\theta}_2) - (-f_4\dot{\theta}_2 + f_2\dot{\theta}_2)\sin(\theta_1)(\dot{\theta}_1) \\
& -(f_1)\cos(\theta_1)(\dot{\theta}_1\dot{\theta}_1) - (f_1)\sin(\theta_1)\ddot{\theta}_1))
\end{aligned}
$$

Therefore,

$$\boldsymbol{\tau}_i = \frac{d}{dt}\left(\frac{\partial \mathcal{L}_i}{\partial \dot{\mathbf{q}}_i}\right) - \frac{\partial \mathcal{L}_i}{\partial \mathbf{q}_i} \tag{11.48}$$

11.5   Quadruped kinematic motion

The robot's global motion is an averaged value on its chassis structure provided by the contributions of each limb's motion. The robot's motion is represented in terms of its pose, velocity, and acceleration; either in fixed inertial frame, or in a global coordinates frame (figure 11.5).

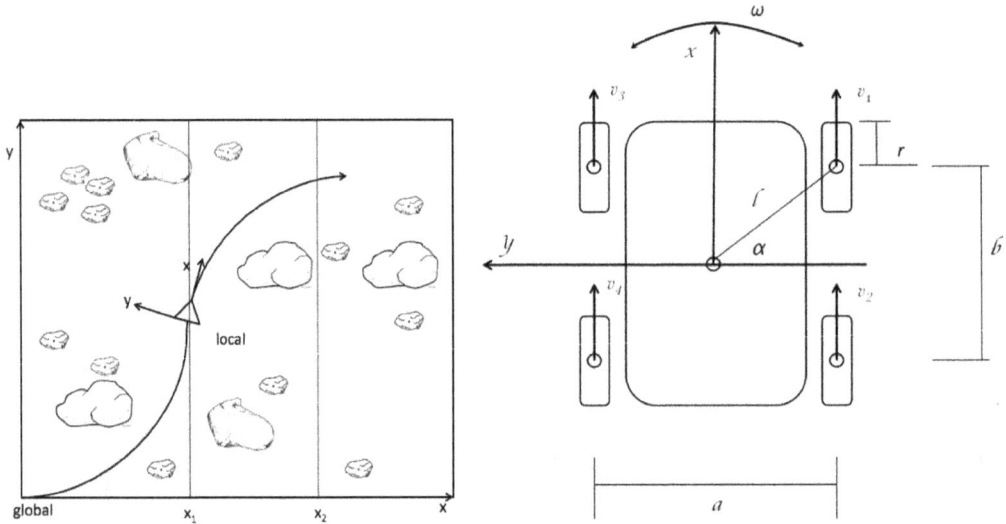

Figure 11.5:   Robot's trajectory control w.r.t. a global inertial system (left). Quadruped general kinematic configuration (right).

The robot's instantaneous linear velocity v_t is approached by an averaged model of the four limbs' speed v_i, as next equation (depicted in figure 11.5),

$$v = \frac{1}{4}(v_1 + v_2 + v_3 + v_4) \tag{11.49}$$

Furthermore, the robot's instantaneous angular velocity ω_t is modelled approaching the robot's differences of lateral velocities. The velocities are yielded by the four limbs w.r.t. the robot's centroid, which is the origin of the coordinate system fixed to the robot's body.

$$\omega = \frac{2a}{a^2 + b^2}(v_1 + v_2 - v_3 - v_4) \tag{11.50}$$

In previous equation $v_{1,2}$, are positive because move counter-clockwise w.r.t. to the robot's centroid. Likewise, $v_{3,4}$ are negative as they normal sense is clockwise.

For instance, let us assume that the limbs are configured as wheels, thus it follows from vector equation (11.7) $\mathbf{p}_i = (x_i, y_i, z_i)^\top$ that the kinematic description of one wheel is provided by

$$x_i = \cos\theta_1 (la + l_b \sin(\theta_2) + l_c \cos\theta_2), \tag{11.51}$$

and

$$y_i = \sin\theta_1 (la + l_b \cos(\theta_2) + l_c \cos(\theta_2)) \tag{11.52}$$

and

$$z_i = l_b \cos\theta_2 + l_c \sin\theta_2. \tag{11.53}$$

To solve the ith limb's tangential velocity the norm of the three Cartesian components is obtained

$$v_i = \left(\dot{x}_i^2 + \dot{y}_i^2 + \dot{z}_i^2 \right)^{\frac{1}{2}} \tag{11.54}$$

which is equivalent to the following expression,

$$v = \left(\left(\frac{d}{dt} x_{p_i} \right)^2 + \left(\frac{d}{dt} y_{p_i} \right)^2 + \left(\frac{d}{dt} z_{p_i} \right)^2 \right)^{\frac{1}{2}} \tag{11.55}$$

Therefore, the whole expression is deduced as provided by the next equation,

Corollary 11.5.1 (Quadruped's instantaneous linear velocity). *The quadruped's instantaneous linear speed model regardless any locomotive configuration is stated by*

$$
\begin{aligned}
v = \frac{r}{4} \Bigg(& \left((\frac{d}{dt} x_{p_1})^2 + (\frac{d}{dt} y_{p_1})^2 + (\frac{d}{dt} z_{p_1})^2 \right)^{\frac{1}{2}} + \left((\frac{d}{dt} x_{p_2})^2 + (\frac{d}{dt} y_{p_2})^2 + (\frac{d}{dt} z_{p_2})^2 \right)^{\frac{1}{2}} \\
& + \left((\frac{d}{dt} x_{p_3})^2 + (\frac{d}{dt} y_{p_3})^2 + (\frac{d}{dt} z_{p_3})^2 \right)^{\frac{1}{2}} + \left((\frac{d}{dt} x_{p_4})^2 + (\frac{d}{dt} y_{p_4})^2 + (\frac{d}{dt} z_{p_4})^2 \right)^{\frac{1}{2}} \Bigg)
\end{aligned}
\tag{11.56}
$$

In addition, for the robot's angular expression a similar expression is deduced,

Corollary 11.5.2 (Quadruped's instantaneous angular velocity). *The quadruped's instantaneous angular velocity model regardless any locomotive configuration is stated by*

$$
\omega = \frac{2ra}{a^2+b^2}\left(\left((\frac{d}{dt}x_{p_1})^2+(\frac{d}{dt}y_{p_1})^2+(\frac{d}{dt}z_{p_1})^2\right)^{\frac{1}{2}}+\left((\frac{d}{dt}x_{p_2})^2+(\frac{d}{dt}y_{p_2})^2+(\frac{d}{dt}z_{p_2})^2\right)^{\frac{1}{2}}\right.
$$
$$
\left.-\left((\frac{d}{dt}x_{p_3})^2+(\frac{d}{dt}y_{p_3})^2+(\frac{d}{dt}z_{p_3})^2\right)^{\frac{1}{2}}-\left((\frac{d}{dt}x_{p_4})^2+(\frac{d}{dt}y_{p_4})^2+(\frac{d}{dt}z_{p_4})^2\right)^{\frac{1}{2}}\right)
$$
$$(11.57)$$

From previous speed models, the robot's first and second derivative are represented in a global inertial system I. Thus, for the global velocity vector,

$$
\mathbf{v}^I = -v_t \begin{pmatrix} \cos(\theta + \psi) \\ \sin(\theta + \psi) \end{pmatrix} \tag{11.58}
$$

by deriving w.r.t. to time and algebraically arranging, the acceleration vector w.r.t. a global coordinate system is obtained,

$$
\mathbf{a}^I = \dot{v}_t \begin{pmatrix} \cos(\theta + \psi) \\ \sin(\theta + \psi) \end{pmatrix} + v \begin{pmatrix} -\sin(\theta + \psi) \\ \cos(\theta + \psi) \end{pmatrix} (\omega_t + \dot{\psi}) \tag{11.59}
$$

Bibliography

[1] SunSpiral V., Wheeler D.W., Chavez-Clemente D., Mittman D., *Development and field testing of the FootFall planning system for the ATHLETE robots*, Journal of Field Test, vol.29(3), pp. 483–505, 2012, doi: 10.1002/rob.20410.

[2] Tadakuma K., Tadakuma R., *Mechanical Design of the Wheel-Leg Hybrid Mobile Robot to Realize a Large Wheel Diameter*, IEEE/RSJ Intl. Conf. on Intelligent Robots and Systems, 2010, doi: http://dx.doi.org/10.1109/iros.2010.5651912.

[3] Azevedoa C., Poignetb P., Espiauc B., *Artificial locomotion control: from human to robots*, Robotics and autonomous Systems, vol. 47(4), pp. 203–223, 2004, doi:10.1016/j.robot.2004.03.013.

[4] Martínez-García E.A., Torres-Mendez L.A. (Eds.), *Autonomous Robots: Control, Sensing and Perception*, Cuvillier Verlag Publishing House, Germany, 2011, isbn: 978-3-86955-866-0.

[5] Beckman, B. Thomson, T., *Kinematic control and posture optimization of a redundantly actuated quadruped robot*, IEEE Intl. Conf. Robotics and Automation, pp. 1895-1900, 2012.

[6] Mutka A., Kovacic Z., *A Leg-wheel Robot-based Approach to the Solution of Flipper-track Robot Kinematics*, Multi-Conference on Systems and Control, USA, 2011.

[7] Watanabe Y., Nonaka k., *Slip Measurement and Vehicle Control for Leg/Wheel Mobile Robots using Caster Type Odometers*, Multi-Conference on Systems and Control, USA., 2011.

[8] Hashimoto K., Hosobata T., Sugahara Y., Mikuriya y., Sunazuka H., Kawase M., *Realization by Biped Leg-wheeled Robot of Biped Walking and Wheel-driven Locomotion*, IEEE ICRA, Spain, 2005.

[9] Tanaka T., Hirose S., *Development of Leg-wheel Hybrid Quadruped "AirHopper" Design of Powerful Light-weight Leg with Wheel*, Intl. Conf. on Intelligent Robots and Systems, France, 2008.

Part IV

Modelling Walking Robots

Chapter 12

DIRECT/INVERSE ANALYSIS OF REDUNDANT WALKING ROBOTS

Diana R. Uribe Escalera and Edgar A. Martínez García

Laboratorio de Robótica, Institute of Engineering and Technology
Universidad Autónoma de Ciudad Juárez, Mexico.

This chapter treats a linearised navigation control law for multi-legged walking robots. The proposed model is stated in terms of robot's global acceleration, and formulated as an average of the Cartesian speeds of n-extremities of k-DOFs each. The state vector is defined as a general solution scoping three cases of robot's tangential acceleration: uniform, non-uniform, and constant speed. Leg's Cartesian velocities are described by their first order Jacobian, which result in redundant kinematics systems. As particular cases of study, two different biological kinematic configurations were analysed in order to be adapted (DOFs reductions) as potential kinematic functions of the navigation control law. Although, the research interest is centralised on walking systems, the *Praying-Mantis* raptorial legs, as well as the *Smithi ant*'s legs are analysed. Because of the kinematic redundancy, by using pseudo-inverse numerical methods, the solution near a singularity region is unstable about these values. It was obtained the first-order derivative pseudo-inverse Jacobian matrix using two different numerical methods for multi-joint legs: the right pseudo-inverse, and by singularity properties using the singular value decomposition approach. Furthermore, *Euler-Lagrange* motion equations are defined.

Hyper-static balanced multi-legged walking robots are mechatronic vehicles capable to walk on multi-joint legs (see figure 12.1). Multi-legged robots with three or more extremities are statically stable when walking, . However, depending on its gait configuration in use, legs must correctly be synchronized while developing free-walking over all-terrain. If some legs become disabled, the robot may still be able to walk, since not all legs might be needed to accomplish stability. Giving other legs the ability to reach new ground placements. Looking into the biology literature , , , one can find an amazingly rich variety of insect's combination of joint legs. Arachnids, crickets, ants and so forth, which are invertebrate animals with eight or more degrees-of-freedom (DOF) in each leg. Much attention has been paid to develop algorithms for gaits control strategies, research on insects biology discloses interesting information that may enrich kinematic and dynamic schemes for gaits control for insect-like artificial walking machines; . Figure 12.1 depicts kinematic combinations of walking robot's legs.

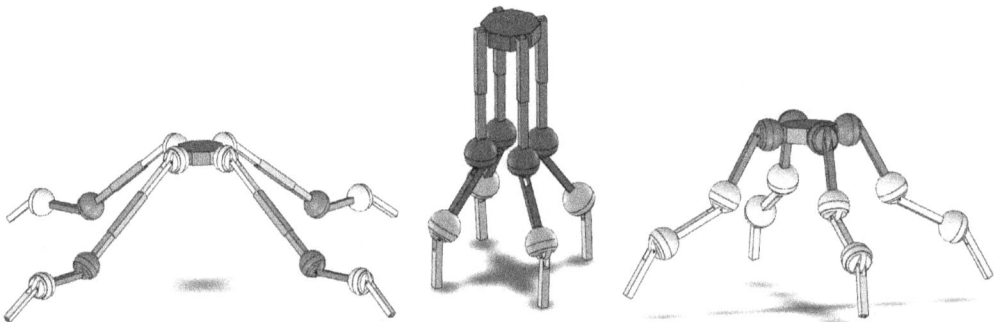

Figure 12.1:   Quadruped machines with different types of legs. (a) type RRPRRRR ; (b) type PRRRRR; (c) type RRRRRR.

This chapter is purposed to provide a generalised velocity-based navigation model of redundantly kinematic control law of n-leg and k-joint. Other similar approaches did not consider redundant Jacobians , . The state vector is defined as a function of robot's tangential acceleration, and leg's Cartesian velocities are described by their first order Jacobian, which results in redundant kinematics systems . A particular interest of this study is to include into the general navigation control law Jacobian matrices of different kinematic configurations. The direct/inverse kinematic analysis for each leg is considered in order to develop an overall position control model for a robot, as to have an integral functional form of variables about the control law .

12.1 Kinematics control law

A combination of synchronized motions of mechanisms, comprised of open-loop serial chains are configured and coordinated to walk yielding controlled trajectories. The fixed-frame Cartesian reference of any walking machine is ideally its centre of mass. Kinematic-based control is critically important to describe the geometry of motion of a body , and its kinematics gives a description of the leg's configuration spaces. Equation (12.1) is a linearised state model to control the Cartesian speed,

$$\dot{\boldsymbol{\xi}} = \mathbf{A}(t) \cdot \boldsymbol{\xi}(t) + \mathbf{B}(t) \cdot \mathbf{u}(t) \tag{12.1}$$

Let us define $\boldsymbol{\xi}(t) = (x, y, \theta)^T$ as the state vector (Cartesian position and angle orientation) . The first derivative state vector w.r.t. time is defined by $\dot{\boldsymbol{\xi}}$. The matrices $\mathbf{A}$ and $\mathbf{B}$ are two transition matrices that connect two different system states, time and Cartesian components, respectively. The input vector $\mathbf{u} = (v, \omega)^T$ is compounded by the linear and angular velocities. Furthermore, the robot's global motion direction is given by $\theta(t)$. The equation (12.2) is the robot's angular velocity, having Cartesian components as first and second orders derivative functions of time.

$$\omega(t) = \left(1 + \left(\frac{\dot{y}}{\dot{x}}\right)^2\right)^{-1} (\ddot{y}\dot{x} - \dot{x}\ddot{y})\left(\dot{x}^2\right)^{-1} \tag{12.2}$$

Therefore, by having a functional form of $\omega(t)$, the linearised first order derivative control law is described by expression (12.3),

$$\dot{\boldsymbol{\xi}} = \begin{pmatrix} \dot{x} \\ \dot{y} \\ \dot{\theta} \end{pmatrix} = \begin{pmatrix} \frac{1}{t} & 0 & 0 \\ 0 & \frac{1}{t} & 0 \\ 0 & 0 & \frac{1}{t} \end{pmatrix} \begin{pmatrix} x \\ y \\ \theta \end{pmatrix} + \begin{pmatrix} \cos\theta & 0 \\ \sin\theta & 0 \\ 0 & 1 \end{pmatrix} \begin{pmatrix} v \\ \omega \end{pmatrix} \tag{12.3}$$

The model of instantaneous robot's linear velocity, for any multi-legged walking machine is expressed by the norm of its legs' Cartesian component speeds, given by equation (12.4),

Corollary 12.1.1. *The robot's instantaneous velocity is an averaged value of the n-leg Cartesian speeds.*

$$v(t) = \left\| \frac{1}{n} \sum_{i=1}^{n} \mathbf{J}_i \cdot \dot{\mathbf{q}}_i \right\| \tag{12.4}$$

Where legs' first derivative Jacobian $\mathbf{J}_i$ is involved for each ith leg. The joint angular velocities vector is defined by $\mathbf{q}_i = (\dot{\phi}_0, \dot{\phi}_0, \ldots, \dot{\phi}_k)^\top$ of k rotational joints. Likewise, each leg's Cartesian velocity vector is defined by,

$$\dot{\mathbf{p}}_i = \mathbf{J}_i \cdot \dot{\mathbf{q}}_i \tag{12.5}$$

where Cartesian velocity vector $\dot{\mathbf{p}}_i = (\dot{x}, \dot{y}, \dot{z})^\top$. With no lost of generality, the state vector $\xi(t)$ is constrained by three controlled equilibrium conditions given by the function $\mathbf{g}$,

$$\xi(t) = \mathbf{g}\,(\dot{\mathbf{x}}, \ddot{\mathbf{x}}, t) \tag{12.6}$$

Hence,

Definition 12.1.2 (Equilibrium conditions). The functional form of $\mathbf{g}(.)$ poses three equilibrium conditions:

$$\mathbf{g}\,(\dot{\mathbf{x}}, \ddot{\mathbf{x}}, t) = \begin{cases} \mathbf{x}_0 + \dot{\mathbf{x}}t - (\ddot{\mathbf{x}})\frac{t^2}{2}, & a = const \\ \mathbf{x}_0 + \frac{t}{2}(\dot{\mathbf{x}}_2 - \dot{\mathbf{x}}_1), & a = f(t) \\ \left(\dot{\mathbf{x}} + \gamma(\dot{\mathbf{x}}^{ref} - \dot{\mathbf{x}})\right)t, & v = const, a = 0 \end{cases} \tag{12.7}$$

Where $\mathbf{x}_0 = (x_0, y_0, \theta_0)^\top$ is the robot's initial position vector.

Thus, from (12.7) in previous theorem, its three constraints are described as,

1. The condition for constant acceleration means the multi-legged machine navigates at open obstacle-free terrains to keep increased its velocity usually to reach large distances.

2. the condition for varying $a(t)$ to slow down, or speed up when dealing with obstacles (this condition allows path planning formulation).

3. The third condition allows keeping a controlled velocity v, with no speed changes under slopes where gravitational effects take place.

Second and third conditions are nearly linear in the presence of inherent speed perturbations. Therefore,

Theorem 12.1.3 (Kinematic control law). *The linearised kinematic control model is given by* (12.8)

$$\dot{\xi} = \begin{pmatrix} \frac{1}{t}00 \\ 0\frac{1}{t}0 \\ 00\frac{1}{t} \end{pmatrix} \cdot \mathbf{g}\,(\dot{\mathbf{x}}, \ddot{\mathbf{x}}, t) + \begin{pmatrix} \cos\theta 0 \\ \sin\theta 0 \\ 0\ \ 1 \end{pmatrix} \cdot \begin{pmatrix} \left\| \frac{1}{n}\sum_{i=1}^{n} \mathbf{J}_i \cdot \dot{\mathbf{q}}_i \right\| \\ \left(1 + \left(\dot{y}^2\dot{x}^{-2}\right)\right)^{-1} \cdot (\ddot{y}\dot{x} - \dot{x}\ddot{y}) \cdot \left(\dot{x}^{-2}\right) \end{pmatrix} \qquad (12.8)$$

12.2 Kinematic analysis

An extremity is said to be redundant when its number of DOFs is greater than the dimension of its task space. For a 3D position task, a leg with more than six joints would be redundant . A definition of what is meant by the term redundant requires that it specifies the number of degrees-of-freedom required to perform a task. Figure 12.2 (b), (d) depict two biological extremities (Praying-mantis and ant) . Figure 12.2 (a), (c) are generic drawings of their reduced DOFs, tarsus in both legs are not considered because of their passive DOFs given for supportive stability, rather than significant rotatory movements. The figure 12.2 depicts six rotative motion variables to represent leg's contact point (farthest Cartesian position from base joint). According to figure 12.2, legs' parameter ℓ_i and variable d_i are links length and prismatic displacement, respectively. Besides, ϕ_i are the joint's rotation angles.

Hereafter, formulation given along this paper are just simplified to shorter mathematical expressions, adopting next equivalences for trigonometric functions due to limits of paper space. Let us assume that for instance $\sin(\phi_0)$ is equivalent to s_0, and $\cos(\phi_1 + \phi_2)$ is equivalent to c_{12}.

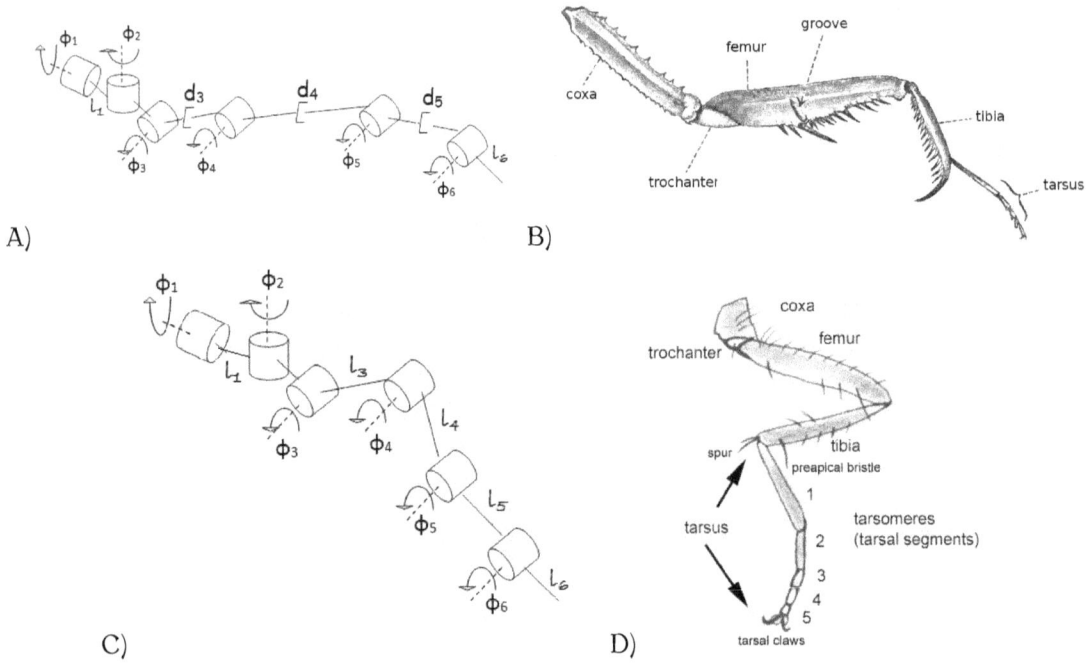

Figure 12.2: Biological limbs and equivalent reduced DOFs. (a) Praying mantis-like leg reduced to six joints and three links; (b) biological Mantis Leg; (c) Ant-like leg with six rotative joints; (d) biological ant Leg.

12.2.1   Mantis-like leg kinematics

Thus, the forward kinematics calculate the contact point Cartesian instantaneous position for the Mantis-like leg by $\mathbf{p}(t) = (x_m, y_m, z_m)^T$ in task space. So that, through direct analysis, its forward kinematics is given by the three equations of components position x_m, y_m and z_m, then,

$$x_m = l_1 s_1 + d_3 c_{13} s_2 + d_4 c_{134} s_2 + d_5 c_{1345} s_2 + l_6 c_{13456} s_2 \tag{12.9}$$

$$y_m = l_1 s_1 + d_3 s_{13} + d_4 s_{134} + d_5 s_{1345} + l_6 s_{13456} \tag{12.10}$$

$$z_m = l_1 c_1 + d_3 c_{13} c_2 + d_4 c_{134} c_2 + d_5 c_{1345} c_2 + l_6 c_{13456} c_2 \tag{12.11}$$

Homogenising trigonometric functions in such expressions by substituting identities, factorizing and algebraically rearranging, equations become further simplified,

$$x_m = l_1 s_1 + \frac{1}{2} d_3 [s_{123} - s_{1-23}] + \frac{1}{2} d_4 [s_{1234} - s_{1-234}] + \frac{1}{2} d_5 [s_{12345} - s_{1-2345}] + l_6 c_{13456} s_2 \tag{12.12}$$

$$y_m = l_1 s_1 + d_3 s_{13} + d_4 s_{134} + d_5 s_{1345} + l_6 s_{13456} \tag{12.13}$$

$$z_m = l_1 c_1 + \frac{1}{2} d_3 [c_{1-23} + c_{123}] + \frac{1}{2} d_4 [c_{1-234} + c_{1234}] + \frac{1}{2} d_5 [c_{1-2345} + c_{12345}] + l_6 c_{13456} c_2 \tag{12.14}$$

Hence, representing in vector notation form, previous expressions are now given by,

$$\mathbf{p}_m(t) = \begin{pmatrix} l_1 s_1 + \frac{1}{2} d_3 [s_{123} - s_{1-23}] + \frac{1}{2} d_4 [s_{1234} - s_{1-234}] \\ \quad + \frac{1}{2} d_5 [s_{12345} - s_{1-2345}] + l_6 c_{13456} s_2 \\ \\ l_1 s_1 + d_3 s_{13} + d_4 s_{134} + d_5 s_{1345} + l_6 s_{13456} \\ \\ l_1 c_1 + \frac{1}{2} d_3 [c_{1-23} + c_{123}] + \frac{1}{2} d_4 [c_{1-234} + c_{1234}] \\ \quad + \frac{1}{2} d_5 [c_{1-2345} + c_{12345}] + l_6 c_{13456} c_2 \end{pmatrix} \tag{12.15}$$

Without loss of generality, the first-order derivative vector $\dot{\mathbf{p}}$ of tangential velocities components are subsequently described. Firstly, the $\dot{x}_m$ component given by,

$$\dot{x}_m = l_1 c_1 \left(\dot{\phi}_1 \right) + \frac{1}{2} d_3 c_{123} \left(\dot{\phi}_1 + \dot{\phi}_2 + \dot{\phi}_3 \right) + \frac{1}{2} \dot{d}_3 s_{123} - \frac{1}{2} d_3 c_{1-23} \left(\dot{\phi}_1 - \dot{\phi}_2 + \dot{\phi}_3 \right) -$$
$$\frac{1}{2} \dot{d}_3 s_{1-23} + \frac{1}{2} d_4 c_{1234} \left(\dot{\phi}_1 + \dot{\phi}_2 + \dot{\phi}_3 + \dot{\phi}_4 \right) + \frac{1}{2} \dot{d}_4 s_{1234} - \frac{1}{2} d_4 c_{1-234} \left(\dot{\phi}_1 - \dot{\phi}_2 + \dot{\phi}_3 + \dot{\phi}_4 \right) -$$
$$\frac{1}{2} \dot{d}_4 s_{1-234} + \frac{1}{2} d_5 c_{12345} \left(\dot{\phi}_1 + \dot{\phi}_2 + \dot{\phi}_3 + \dot{\phi}_4 + \dot{\phi}_5 \right) \tag{12.16}$$
$$+ \frac{1}{2} \dot{d}_5 s_{12345} - \frac{1}{2} d_5 c_{1-2345} \left(\dot{\phi}_1 - \dot{\phi}_2 + \dot{\phi}_3 + \dot{\phi}_4 + \dot{\phi}_5 \right) -$$
$$\frac{1}{2} \dot{d}_5 s_{1-2345} + l_6 c_{13456} \left(\dot{\phi}_1 + \dot{\phi}_3 + \dot{\phi}_4 + \dot{\phi}_5 + \dot{\phi}_6 \right) c_2 - l_6 s_{13456} s_2 \left(\dot{\phi}_2 \right)$$

Secondly, the $\dot{y}_m$ component

$$\dot{y}_m = l_1 c_1 \left(\dot{\phi}_1 \right) + d_3 c_{13} \left(\dot{\phi}_1 + \dot{\phi}_3 \right) + \dot{d}_3 s_{13} - d_4 c_{134} \left(\dot{\phi}_1 + \dot{\phi}_3 + \dot{\phi}_4 \right) + \dot{d}_4 s_{134} +$$
$$d_5 c_{1345} \left(\dot{\phi}_1 + \dot{\phi}_3 + \dot{\phi}_4 + \dot{\phi}_5 \right) + \dot{d}_5 s_{1345} + l_6 c_{13456} \left(\dot{\phi}_1 + \dot{\phi}_3 + \dot{\phi}_4 + \dot{\phi}_5 + \dot{\phi}_6 \right) \tag{12.17}$$

and then, the $\dot{z}_m$ component,

$$\dot{z}_m = -l_1 s_1\left(\dot{\phi}_1\right) - \frac{1}{2}d_3 s_{1-23}\left(\dot{\phi}_1 - \dot{\phi}_2 + \dot{\phi}_3\right) + \frac{1}{2}d_3 c_{1-23} - \frac{1}{2}d_3 s_{123}\left(\dot{\phi}_1 + \dot{\phi}_2 + \dot{\phi}_3\right) + \frac{1}{2}d_3 c_{123} -$$

$$\frac{1}{2}d_4 s_{1-234}\left(\dot{\phi}_1 - \dot{\phi}_2 + \dot{\phi}_3 + \dot{\phi}_4\right) + \frac{1}{2}d_4 c_{1-234} - \frac{1}{2}d_4 s_{1234}\left(\dot{\phi}_1 + \dot{\phi}_2 + \dot{\phi}_3 + \dot{\phi}_4\right) + \frac{1}{2}d_4 c_{1234} -$$

$$\frac{1}{2}d_5 s_{1-2345}\left(\dot{\phi}_1 - \dot{\phi}_2 + \dot{\phi}_3 + \dot{\phi}_4 + \dot{\phi}_5\right) + \frac{1}{2}d_5 c_{1-2345} - \frac{1}{2}d_5 s_{12345}\left(\dot{\phi}_1 + \dot{\phi}_2 + \dot{\phi}_3 + \dot{\phi}_4 + \dot{\phi}_5\right) +$$

$$\frac{1}{2}d_5 c_{12345} - l_6 s_{13456}\left(\dot{\phi}_1 + \dot{\phi}_3 + \dot{\phi}_4 + \dot{\phi}_5 + \dot{\phi}_6\right)c_2 - l_6 c_{13456} s_2\left(\dot{\phi}_2\right)$$

$$(12.18)$$

Then, factorizing common terms, the joints angular velocity vector is $\dot{\boldsymbol{\phi}} = (\dot{\phi}_0, \dot{\phi}_1, \ldots \dot{\phi}_6)^\top$, and by simplifying the forward kinematics model for 3D, the equation (12.19) resulted,

$$\dot{\mathbf{p}} = \mathbf{J}\cdot\dot{\mathbf{q}} + \dot{d}_3\begin{pmatrix}\frac{1}{2}s_{123} - \frac{1}{2}c_{1-23} \\ s_{13} \\ \frac{1}{2}c_{1-23} + \frac{1}{2}c_{123}\end{pmatrix} + \dot{d}_4\begin{pmatrix}\frac{1}{2}s_{1234} - \frac{1}{2}s_{1-234} \\ s_{134} \\ \frac{1}{2}c_{1-234} + \frac{1}{2}c_{1234}\end{pmatrix} + \dot{d}_5\begin{pmatrix}\frac{1}{2}s_{12345} - \frac{1}{2}s_{1-2345} \\ s_{1345} \\ \frac{1}{2}c_{1-2345} + \frac{1}{2}c_{12345}\end{pmatrix} \quad (12.19)$$

which is the equation of direct kinematics presented as a first order derivative, which are the Cartesian velocities of the leg's contact point.

12.2.2 Ant-like leg kinematics

Figure 12.2-right shows a generic drawing of a leg's Cartesian framework and its DOFs, inspired by the Ant Smithi . It poses nine real DOF, but it was adopted only six DOF because of the rest of them are tarsus specifically used to keep adapted to ground texture as if they were passive joints. The first three DOF are embedded in the same joint, so that, the equations of direct kinematics for the three Cartesian components are as follow,

$$x_a = l_1 s_1 + l_3 c_{13} s_2 + l_4 c_{134} s_2 + l_5 c_{1345} s_2 + l_6 c_{13456} s_2 \quad (12.20)$$

and

$$y_a = l_1 s_1 + l_3 s_{13} + l_4 s_{134} + l_5 s_{1345} + l_6 s_{13456} \quad (12.21)$$

as well as

$$z_a = l_1 c_1 + l_3 c_{13} c_2 + l_4 c_{134} c_2 + l_5 c_{1345} c_2 + l_6 c_{13456} c_2 \tag{12.22}$$

Therefore, in vector form the leg position is given by equation (12.23),

$$\mathbf{p}_a(t) = \begin{pmatrix} l_1 s_1 + l_3 c_{13} s_2 + l_4 c_{134} s_2 + l_5 c_{1345} s_2 + l_6 c_{13456} s_2 \\ l_1 s_1 + l_3 s_{13} + l_4 s_{134} + l_5 s_{1345} + l_6 s_{13456} \\ l_1 c_1 + l_3 c_{13} c_2 + l_4 c_{134} c_2 + l_5 c_{1345} c_2 + l_6 c_{13456} c_2 \end{pmatrix} \tag{12.23}$$

In addition, by deriving w.r.t. time, our three first-order derivative Cartesian components,

$$\dot{x}_a = l_1 c_1 \left(\dot{\phi}_1\right) - l_3 s_{13} \left(\dot{\phi}_1 + \dot{\phi}_3\right) s_2 + l_3 c_{13} c_2 \left(\dot{\phi}_2\right) - l_4 s_{134} \left(\dot{\phi}_1 + \dot{\phi}_3 + \dot{\phi}_4\right) s_2 + l_4 c_{134} c_2 \left(\dot{\phi}_2\right) s_0 -$$
$$l_5 s_{1345} \left(\dot{\phi}_1 + \dot{\phi}_3 + \dot{\phi}_4 + \dot{\phi}_5\right) c_2 + l_5 c_{1345} s_2 \left(\dot{\phi}_2\right) - l_6 s_{13456} \left(\dot{\phi}_1 + \dot{\phi}_3 + \dot{\phi}_4 + \dot{\phi}_5 + \dot{\phi}_6\right) s_2 + l_6 c_{13456} c_2 \left(\dot{\phi}_2\right) \tag{12.24}$$

$$\dot{y}_a = l_1 c_1 \left(\dot{\phi}_1\right) + l_3 c_{13} \left(\dot{\phi}_1 + \dot{\phi}_3\right) + l_4 c_{134} \left(\dot{\phi}_1 + \dot{\phi}_3 + \dot{\phi}_4\right) + l_5 c_{1345} \left(\dot{\phi}_1 + \dot{\phi}_3 + \dot{\phi}_4 + \dot{\phi}_5\right) +$$
$$l_6 c_{13456} \left(\dot{\phi}_1 + \dot{\phi}_3 + \dot{\phi}_4 + \dot{\phi}_5 + \dot{\phi}_6\right) \tag{12.25}$$

$$\dot{z}_a = -l_1 s_1 \left(\dot{\phi}_1\right) - l_3 s_{13} \left(\dot{\phi}_1 + \dot{\phi}_3\right) c_2 - l_3 c_{13} s_2 \left(\dot{\phi}_2\right) - l_4 s_{134} \left(\dot{\phi}_1 + \dot{\phi}_3 + \dot{\phi}_4\right) c_2 - l_4 c_{134} s_2 \left(\dot{\phi}_2\right) -$$
$$l_5 c_{1345} s_2 \left(\dot{\phi}_2\right) - l_5 s_{1345} \left(\dot{\phi}_1 + \dot{\phi}_3 + \dot{\phi}_4 + \dot{\phi}_5\right) c_2 - l_6 s_{13456} \left(\dot{\phi}_1 + \dot{\phi}_3 + \dot{\phi}_4 + \dot{\phi}_5 + \dot{\phi}_6\right) c_2 - l_6 c_{13456} s_2 \left(\dot{\phi}_2\right) \tag{12.26}$$

Therefore, the forward kinematics model is formulated. For the ant-like leg, only rotative joints were mathematically described, and no prismatic variables are involved.

12.3 Jacobian matrix analysis

Although the general problem of inverse kinematics is not straightforward, it turns out that for extremities having six joints, with their last three DOF intersecting at a point, it is possible to decouple the inverse kinematics problem into two simpler problems . It is known respectively, as inverse position kinematics, and inverse orientation kinematics. Thus, from the required algebraic solution for joint velocities vector,

$$\dot{\mathbf{q}} = \mathbf{J}^{-1} \cdot \dot{\mathbf{p}} \tag{12.27}$$

The Jacobian matrix for the Mantis-like extremity in terms of first-order derivative is disclosed next,

$$
\begin{pmatrix} \dot{x} \\ \\ \dot{y} \\ \\ \dot{z} \end{pmatrix} =
\begin{pmatrix} a_1 a_2 a_3 a_4 a_5 a_6 \\ b_1 b_2 b_3 b_4 b_5 b_6 \\ c_1 c_2 c_3 c_4 c_5 c_6 \end{pmatrix} \cdot
\begin{pmatrix} \dot{\phi}_1 \\ \dot{\phi}_2 \\ \dot{\phi}_3 \\ \dot{\phi}_4 \\ \dot{\phi}_5 \\ \dot{\phi}_6 \end{pmatrix}
+ \dot{d}_3 \begin{pmatrix} \frac{1}{2}s_{123} - \frac{1}{2}c_{1-23} \\ s_{13} \\ \frac{1}{2}c_{1-23} + \frac{1}{2}c_{123} \end{pmatrix} +
$$

$$
d_4 \begin{pmatrix} \frac{1}{2}s_{1234} - \frac{1}{2}s_{1-234} \\ s_{134} \\ \frac{1}{2}c_{1-234} + \frac{1}{2}c_{1234} \end{pmatrix}
+ \dot{d}_5 \begin{pmatrix} \frac{1}{2}s_{12345} - \frac{1}{2}s_{1-2345} \\ s_{1345} \\ \frac{1}{2}c_{1-2345} + \frac{1}{2}c_{12345} \end{pmatrix} \tag{12.28}
$$

The Jacobian matrix terms are defined next in their functional form by,

$a_1 = l_1 c_1 + \frac{1}{2}d_3 c_{123} - \frac{1}{2}d_3 c_{1-23} + \frac{1}{2}d_4 c_{1234} - \frac{1}{2}d_4 c_{1-234} + \frac{1}{2}d_5 c_{12345} - \frac{1}{2}d_5 c_{1-2345} - l_6 s_{13456} s_2$

$a_2 = +\frac{1}{2}d_3 c_{123} - \frac{1}{2}d_3 c_{1-23} + \frac{1}{2}d_4 c_{1234} - \frac{1}{2}d_4 c_{1-234} + \frac{1}{2}d_5 c_{12345} - \frac{1}{2}d_5 c_{1-2345} + l_6 c_{13456} c_2$

$a_3 = +\frac{1}{2}d_3 c_{123} - \frac{1}{2}d_3 c_{1-23} + \frac{1}{2}d_4 c_{1234} - \frac{1}{2}d_4 c_{1-234} + \frac{1}{2}d_5 c_{12345} - \frac{1}{2}d_5 c_{1-2345} - l_6 s_{13456} s_2$

$a_4 = +\frac{1}{2}d_4 c_{1234} - \frac{1}{2}d_4 c_{1-234} + \frac{1}{2}d_5 c_{12345} - \frac{1}{2}d_5 c_{1-2345} - l_6 s_{13456} s_2$

$a_5 = +\frac{1}{2}d_5 c_{12345} - \frac{1}{2}d_5 c_{1-2345} - l_6 s_{13456} s_2$

$a_6 = -l_6 s_{13456} s_2$

$b_1 = l_1 + d_3 c_{13} + d_4 c_{134} + d_5 c_{1345} + l_6 c_{13456}$

$b_2 = 0$

$b_3 = +d_3 c_{13} + d_4 c_{134} + d_5 c_{1345} + l_6 c_{13456}$

$b_4 = +d_4 c_{134} + d_5 c_{1345} + l_6 c_{13456}$

$b_5 = +d_5 c_{1345} + l_6 c_{13456}$

$b_6 = +l_6 c_{13456}$

$c_1 = -l_1 s_1 - \frac{1}{2}d_3 s_{1-23} - \frac{1}{2}d_3 s_{123} - \frac{1}{2}d_4 s_{1-234} - \frac{1}{2}d_4 s_{1234} - \frac{1}{2}d_5 s_{1-2345} - \frac{1}{2}d_5 s_{12345} - l_6 s_{13456} c_2$

$c_2 = -\frac{1}{2}d_3 s_{1-23} - \frac{1}{2}d_3 s_{123} - \frac{1}{2}d_4 s_{1-234} - \frac{1}{2}d_4 s_{1234} - \frac{1}{2}d_5 s_{1-2345} - \frac{1}{2}d_5 s_{12345} - l_6 c_{13456} s_2$

$c_3 = -\frac{1}{2}d_3 s_{1-23} - \frac{1}{2}d_3 s_{123} - \frac{1}{2}d_4 s_{1-234} - \frac{1}{2}d_4 s_{1234} - \frac{1}{2}d_5 s_{1-2345} - \frac{1}{2}d_5 s_{12345} - l_6 s_{13456} c_2$

$c_4 = -\frac{1}{2}d_4 s_{1-234} - \frac{1}{2}d_4 s_{1234} - \frac{1}{2}d_5 s_{1-2345} - \frac{1}{2}d_5 s_{12345} - l_6 s_{13456} c_2$

$c_5 = -\frac{1}{2}d_5 s_{1-2345} - \frac{1}{2}d_5 s_{12345} - l_6 s_{13456} c_2$

$c_6 = -l_6 s_{13456} c_2$

In addition, for the case of the ant-like Jacobian expression of first-order derivative, equation (12.29) is the analytical solution,

$$\begin{pmatrix} \dot{x} \\ \dot{y} \\ \dot{z} \end{pmatrix} = \begin{pmatrix} a_1 a_2 a_3 a_4 a_5 a_6 \\ b_1 b_2 b_3 b_4 b_5 b_6 \\ c_1 c_2 c_3 c_4 c_5 c_6 \end{pmatrix} \cdot \begin{pmatrix} \dot{\phi}_1 \\ \dot{\phi}_2 \\ \dot{\phi}_3 \\ \dot{\phi}_4 \\ \dot{\phi}_5 \\ \dot{\phi}_6 \end{pmatrix} \qquad (12.29)$$

Similarly, with its Jacobian matrix terms given in their functional representations,

$a_1 = l_1 c_1 - l_3 s_{13} s_2 - l_4 s_{134} s_2 - l_5 s_{1345} s_2 - l_6 s_{13456} s_2$

$a_2 = +l_3 c_{13} c_2 - l_4 c_{134} c_2 - l_5 c_{1345} c_2 - l_6 c_{13456} c_2$

$a_3 = -l_3 s_{13} s_2 - l_4 s_{134} s_2 - l_5 s_{1345} s_2 - l_6 s_{13456} s_2$

$a_4 = -l_4 s_{134} s_2 - l_5 s_{1345} s_2 - l_6 s_{13456} s_2$

$a_5 = -l_5 s_{1345} s_2 - l_6 s_{13456} s_2$

$a_6 = -l_6 s_{13456} s_2$

$b_1 = l_1 c_1 + l_3 c_{13} + l_4 c_{134} + l_5 c_{1345} + l_6 c_{13456}$

$b_2 = 0$

$b_3 = l_3 c_{13} + l_4 c_{134} + l_5 c_{1345} + l_6 c_{13456}$

$b_4 = +l_4 c_{134} + l_5 c_{1345} + l_6 c_{13456}$

$b_5 = +l_5 c_{1345} + l_6 c_{13456}$

$b_6 = +l_6 c_{13456}$

$c_1 = -l_1 s_1 + l_3 s_{13} c_2 + l_4 s_{134} c_2 + l_5 s_{1345} c_2 + l_6 s_{13456} c_2$

$c_2 = -l_3 c_{13} s_2 - l_4 c_{134} s_2 - l_5 c_{1345} s_2 - l_6 c_{13456} s_2$

$c_3 = +l_3 s_{13} c_2 + l_4 s_{134} c_2 + l_5 s_{1345} c_2 + l_6 s_{13456} c_2$

$c_4 = +l_4 s_{134} c_2 + l_5 s_{1345} c_2 + l_6 s_{13456} c_2$

$c_5 = +l_5 s_{1345} c_2 + l_6 s_{13456} c_2$

$c_6 = +l_6 s_{13456} c_2$

Nevertheless, when Jacobian is not square, as the case of redundant multi-joint extremities, the method is numerically solved since for a non-squared matrix there is no determinant and therefore cannot be directly inverted.

12.3.1   Right pseudoinverse

The inverse kinematic problem is straightforward solved when Jacobian is square with non zero determinant. Nevertheless, when Jacobian is not square, as is the case for redundant multi-joint mechanism, the method is numerically solved since for a non-squared matrix there is no determinant and therefore cannot be inverted. Thus, to deal with the case $m < n$, we use the following resilt from linear algebra.

Proposition 12.3.1. *For* $\mathbf{J} \in \Re^{m \times n}$*, and rank* $\mathbf{J} = m$*, then* $(\mathbf{JJ})^{-1}$ *exists.*

In this case $(\mathbf{JJ}^T) \in \Re^{m \times m}$ and has rank m. Using this result, we can regroup terms to obtain,

$$(\mathbf{JJ}^T)(\mathbf{JJ}^T)^{-1} = \mathbf{I}$$

it follows that,

$$\mathbf{J}\left[\mathbf{J}^T(\mathbf{JJ}^T)^{-1}\right] = \mathbf{I}$$

and,

$$\mathbf{JJ}^+ = \mathbf{I}$$

Here, $\mathbf{J}^+ = \mathbf{J}^T(\mathbf{JJ}^T)^{-1}$ is called a right pseudoinverse of $\mathbf{J}$, since $\mathbf{JJ}^+ = \mathbf{I}$. Note that, $\mathbf{J}^+\mathbf{J} \in \Re^{n \times n}$, and that in general, $\mathbf{J}^+\mathbf{J} \neq \mathbf{i}$ (recall that matrix multiplication is not commutative). Therefore, the inverse leg's kinematics model is given for either Mantis-like leg or ant-like leg.

Definition 12.3.2. The ant-like leg inverse kinematic model $\dot{\mathbf{q}}$ is formalised by:

$$\dot{\mathbf{q}}(t) = \mathbf{J}^T(\mathbf{J} \cdot \mathbf{J}^T)^{-1} \cdot \dot{\mathbf{p}}(t) \tag{12.30}$$

$$
\begin{pmatrix} \dot{\phi}_1 \\ \dot{\phi}_2 \\ \dot{\phi}_3 \\ \dot{\phi}_4 \\ \dot{\phi}_5 \\ \dot{\phi}_6 \end{pmatrix} = \begin{pmatrix} a_1 b_1 c_1 \\ a_2 b_2 c_2 \\ a_3 b_3 c_3 \\ a_4 b_4 c_4 \\ a_5 b_5 c_5 \\ a_6 b_6 c_6 \end{pmatrix} \cdot \left(\begin{pmatrix} a_1 a_2 a_3 a_4 a_5 a_6 \\ b_1 b_2 b_3 b_4 b_5 b_6 \\ c_1 c_2 c_3 c_4 c_5 c_6 \end{pmatrix} \cdot \begin{pmatrix} a_1 b_1 c_1 \\ a_2 b_2 c_2 \\ a_3 b_3 c_3 \\ a_4 b_4 c_4 \\ a_5 b_5 c_5 \\ a_6 b_6 c_6 \end{pmatrix} \right)^{-1} \cdot \begin{pmatrix} \dot{x} \\ \dot{y} \\ \dot{z} \end{pmatrix} \tag{12.31}
$$

12.3.2 Singular value decomposition

The inverse solution is complemented alternatively with a second numeric method to find the inverse kinematics by decomposing the Jacobian singular values, by the eigenvalues $\lambda_1 \geq \lambda_2 \cdots \geq \lambda_m \geq 0$ of square matrix JJ^T.

The singular values for the Jacobian matrix J are given by the square roots of the eigenvalues of JJ^T through $\sigma_i = \sqrt[2]{\lambda_i}$, where $U = [u_1, u_2, \dots, u_m]$, and $V = [v_1, v_2, \dots, v_n]$ that are orthogonal matrices, and $\Sigma \in \Re^{m \times n}$.

The diagonal matrix $\Sigma_m = \text{diag}(\sigma_0, \sigma_1, \dots, \sigma_k)$ is squared and symmetric.

$$
\Sigma_m = \begin{bmatrix} \sigma_1 & & & \\ & \sigma_2 & & \\ & & \ddots & \\ & & & \sigma_m \end{bmatrix} \tag{12.32}
$$

It is found the singular values, σ_i of J that can be used to find the eigenvectors $u, \dots, u_m$ that satisfy $JJ^T = u_i = \sigma_i u_i$. Such eigenvectors comprise the matrix $U = [u_1, u_2, \cdots, u_m]$. The system is then rewritten as,

$$
JJ^T U = U \Sigma_m^2 \tag{12.33}
$$

Thus, by defining the matrix,

$$
\Sigma_m = \begin{bmatrix} \sigma_1 & & & \\ & \sigma_2 & & \\ & & \ddots & \\ & & & \sigma_1 \end{bmatrix} \tag{12.34}
$$

Hence, it is defined $V_m = J^T U \Sigma_m^{-1}$. and let V be any orthogonal matrix that satisfies $V = [V_m | V_{n-m}]$. Notice that V is an $n \times n$ matrix. Then, constructing the right pseudo-inverse of J using singular value decomposition, the Jacobian pseudo-inverse $J^+ = V \Sigma^{-1} U^T$. Therefore, through SVD the ant-like leg inverse kinematics is given by (12.36), in which Σ_m^+ is the inverse (square) matrix of Σ_m.

$$\Sigma_m^+ = \begin{bmatrix} \sigma_1^{-1} & & & \\ & \sigma_2^{-1} & & \\ & & \ddots & \\ & & & \sigma_1^{-1} \end{bmatrix} \tag{12.35}$$

For the same kinematic example as previous subsection, now the inverse matrix solution is given by,

$$\begin{pmatrix} \dot{\phi}_1 \\ \dot{\phi}_2 \\ \dot{\phi}_3 \\ \dot{\phi}_4 \\ \dot{\phi}_5 \\ \dot{\phi}_6 \end{pmatrix} = \left(\begin{pmatrix} a_1 b_1 c_1 \\ a_2 b_2 c_2 \\ a_3 b_3 c_3 \\ a_4 b_4 c_4 \\ a_5 b_5 c_5 \\ a_6 b_6 c_6 \end{pmatrix} \cdot U \cdot \Sigma_m^{-1} \right) \cdot \Sigma_m^{-1} \cdot$$

$$\left(\left(\begin{pmatrix} a_1 a_2 a_3 a_4 a_5 a_6 \\ b_1 b_2 b_3 b_4 b_5 b_6 \\ c_1 c_2 c_3 c_4 c_5 c_6 \end{pmatrix} \cdot \begin{pmatrix} a_1 b_1 c_1 \\ a_2 b_2 c_2 \\ a_3 b_3 c_3 \\ a_4 b_4 c_4 \\ a_5 b_5 c_5 \\ a_6 b_6 c_6 \end{pmatrix} \right)^{-1} U \cdot \Sigma_m^2 \right)^{T} \cdot \begin{pmatrix} \dot{x} \\ \dot{y} \\ \dot{z} \end{pmatrix} \tag{12.36}$$

The forward kinematics $p(t)$ and its first derivative $\dot{p}(t)$ are analytic solutions. However, their inverse solutions $\Phi(t)$ and first derivative $\dot{\Phi}(t)$ are numeric ones. Either types of equation can be used within the general navigation control law $\dot{\xi}(t)$, which will only depend on its navigational algorithmic convenience.

12.4 Dynamics analysis

By means of energies analysis through Euler-Lagrange motion equations, walking tasks are solely considered over the plane,()two dimensions). The next kinematics equations only

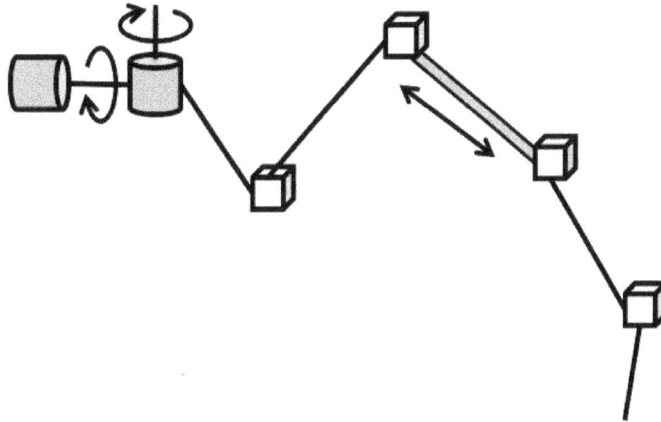

Figure 12.3: Degrees of freedom required for walking in a Mantis-like extremity.

describes the kinematic motion of Mantis-like leg.

$$x = l_1 c_{01} + l_a c_{01a} + d_2(t) c_{01ab} + l_c c_{01abc} + l_d c_{01abcd} \tag{12.37}$$

and,

$$y = l_1 s_{01} + l_a s_{01a} + d_2(t) s_{01ab} + l_c s_{01abc} + l_d s_{01abcd} \tag{12.38}$$

Their, first derivative is given by,

$$
\begin{aligned}
\frac{d}{dt}x = {} & -l_1 s_{01}\left(\frac{d}{dt}\phi_0 + \frac{d}{dt}\phi_1\right) - l_a s_{01a}\left(\frac{d}{dt}\phi_0 + \frac{d}{dt}\phi_1 + a\right) + \frac{d}{dt}d_2(t) c_{01ab} \\
& -d_2(t) s_{01ab}\left(\frac{d}{dt}\phi_0 + \frac{d}{dt}\phi_1 + a + b\right) - l_c s_{01abc}\left(\frac{d}{dt}\phi_0 + \frac{d}{dt}\phi_1 + a + b + c\right) \\
& -l_d s_{01abcd}\left(\frac{d}{dt}\phi_0 + \frac{d}{dt}\phi_1 + a + b + c + d\right)
\end{aligned}
\tag{12.39}
$$

and,

$$
\begin{aligned}
\frac{d}{dt}y &= l_1 c_{01}\left(\frac{d}{dt}\phi_0 + \frac{d}{dt}\phi_1\right) + l_a c_{01a}\left(\frac{d}{dt}\phi_0 + \frac{d}{dt}\phi_1 + a\right) + \frac{d}{dt}d_2(t)s_{01ab} \\
&\quad -d_2(t)c_{01ab}\left(\frac{d}{dt}\phi_0 + \frac{d}{dt}\phi_1 + a + b\right) + l_c c_{01abc}\left(\frac{d}{dt}\phi_0 + \frac{d}{dt}\phi_1 + a + b + c\right) \\
&\quad\quad +l_d c_{01abcd}\left(\frac{d}{dt}\phi_0 + \frac{d}{dt}\phi_1 + a + b + c + d\right)
\end{aligned}
\tag{12.40}
$$

The kinetic energy for joint 0 with (potential energy $p_0 = 0$),

$$
k_0 = \frac{1}{2}I_0(\frac{d}{dt}\phi_0)^2
\tag{12.41}
$$

and its linear velocity model,

$$
v_1^2 = l_1^2\left(\frac{d}{dt}\phi_0 + \frac{d}{dt}\phi_1\right)^2
\tag{12.42}
$$

The kinetic energy for joint 1,

$$
k_1 = \frac{1}{2}m_1 v_1^2 + \frac{1}{2}I_1(\frac{d}{dt}\phi_0)^2
\tag{12.43}
$$

and potential energy p_1,

$$
p_1 = mg\frac{l_1}{2}s_{01}
\tag{12.44}
$$

The linear velocity model for joint-1 is given by

$$
\begin{aligned}
v_2^2 &= l_1^2\left(\frac{d}{dt}\phi_0 + \frac{d}{dt}\phi_1\right)^2 + l_a^2\left(\frac{d}{dt}\phi_0 + \frac{d}{dt}\phi_1 + a\right)^2 + \frac{d}{dt}d_2(t)^2 \\
&\quad +d_2(t)^2\left(\frac{d}{dt}\phi_0 + \frac{d}{dt}\phi_1 + a + b\right)^2 + 2l_1 l_a c_a\left(\frac{d}{dt}\phi_0 + \frac{d}{dt}\phi_1\right)\left(\frac{d}{dt}\phi_0 + \frac{d}{dt}\phi_1 + a\right) \\
&\quad\quad +2\frac{d}{dt}d_2(t)l_1 s_{ab}\left(\frac{d}{dt}\phi_0 + \frac{d}{dt}\phi_1\right) \\
&\quad -2d_2(t)l_1 c_{2\phi_0 2\phi_1 ab}\left(\frac{d}{dt}\phi_0 + \frac{d}{dt}\phi_1\right)\left(\frac{d}{dt}\phi_0 + \frac{d}{dt}\phi_1 + a + b\right) \\
&\quad\quad +2\frac{d}{dt}d_2(t)^2 l_a s_b\left(\frac{d}{dt}\phi_0 + \frac{d}{dt}\phi_1 + a\right) \\
&\quad -2d_2(t)l_a c_{2\phi_0 2\phi_1 2ab}\left(\frac{d}{dt}\phi_0 + \frac{d}{dt}\phi_1 + a\right)\left(\frac{d}{dt}\phi_0 + \frac{d}{dt}\phi_1 + a + b\right) \\
&\quad\quad -2\frac{d}{dt}d_2(t)d_2(t)s_{2(01ab)}\left(\frac{d}{dt}\phi_0 + \frac{d}{dt}\phi_1 + a + b\right)
\end{aligned}
\tag{12.45}
$$

Likewise, the kinetic and potential energy models for the second joint p_2,

$$k_2 - p_2 = \frac{1}{2}m_2 v_2^2 + \frac{1}{2}I_2(\frac{d}{dt}\phi_0)^2 - mg\frac{l_1}{2}s_{01} + \frac{1}{2}mg\frac{d_2(t)}{2}s_{01ab} \tag{12.46}$$

The linear velocity model for joint-2 is defined by,

$$v_w^2 = l_1^2 \left(\frac{d}{dt}\phi_0 + \frac{d}{dt}\phi_1\right)^2 + l_a^2 \left(\frac{d}{dt}\phi_0 + \frac{d}{dt}\phi_1 + a\right)^2 + \frac{d}{dt}d_2(t)^2$$

$$+d_2(t)^2 \left(\frac{d}{dt}\phi_0 + \frac{d}{dt}\phi_1 + a + b\right)^2 + l_c^2 \left(\frac{d}{dt}\phi_0 + \frac{d}{dt}\phi_1 + a + b + c\right)^2$$

$$l_d^2 \left(\frac{d}{dt}\phi_0 + \frac{d}{dt}\phi_1 + a + b + c + d\right)^2 + 2l_1 l_a c_a \left(\frac{d}{dt}\phi_0 + \frac{d}{dt}\phi_1\right)\left(\frac{d}{dt}\phi_0 + \frac{d}{dt}\phi_1 + a\right)$$

$$+2\frac{d}{dt}d_2(t)l_1 s_{ab}\left(\frac{d}{dt}\phi_0 + \frac{d}{dt}\phi_1\right)$$

$$-2d_2(t)l_1 c_{2\phi_0 2\phi_1 ab}\left(\frac{d}{dt}\phi_0 + \frac{d}{dt}\phi_1\right)\left(\frac{d}{dt}\phi_0 + \frac{d}{dt}\phi_1 + a + b\right)$$

$$+2\frac{d}{dt}d_2(t)^2 l_a s_b \left(\frac{d}{dt}\phi_0 + \frac{d}{dt}\phi_1 + a\right)$$

$$-2d_2(t)l_a c_{2\phi_0 2\phi_1 2ab}\left(\frac{d}{dt}\phi_0 + \frac{d}{dt}\phi_1 + a\right)\left(\frac{d}{dt}\phi_0 + \frac{d}{dt}\phi_1 + a + b\right)$$

$$-2\frac{d}{dt}d_2(t)d_2(t)s_{2(01ab)}\left(\frac{d}{dt}\phi_0 + \frac{d}{dt}\phi_1 + a + b\right)$$

$$+2l_1 l_c c_{abc}\left(\frac{d}{dt}\phi_0 + \frac{d}{dt}\phi_1\right)\left(\frac{d}{dt}\phi_0 + \frac{d}{dt}\phi_1 + a + b + c\right) \tag{12.47}$$

$$+2l_a l_c c_{bc}\left(\frac{d}{dt}\phi_0 + \frac{d}{dt}\phi_1 + a\right)\left(\frac{d}{dt}\phi_0 + \frac{d}{dt}\phi_1 + a + b + c\right)$$

$$+2\frac{d}{dt}d_2(t)l_c s_c \left(\frac{d}{dt}\phi_0 + \frac{d}{dt}\phi_1 + a + b + c\right)$$

$$-2d_2(t)l_c c_{2\phi_0 2\phi_1 2a2bc}\left(\frac{d}{dt}\phi_0 + \frac{d}{dt}\phi_1 + a + b + c\right)\left(\frac{d}{dt}\phi_0 + \frac{d}{dt}\phi_1 + a + b + c + d\right)$$

$$+2l_c l_d c_d \left(\frac{d}{dt}\phi_0 + \frac{d}{dt}\phi_1 + a + b + c\right)\left(\frac{d}{dt}\phi_0 + \frac{d}{dt}\phi_1 + a + b + c + d\right)$$

$$+2l_1 l_d c_{abcd}\left(\frac{d}{dt}\phi_0 + \frac{d}{dt}\phi_1\right)\left(\frac{d}{dt}\phi_0 + \frac{d}{dt}\phi_1 + a + b + c + d\right)$$

$$+2l_a l_d c_{bcd}\left(\frac{d}{dt}\phi_0 + \frac{d}{dt}\phi_1 + a\right)\left(\frac{d}{dt}\phi_0 + \frac{d}{dt}\phi_1 + a + b + c + d\right)$$

$$+2\frac{d}{dt}d_2(t)l_d s_{cd}\left(\frac{d}{dt}\phi_0 + \frac{d}{dt}\phi_1 + a + b + c + d\right)$$

$$-2d_3(t)l_d c_{2\phi_0 2\phi_1 2a2bcd}\left(\frac{d}{dt}\phi_0 + \frac{d}{dt}\phi_1 + a + b\right)\left(\frac{d}{dt}\phi_0 + \frac{d}{dt}\phi_1 + a + b + c + d\right)$$

finally, the kinetic energy of the extremity with load w,

$$k_w = \frac{1}{2}m_w v_w^2 + \frac{1}{2}I_w(\frac{d}{dt}\phi_0)^2 \tag{12.48}$$

The potential energy model,

$$p_w = mgA_w s_{\phi_w} \tag{12.49}$$

Then according to Euler-Lagrange equations, and by partially deriving, the force equations are,

$$\tau_0 = I_0\frac{d^2}{dt^2}\phi_0 \tag{12.50}$$

$$\tau_1 = \frac{1}{2}m_1a_1 + I_1\frac{d^2}{dt^2}\phi_0 - mg\frac{l_1}{2}s_{01} \tag{12.51}$$

$$\tau_2 = \frac{1}{2}m_2a_2 + I_2\frac{d^2}{dt^2}\phi_0 - mg\frac{l_1}{2}s_{01} - \frac{1}{2}mg\frac{d_2(t)}{2}s_{01ab} \tag{12.52}$$

$$\tau_w = \frac{1}{2}m_w a_w + I_w\frac{d^2}{dt^2}\phi_0 - mgA_w s_w \tag{12.53}$$

Furthermore, by developing the same systematic derivation and algebraic process, now on we state the set of dynamic equations for the ant-like extremities for a process of walking kinematic configuration.

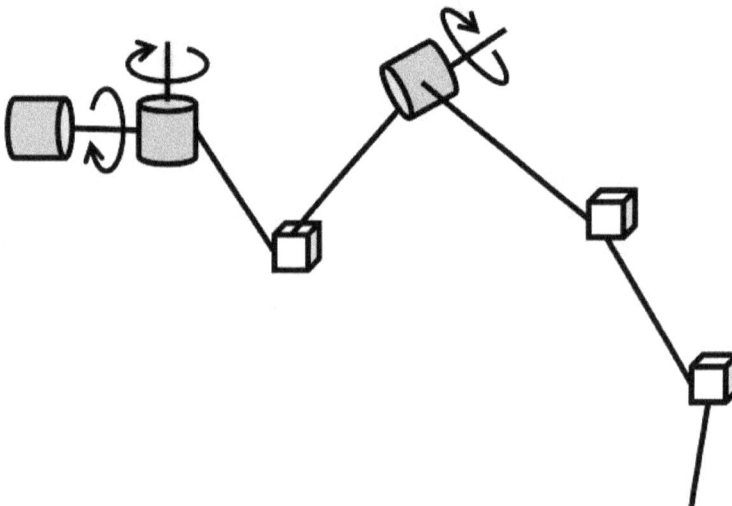

Figure 12.4: Degrees of freedom required for walking of an ant-like extremity.

Then, the walking kinematic configuration is given by next equations of planar positions $(x, y)^\top$,

$$x = l_1 c_{01} + l_a c_{01a} + l_3 c_{01a3} + l_b c_{01a3b} + l_c c_{01a3bc} \tag{12.54}$$

and

$$y = l_1 s_{01} + l_a s_{01a} + l_3 s_{01a3} + l_b s_{01a3b} + l_c s_{01a3bc} \tag{12.55}$$

Thus, the first order derivative for the X component

$$
\begin{aligned}
\frac{d}{dt} x = &-l_1 s_{01} \left(\frac{d}{dt} \phi_0 + \frac{d}{dt} \phi_1 \right) - l_a s_{01a} \left(\frac{d}{dt} \phi_0 + \frac{d}{dt} \phi_1 + a \right) \\
&-l_3 s_{01a3} \left(\frac{d}{dt} \phi_0 + \frac{d}{dt} \phi_1 + a + \frac{d}{dt} \phi_3 \right) - l_b s_{01a3b} \left(\frac{d}{dt} \phi_0 + \frac{d}{dt} \phi_1 + a + \frac{d}{dt} \phi_3 + b \right) \\
&-l_c s_{01a3bc} \left(\frac{d}{dt} \phi_0 + \frac{d}{dt} \phi_1 + a + \frac{d}{dt} \phi_3 + b + c \right)
\end{aligned}
\tag{12.56}
$$

likewise, the first order derivative for the Y component

$$
\begin{aligned}
\frac{d}{dt} y = &\ l_1 c_{01} \left(\frac{d}{dt} \phi_0 + \frac{d}{dt} \phi_1 \right) + l_a c_{01a} \left(\frac{d}{dt} \phi_0 + \frac{d}{dt} \phi_1 + a \right) \\
&+l_3 c_{01a3} \left(\frac{d}{dt} \phi_0 + \frac{d}{dt} \phi_1 + a + \frac{d}{dt} \phi_3 \right) + l_b c_{01a3b} \left(\frac{d}{dt} \phi_0 + \frac{d}{dt} \phi_1 + a + \frac{d}{dt} \phi_3 + b \right) \\
&+l_c c_{01a3bc} \left(\frac{d}{dt} \phi_0 + \frac{d}{dt} \phi_1 + a + \frac{d}{dt} \phi_3 + b + c \right)
\end{aligned}
\tag{12.57}
$$

the kinetic energy for the active joint 0,

$$k_0 = \frac{1}{2} I_0 \left(\frac{d}{dt} \phi_0 \right)^2 \tag{12.58}$$

The model for linear velocity of active joint 1,

$$v_1^2 = l_1^2 \left(\frac{d}{dt} \phi_0 + \frac{d}{dt} \phi_1 \right)^2 \tag{12.59}$$

The kinetic and potential energy models for active joint 1,

$$k_1 = \frac{1}{2} m_1 v_1^2 + \frac{1}{2} I_1 \left(\frac{d}{dt} \phi_0 \right)^2 \tag{12.60}$$

and

$$p_1 = mg\frac{l_1}{2}s_{01} \tag{12.61}$$

Likewise,

$$
\begin{aligned}
v_2^2 = {} & l_1\left(\frac{\mathrm{d}}{\mathrm{dt}}\phi_0 + \frac{\mathrm{d}}{\mathrm{dt}}\phi_1\right)^2 + l_a^2\left(\frac{\mathrm{d}}{\mathrm{dt}}\phi_0 + \frac{\mathrm{d}}{\mathrm{dt}}\phi_1 + a\right)^2 + l_3^2\left(\frac{\mathrm{d}}{\mathrm{dt}}\phi_0 + \frac{\mathrm{d}}{\mathrm{dt}}\phi_1 + a + \frac{\mathrm{d}}{\mathrm{dt}}\phi_3\right)^2 \\
& + 2l_1l_3c_{a3}\left(\frac{\mathrm{d}}{\mathrm{dt}}\phi_0 + \frac{\mathrm{d}}{\mathrm{dt}}\phi_1\right)\left(\frac{\mathrm{d}}{\mathrm{dt}}\phi_0 + \frac{\mathrm{d}}{\mathrm{dt}}\phi_1 + a + \frac{\mathrm{d}}{\mathrm{dt}}\phi_3\right) \\
& + 2l_1l_ac_a\left(\frac{\mathrm{d}}{\mathrm{dt}}\phi_0 + \frac{\mathrm{d}}{\mathrm{dt}}\phi_1\right)\left(\frac{\mathrm{d}}{\mathrm{dt}}\phi_0 + \frac{\mathrm{d}}{\mathrm{dt}}\phi_1 + a\right) \\
& + 2l_al_3c_3\left(\frac{\mathrm{d}}{\mathrm{dt}}\phi_0 + \frac{\mathrm{d}}{\mathrm{dt}}\phi_1 + a\right)\left(\frac{\mathrm{d}}{\mathrm{dt}}\phi_0 + \frac{\mathrm{d}}{\mathrm{dt}}\phi_1 + a + \frac{\mathrm{d}}{\mathrm{dt}}\phi_3\right)
\end{aligned}
\tag{12.62}
$$

The kinetic and potential energy models for active joint 2,

$$k_2 = \frac{1}{2}m_2v_2^2 + \frac{1}{2}I_2(\frac{\mathrm{d}}{\mathrm{dt}}\phi_0)^2 \tag{12.63}$$

and

$$p_2 = mg\frac{l_1}{2}s_{01} + \frac{1}{2}mg\frac{l_3}{2}s_{01a3} \tag{12.64}$$

Thus, the model for the linear speed,

$$
\begin{aligned}
v_w^2 &= l_1 \left(\frac{\mathrm{d}}{\mathrm{dt}} \phi_0 + \frac{\mathrm{d}}{\mathrm{dt}} \phi_1 \right)^2 + l_a^2 \left(\frac{\mathrm{d}}{\mathrm{dt}} \phi_0 + \frac{\mathrm{d}}{\mathrm{dt}} \phi_1 + a \right)^2 + l_3^2 \left(\frac{\mathrm{d}}{\mathrm{dt}} \phi_0 + \frac{\mathrm{d}}{\mathrm{dt}} \phi_1 + a + \frac{\mathrm{d}}{\mathrm{dt}} \phi_3 \right)^2 \\
&+ l_b^2 \left(\frac{\mathrm{d}}{\mathrm{dt}} \phi_0 + \frac{\mathrm{d}}{\mathrm{dt}} \phi_1 + a + \frac{\mathrm{d}}{\mathrm{dt}} \phi_3 + b \right)^2 + l_c^2 \left(\frac{\mathrm{d}}{\mathrm{dt}} \phi_0 + \frac{\mathrm{d}}{\mathrm{dt}} \phi_1 + a + \frac{\mathrm{d}}{\mathrm{dt}} \phi_3 + b + c \right)^2 \\
&+ 2 l_1 l_3 c_{a3} \left(\frac{\mathrm{d}}{\mathrm{dt}} \phi_0 + \frac{\mathrm{d}}{\mathrm{dt}} \phi_1 \right) \left(\frac{\mathrm{d}}{\mathrm{dt}} \phi_0 + \frac{\mathrm{d}}{\mathrm{dt}} \phi_1 + a + \frac{\mathrm{d}}{\mathrm{dt}} \phi_3 \right) \\
&+ 2 l_1 l_a c_a \left(\frac{\mathrm{d}}{\mathrm{dt}} \phi_0 + \frac{\mathrm{d}}{\mathrm{dt}} \phi_1 \right) \left(\frac{\mathrm{d}}{\mathrm{dt}} \phi_0 + \frac{\mathrm{d}}{\mathrm{dt}} \phi_1 + a \right) \\
&+ 2 l_a l_3 c_3 \left(\frac{\mathrm{d}}{\mathrm{dt}} \phi_0 + \frac{\mathrm{d}}{\mathrm{dt}} \phi_1 + a \right) \left(\frac{\mathrm{d}}{\mathrm{dt}} \phi_0 + \frac{\mathrm{d}}{\mathrm{dt}} \phi_1 + a + \frac{\mathrm{d}}{\mathrm{dt}} \phi_3 \right) \\
&+ 2 l_1 l_b c_{a3b} \left(\frac{\mathrm{d}}{\mathrm{dt}} \phi_0 + \frac{\mathrm{d}}{\mathrm{dt}} \phi_1 \right) \left(\frac{\mathrm{d}}{\mathrm{dt}} \phi_0 + \frac{\mathrm{d}}{\mathrm{dt}} \phi_1 + a + \frac{\mathrm{d}}{\mathrm{dt}} \phi_3 + b \right) \\
&+ 2 l_a l_b c_{3b} \left(\frac{\mathrm{d}}{\mathrm{dt}} \phi_0 + \frac{\mathrm{d}}{\mathrm{dt}} \phi_1 + a \right) \left(\frac{\mathrm{d}}{\mathrm{dt}} \phi_0 + \frac{\mathrm{d}}{\mathrm{dt}} \phi_1 + a + \frac{\mathrm{d}}{\mathrm{dt}} \phi_3 + b \right) \\
&+ 2 l_3 l_b c_b \left(\frac{\mathrm{d}}{\mathrm{dt}} \phi_0 + \frac{\mathrm{d}}{\mathrm{dt}} \phi_1 + a + \frac{\mathrm{d}}{\mathrm{dt}} \phi_3 \right) \left(\frac{\mathrm{d}}{\mathrm{dt}} \phi_0 + \frac{\mathrm{d}}{\mathrm{dt}} \phi_1 + a + \frac{\mathrm{d}}{\mathrm{dt}} \phi_3 + b \right) \\
&+ 2 l_b l_c c_c \left(\frac{\mathrm{d}}{\mathrm{dt}} \phi_0 + \frac{\mathrm{d}}{\mathrm{dt}} \phi_1 + a + \frac{\mathrm{d}}{\mathrm{dt}} \phi_3 + b \right) \left(\frac{\mathrm{d}}{\mathrm{dt}} \phi_0 + \frac{\mathrm{d}}{\mathrm{dt}} \phi_1 + a + \frac{\mathrm{d}}{\mathrm{dt}} \phi_3 + b + c \right) \\
&+ 2 l_1 l_c c_{a3bc} \left(\frac{\mathrm{d}}{\mathrm{dt}} \phi_0 + \frac{\mathrm{d}}{\mathrm{dt}} \phi_1 \right) \left(\frac{\mathrm{d}}{\mathrm{dt}} \phi_0 + \frac{\mathrm{d}}{\mathrm{dt}} \phi_1 + a + \frac{\mathrm{d}}{\mathrm{dt}} \phi_3 + b + c \right) \\
&+ 2 l_a l_c c_{3bc} \left(\frac{\mathrm{d}}{\mathrm{dt}} \phi_0 + \frac{\mathrm{d}}{\mathrm{dt}} \phi_1 + a \right) \left(\frac{\mathrm{d}}{\mathrm{dt}} \phi_0 + \frac{\mathrm{d}}{\mathrm{dt}} \phi_1 + a + \frac{\mathrm{d}}{\mathrm{dt}} \phi_3 + b + c \right) \\
&+ 2 l_3 l_c c_{bc} \left(\frac{\mathrm{d}}{\mathrm{dt}} \phi_0 + \frac{\mathrm{d}}{\mathrm{dt}} \phi_1 + a + \frac{\mathrm{d}}{\mathrm{dt}} \phi_3 \right) \left(\frac{\mathrm{d}}{\mathrm{dt}} \phi_0 + \frac{\mathrm{d}}{\mathrm{dt}} \phi_1 + a + \frac{\mathrm{d}}{\mathrm{dt}} \phi_3 + b + c \right)
\end{aligned} \tag{12.65}
$$

The models for the kinetic and potential energies,

$$
k_w = \frac{1}{2} m_w v_w^2 + \frac{1}{2} I_w \left(\frac{\mathrm{d}}{\mathrm{dt}} \phi_0 \right)^2 \tag{12.66}
$$

and

$$
p_w = m g A_w s_{\phi_w} \tag{12.67}
$$

The rotational joints angular moments are described by

$$
\tau_0 = I_0 \frac{\mathrm{d}^2}{\mathrm{dt}^2} \phi_0 \tag{12.68}
$$

$$\tau_1 = \frac{1}{2}m_1a_1 + I_1\frac{d^2}{dt^2}\phi_0 - mg\frac{l_1}{2}s_{01} \tag{12.69}$$

$$\tau_2 = \frac{1}{2}m_2a_2 + I_2\frac{d^2}{dt^2}\phi_0 - mg\frac{l_1}{2}s_{01} - \frac{1}{2}mg\frac{l_3}{2}s_{01a3} \tag{12.70}$$

and

$$\tau_w = \frac{1}{2}m_wa_w + I_w\frac{d^2}{dt^2}\phi_0 - mgA_ws_w \tag{12.71}$$

12.5 Simulation results

Figure 12.5 depicts work spaces of both extremities Mantis-like (12.5-(a)), and ant-like (12.5-(b)) legs. The coordinate $(0,0)^\top$ correspond to each leg's first joint dubbed ϕ_0. A transversal cut of either plots represent an approximation of a single leg step. The workspace is a Cartesian plot showing the set of possible 3D locations a leg is able to reach in terms of its kinematic restrictions. The vector models first derivative $\dot{p}_m$ and $\dot{p}_a$ are the Mantis-like and ant-like forward kinematic equations respectively. For the case of first order legs kinematics $\dot{p}(t)$,

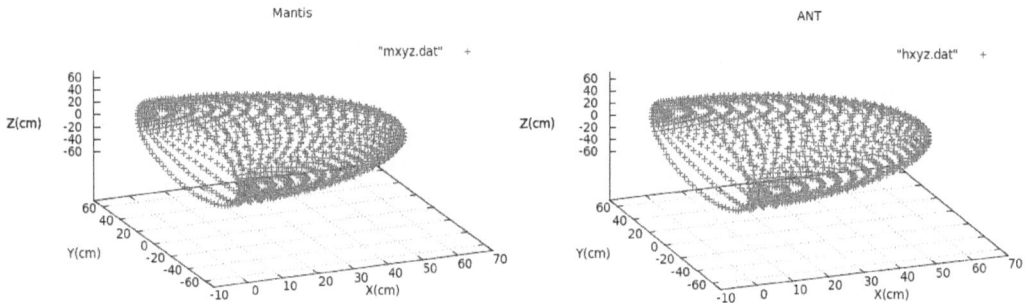

Figure 12.5: (a) Mantis-like walking-step space; (b) Ant-like 3D walking-step space.

the figure 12.6 illustrates the Cartesian speeds $\dot{x}$, $\dot{y}$, $\dot{z}$ evolution in reference to a single leg-step trajectory, figure 12.6-(a) for Mantis-like leg step, and 12.6-(c) for ant-like leg step. Their magnitudes in velocity given by the norm $\|\dot{p}_m\|$ and $\|\dot{p}_a\|$ are depicted by figure 12.6-(b)(d), respectively. For these results, a step given by a Mantis-like leg, the joints ϕ_0, ϕ_3, and ϕ_5 were controlled, according to empirical observations of the insect's walking movements.

In addition, for the ant-like leg, a step trajectory was emulated by controlling the joints ϕ_1, ϕ_2, and ϕ_4 only, according to insects empirical observations. The small differences of speed magnitudes between Mantis and ant, resulted because of links lengths configuration

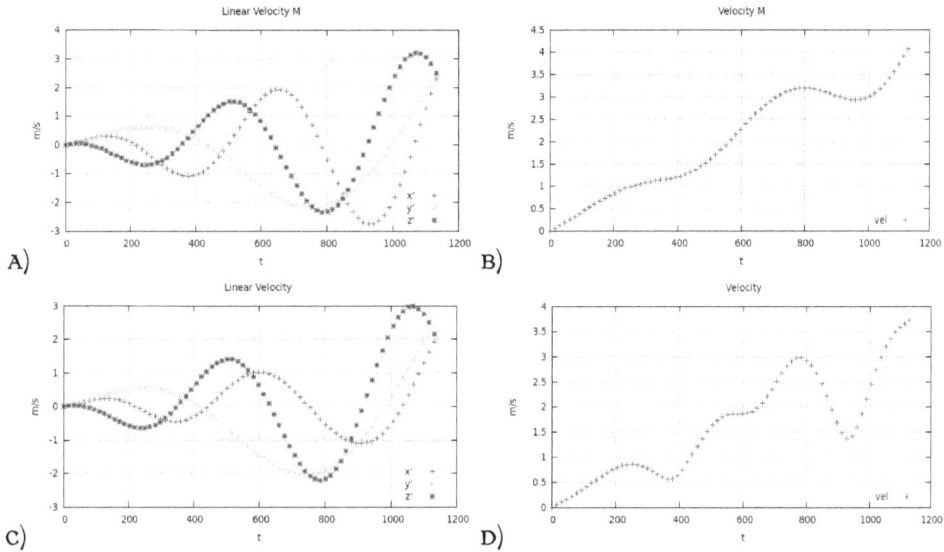

Figure 12.6:   Robot's one step Cartesian velocities. (a) Mantis-like leg; (b) Mantis-like magnitude; (c) Ant-like leg; (d) ant-like leg magnitude.

were arbitrary taken considering insects' size rates. The presented formulation in this work stated that the joints may be adjusted and configured in accordance to the expression $\|\mathbf{J} \cdot \mathbf{\Phi}\|$, which is the equation (12.4) that allow to configure numerous gaits patterns.

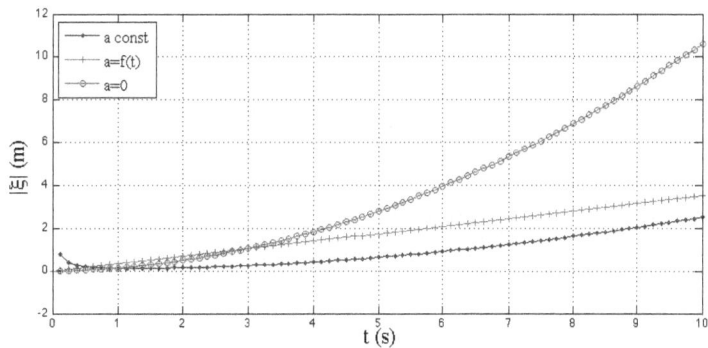

Figure 12.7:   Robot's position given by the accelerative restrictions $\mathbf{g}(\mathbf{x}, \dot{\mathbf{x}}, \ddot{\mathbf{x}})$.

The general acceleration restrictions given by function $\mathbf{g}(\mathbf{x}, \dot{\mathbf{x}}, \ddot{\mathbf{x}})$ as the motion state vector in the navigation control law is illustrated by figure 12.7. It depicts the instantaneous robot's tangential velocity evolution, according to equation (12.7). At constant acceleration $0.25ms^{-2}$ for the first equation condition. However, for second restriction, despite is a linear function of time, its behaviour is prone to slope changes affecting displacements. It is because the instantaneous acceleration gives such slope value. The acceleration depends on differences in time of two successive instantaneous velocities. The third acceleration restriction was set at a constant reference velocity of $v^{ref} = 0.25$m/s, and gamma (attenuation factor) $\gamma = 0.25$. Although $a = 0$, its behaviour in time is gradually increased or decreased (if the case) because it is velocity controlled equation at constant speed change overtime gives stability of rate displacement. The linearised state equation determines the global robot's motion behaviour, and it was based on observable conditions that depend on any of the acceleration condition states: uniform, non-uniform and magnitude zero. The navigation control relies upon an averaged Cartesian norm of legs' speed components. It allows to adjust and configure the walking gaits since the control law's input vector is defined in terms of the norm of Jacobians $\frac{1}{n} \| \sum_i^n J_i \cdot \dot{\Phi}_i \|$, where n is the number of legs, and joints vector $\dot{\Phi}$ is explicitly defined in accordance to a given walking gait. Instead, a first order derivative analysis of kinematic equations system was stated. Two different bio-inspired six-DOF extremities with redundant kinematics were studied and analysed through simulation results; the Praying-Mantis, and the Ant Smithi. Redundant kinematic extremities were inversely solved by two solutions: the right pseudo-inverse, and decomposition the Jacobian's singular values through SVD.

Bibliography

[1] Martínez-García E., Torres D., Ortega A., Zamora A., Torres-Cordoba R., Martínez-Villafañe A., *Chap. 4, Modeling Dynamics and Navigation Control of an Explorer Hexapod*, Autonomous Robots: Control, Sensing and Perception, Cuvillier Verlag, pp. 82-115, 2011.

[2] Roy S.S., Singh A.K., and Pratihar D.K., *Analysis of six-legged walking robots*, 14th Conf. Machines and Mechanisms, Durgapur India, Dec 17- 18, pp. 259-265, 2009.

[3] Martínez-García E., Villalobos E., Lara M.C., Aquino E., Mohan R.E., *Six-legged robot modeling with autonomous visual navigation control*, 8th Intl. Conf. on Intelligent Unmanned Systems, 2012, Oct 22-24, 2012.

[4] Umar A., Javaid I., *Comparative Study of Biologically Inspired Walking Gaits through Waypoint Navigation*, Advances in Mechanical Engineering, Vol. 2011, Hindawi, pp. 1-9.

[5] Figliolini G., Rea P., *Mechanics and Simulation of Six-Legged Walking Robots*, Climbing and Walking Robots: towards New Applications, Houxi ang Zhang (Ed.), InTech, 2007.

[6] Nakamura Y., *Advanced Robotics: Redundancy and Optimization*, Addison Wesley, 1991.

[7] Spong M.W. and Vidyasagar M., *Robot Dynamics and Control*, Wiley India, 2008.

[8] Mackay W. P., Maes J. M., Fernandez P.R., Luna G., *The ants of North and Central America: the genus Mycocepurus (Hymenoptera: Formicidae*, Journal of Science, Vol. 4, 2004.

[9] Zomeren L.V., *Keeping Insects, Praying mantis, butterflies, stick insects and beetles as pets*, http://www.keepinginsects.com/praying-mantis 2013.

[10] Bolton B., *Species: Mycocepurus smithii*, http://www.antweb.org, 2013.

[11] Umar A., *Virtual Reality-Human Computer Interaction, Chap. 5*, Virtual Reality to Simulate Adaptive Walking in Unstructured Terrains for Multi-Legged Robots, IntechOpen, pp. 79–102, 2012.

[12] Roennau, A. Kerscher, T. ; Dillmann, R., *Design and kinematics of a biologically-inspired leg for a six-legged walking machine*, 3rd IEEE RAS and EMBS Intl. Conf. on Biomedical Robotics and Biomechatronics, pp. 626–631, 2010.

Chapter 13

Under-actuated Jansen-based Robot Control

Jaime Candelaria Solís and Edgar A. Martínez García

Laboratorio de Robótica, Institute of Engineering and Technology
Universidad Autónoma de Ciudad Juárez, Mexico.

Latest advances in science and technology have incredibly imposed robotic walkers, which is becoming a very important aspect in human society. A few examples of this role are robots employed in industrial, space fields, all-terrain exploration, rehabilitation, and even entertainment ⁻ . As these kind of robotic systems technologically grow, they require more sensors and devices in order to make them behave in a more realistic manner. Therefore, achieving a better interaction with their environment. Wheeled mobile robots are characterized by their simplicity, but high performance to travel on even surfaces. Nevertheless their movement over complex terrains becomes limited. Unlike wheeled robots, the robots with extremities capable to walk over all types of grounds, and they consume less energy than wheeled robots. All these advantages make these systems to depend on the number of actuators to travel and complete tasks, since they regularly require one independent variable to be controlled per joint. As a consequence, they gain more weight, use more energy and computational complexity gets incremented. As a result, the systems known as under-actuated ones have been developed, and represent a reduction in their driven joints to perform dynamic walking employing more efficiently potential and kinetic energies . As this reduction may constraint the robot's movement, but it provides the benefit of reducing the number of actuators needed for working. This chapter presents the design and kinematic control of an octapod walking robot with foundations on the study of the under-actuated Theo Jansen mechanism . The analysis departs from a study of all passive joints present in Jansen linkage. The position equations of Jansen-based limbs' interacting as an octapod are analysed, involving mathematical modelling of both walking patterns: for one limb, and for the eight limbs. The Jansen's mechanism is a planar mechanical linkage that consists of a frame, a crankshaft, and 11 links; shown in figure 13.1-a). The mechanism is based on the next idea: if the crankshaft's rotary movement is controlled, all mechanism's linkages are also easily controlled.

13.1   Passive joints analysis

In this section, an algebraic analysis of the Jansen mechanism of figure 13.1-a) is discussed. The main strategy of this chapter is to gain understanding of the mechanism's behaviour by mathematically treat partial closed chains of links (figures 13.1-b)–f). The solution of the kinematic system is provided by analysing all rotational variables (passive joints) that are controlled through the driven input, namely crankshaft angle (denoted by link l_2) (see chapter 1.1). The kinematic analysis of the limb, is by following a way of interconnected links from the end of l_2 or node $\overrightarrow{jk}$ until node $\overrightarrow{ih}$ (figure 13.1-a)). Any linkage relationship may correctly be used to obtain a solution about the planar position of the limb's contact (node $\overrightarrow{ih}$). Therefore, considering the rigid links L_i and passive angles θ_j, the following position equations are stated,

Axiom 13.1.1 (Closed link kinematic equations). *A formal statement of two equations to solve a planar four links closed chain is postulated. Thus, equation for the component X,*

$$L_2 \cos \theta + L_3 \cos \theta_3 = L_1 \cos \theta_1 + L_4 \cos \theta_4 \tag{13.1}$$

and for the component Y,

$$L_2 \sin \theta + L_3 \sin \theta_3 = L_1 \sin \theta_1 + L_4 \sin \theta_4 \tag{13.2}$$

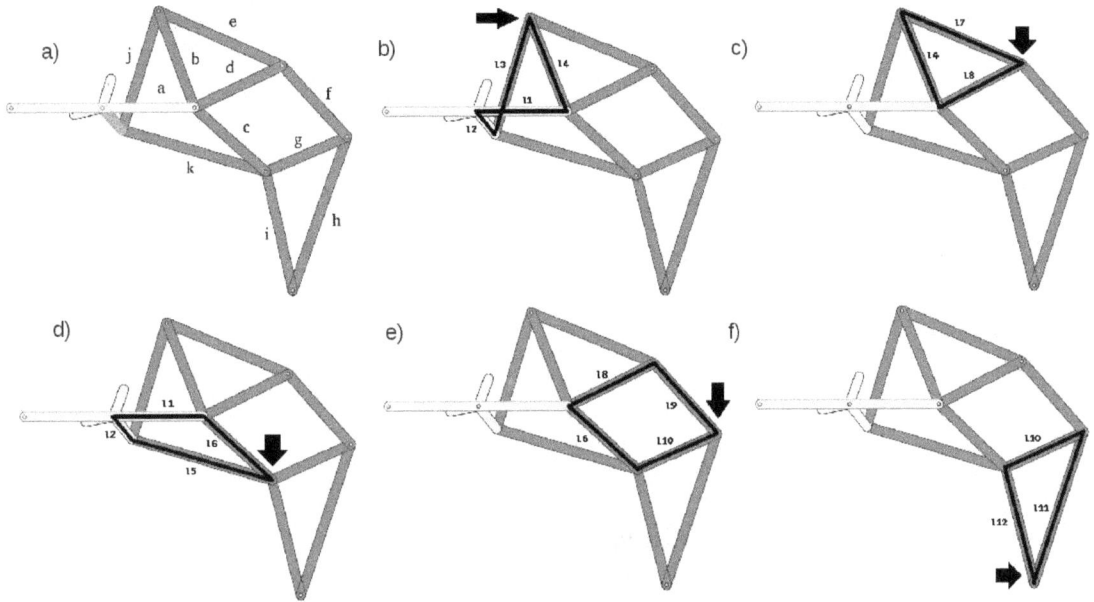

Figure 13.1:   a)Jansen links. Closed chain of links: b) linkage l_1, l_2, l_3, l_4; c) linkage l_4, l_7, l_8; d) linkage l_1, l_2, l_5, l_6; e) linkage l_6, l_8, l_9, l_{10}; and f) linkage l_{10}, l_{11}, l_{12}.

In this case, the input angle θ is the only independent driven variable known, as well as the angle of passive joint $\theta_1 = 0$ for motion of link L_1. In order to simplify equations, the value of θ_1 is algebraically substituted from the previous equations:

$$L_2 \cos \theta + L_3 \cos \theta_3 = L_1 + L_4 \cos \theta_4 \tag{13.3}$$

and

$$L_2 \sin \theta + L_3 \sin \theta_3 = L_4 \sin \theta_4 \tag{13.4}$$

Thus, a known angle is obtained temporally called θ, and two unknowns θ_3 and θ_4. Hence, the last expression is rearranged as

$$L_3 \cos \theta_3 = L_1 + L_4 \cos \theta_4 - L_2 \cos \theta \tag{13.5}$$

and

$$L_3 \sin \theta_3 = L_4 \sin \theta_4 - L_2 \sin \theta \tag{13.6}$$

Then, the two equations are squared in both sides of equality, and algebraically and simplified,

$$(L_3 \cos \theta_3)^2 = (L_1 + L_4 \cos \theta_4 - L_2 \cos \theta)^2 \tag{13.7}$$

as well as

$$L_3^2 \cos^2 \theta_3 = L_1^2 + 2L_1 L_4 \cos \theta_4 + L_4^2 \cos^2 \theta_4 - 2(L_1 + L_4 \cos \theta_4)(L_2 \cos \theta) + L_2^2 \cos^2 \theta \tag{13.8}$$

It follow that,

$$(L_3 \sin \theta_3)^2 = (L_4 \sin \theta_4 - L_2 \sin \theta)^2 \tag{13.9}$$

and

$$L_3^2 \sin^2 \theta_3 = L_4^2 \sin^2 \theta_4 - 2(L_4 \sin \theta_4)(L_2 \sin \theta) + L_2^2 \sin^2 \theta \tag{13.10}$$

by adding $(L_3 \cos \theta_3)^2$ with $(L_3 \sin \theta_3)^2$, we have:

$$L_3^2 = L_1^2 + L_2^2 + L_4^2 + 2L_1 L_4 \cos \theta_4 - 2L_1 L_2 \cos \theta - 2L_2 L_4 (\cos \theta \cos \theta_4 + \sin \theta \sin \theta_4) \tag{13.11}$$

Algebraically rearranging

$$(2L_1 L_4 - 2L_2 L_4 \cos \theta) \cos \theta_4 - (2L_2 L_4 \sin \theta) \sin \theta_4 + (L_1^2 + L_2^2 - L_3^2 + L_4^2 - 2L_1 L_2 \cos \theta) = 0 \tag{13.12}$$

Thus, the double-angle formulas are then used according to the following manner,

$$\cos \theta_4 = \frac{1 - \tan^2(\frac{\theta_4}{2})}{1 + \tan^2(\frac{\theta_4}{2})} \tag{13.13}$$

and

$$\sin \theta_4 = \frac{2 \tan(\frac{\theta_4}{2})}{1 + \tan^2(\frac{\theta_4}{2})} \tag{13.14}$$

Hence, the following expression is obtained,

$$A \left(\frac{1 - t^2}{1 + t^2} \right) + B \left(\frac{2t}{1 + t^2} \right) + C = 0 \tag{13.15}$$

Where the following terms are defined as,

$$A = 2L_1 L_4 - 2L_2 L_4 \cos \theta \tag{13.16}$$

$$B = -2L_2 L_4 \sin \theta \tag{13.17}$$

$$C = L_1^2 + L_2^2 - L_3^2 + L_4^2 - 2L_1 L_2 \cos \theta \tag{13.18}$$

and

$$t = \tan(\frac{\theta_4}{2}) \tag{13.19}$$

From previous equation (13.15), the following process is developed,

$$\frac{A - At^2 + 2Bt}{1 + t^2} + C = 0 \tag{13.20}$$

$$\frac{A - At^2 + 2Bt + C + Ct^2}{1 + t^2} = 0 \tag{13.21}$$

$$A - At^2 + 2Bt + C + Ct^2 = 0 \tag{13.22}$$

thus,

$$(C - A)t^2 + (2B)t + (A + C) = 0 \tag{13.23}$$

It is observed that (13.23) is a quadratic equation of the general quadratic form $ax^2 + bx + c = 0$. Thereby, a second degree equation is solved by the general form,

$$t = \frac{-2B - \sqrt{4B^2 - 4(C - A)(A + C)}}{2(C - A)} = \frac{-B - \sqrt{B^2 - C^2 + A^2}}{C - A} \tag{13.24}$$

Thus, let us consider that $t = \tan(\frac{\theta_4}{2})$, and θ_4 is drop-off,

$$\theta_4 = 2 \tan^{-1}(t) \tag{13.25}$$

Since θ_4 is already known, $L_3 \sin \theta_3$ is divided by $L_3 \cos \theta_3$ to obtain a solution for θ_3

$$\tan \theta_3 = (\frac{L_4 \sin \theta_4 - L_2 \sin \theta}{L_1 + L_4 \cos \theta_4 - L_2 \cos \theta}) \tag{13.26}$$

and subsequently,

$$\theta_3 = \tan^{-1}\left(\frac{L_4 \sin \theta_4 - L_2 \sin \theta}{L_1 + L_4 \cos \theta_4 - L_2 \cos \theta}\right) \tag{13.27}$$

It follows to solve for the a second linkage formed by L_4, L_7, L_8 that is depicted in figure 13.1-c). Thus, the following kinematic equations of position are stated, for the X component,

$$L_4 \cos \theta_4 + L_7 \cos \theta_7 = L_8 \cos \theta_8 \tag{13.28}$$

and then for the Y component,

$$L_4 \sin \theta_4 + L_7 \sin \theta_7 = L_8 \sin \theta_8 \tag{13.29}$$

It is worth noting that an analytic solution for θ_4 has already been obtained, and it will be used as a known variable to solve for the next chain of links where angles θ_7 and θ_8 are involved.

$$L_7 \cos \theta_7 = L_8 \cos \theta_8 - L_4 \cos \theta_4 \tag{13.30}$$

and

$$L_7 \sin \theta_7 = L_8 \sin \theta_8 - L_4 \sin \theta_4 \tag{13.31}$$

Similarly, both equations are squared and algebraically simplified to obtain

$$(L_7 \cos \theta_7)^2 = (L_8 \cos \theta_8 - L_4 \cos \theta_4)^2 \tag{13.32}$$

and developing square binomial,

$$L_7^2 \cos^2 \theta_7 = L_8^2 \cos^2 \theta_8 - 2L_4 L_8 \cos \theta_4 \cos \theta_8 + L_4^2 \cos^2 \theta_4 \tag{13.33}$$

applying same algebraic procedure to the Y component,

$$(L_7 \sin \theta_7)^2 = (L_8 \sin \theta_8 - L_4 \sin \theta_4)^2 \tag{13.34}$$

and expanding the squared binomial,

$$L_7^2 \sin^2 \theta_7 = L_8^2 \sin^2 \theta_8 - 2L_4 L_8 \sin \theta_4 \sin \theta_8 + L_4^2 \sin^2 \theta_4 \tag{13.35}$$

Adding $(L_7 \cos \theta_7)^2$ and $(L_7 \sin \theta_7)^2$ in both sides of equation to reduce by substituting trigonometric identities, it results in

$$L_7^2 = L_8^2 - 2L_4 L_8 (\cos \theta_4 \cos \theta_8 + \sin \theta_4 \sin \theta_8) + L_4^2 \qquad (13.36)$$

In addition, by rearranging

$$- (2L_4 L_8 \cos \theta_4) \cos \theta_8 - (2L_4 L_8 \sin \theta_4) \sin \theta_8 + (L_8^2 - L_7^2 + L_4^2) = 0 \qquad (13.37)$$

And by substituting next trigonometric identities,

$$\cos \theta_8 = \frac{1 - \tan^2(\frac{\theta_8}{2})}{1 + \tan^2(\frac{\theta_8}{2})} \qquad (13.38)$$

as well as

$$\sin \theta_8 = \frac{2 \tan(\frac{\theta_8}{2})}{1 + \tan^2(\frac{\theta_8}{2})} \qquad (13.39)$$

As early stated, similarly to equation (13.15), the following expressions are established,

$$A = -2L_4 L_8 \cos \theta_4, \qquad (13.40)$$

$$B = -2L_4 L_8 \sin \theta_4, \qquad (13.41)$$

$$C = L_8^2 - L_7^2 + L_4^2 \qquad (13.42)$$

and

$$t = \tan(\frac{\theta_8}{2}) \qquad (13.43)$$

As in equation (13.15), a similar algebraic process is developed and the quadratic equation is solved by the a general form

$$t = \frac{-B - \sqrt{B^2 - C^2 + A^2}}{C - A} \qquad (13.44)$$

Let us recall that $t = \tan(\frac{\theta_8}{2})$, and we solve for θ_8,

$$\theta_8 = 2 \tan^{-1}(t) \qquad (13.45)$$

hence,

$$\theta_8 = 2\tan^{-1}\left(\frac{2L_4L_8\sin\theta_4 - \sqrt{(-2L_4L_8\sin\theta_4)^2 - (L_8^2 - L_7^2 + L_4^2)^2 + (-2L_4L_8\cos\theta_4)^2}}{(L_8^2 - L_7^2 + L_4^2) - (-2L_4L_8\cos\theta_4)}\right) \qquad (13.46)$$

Once θ_8 is known, $L_7\sin\theta_7$ is divided by $L_7\cos\theta_7$ to obtain θ_7

$$\tan\theta_7 = \left(\frac{L_8\sin\theta_8 - L_4\sin\theta_4}{L_8\cos\theta_8 - L_4\cos\theta_4}\right) \qquad (13.47)$$

$$\theta_7 = \tan^{-1}\left(\frac{L_8\sin\theta_8 - L_4\sin\theta_4}{L_8\cos\theta_8 - L_4\cos\theta_4}\right) \qquad (13.48)$$

$$\theta_7 = \tan^{-1}\left(\frac{L_8\sin(2\tan^{-1}(\frac{2L_4L_8\sin\theta_4 - \sqrt{(-2L_4L_8\sin\theta_4)^2 - (L_8^2 - L_7^2 + L_4^2)^2 + (-2L_4L_8\cos\theta_4)^2}}{(L_8^2 - L_7^2 + L_4^2) - (-2L_4L_8\cos\theta_4)})) - L_4\sin\theta_4}{L_8\cos(2\tan^{-1}(\frac{2L_4L_8\sin\theta_4 - \sqrt{(-2L_4L_8\sin\theta_4)^2 - (L_8^2 - L_7^2 + L_4^2)^2 + (-2L_4L_8\cos\theta_4)^2}}{(L_8^2 - L_7^2 + L_4^2) - (-2L_4L_8\cos\theta_4)})) - L_4\cos\theta_4}\right) \qquad (13.49)$$

Furthermore, as depicted in figure 13.1-d) the third linkage to be analysed is formed by the rigid links L_1, L_2, L_5, L_6, which form a closed kinematic chain. Thus, the following equations are stated,

$$L_2\cos\theta + L_5\cos\theta_5 = L_1\cos\theta_1 + L_6\cos\theta_6 \qquad (13.50)$$

and

$$L_2\sin\theta + L_5\sin\theta_5 = L_1\sin\theta_1 + L_6\sin\theta_6 \qquad (13.51)$$

In this case, the angle θ is known a priori from the link L_1, and where $\theta_1 = 0$. Thus, the set of equations are simplified due to the value of θ_1,

$$L_2\cos\theta + L_5\cos\theta_5 = L_1 + L_6\cos\theta_6 \qquad (13.52)$$

and

$$L_2\sin\theta + L_5\sin\theta_5 = L_6\sin\theta_6 \qquad (13.53)$$

Rearranging previous expressions,

$$L_5\cos\theta_5 = L_1 + L_6\cos\theta_6 - L_2\cos\theta \qquad (13.54)$$

and

$$L_5 \sin \theta_5 = L_6 \sin \theta_6 - L_2 \sin \theta \qquad (13.55)$$

and both equations are squared, and algebraically simplified,

$$(L_5 \cos \theta_5)^2 = (L_1 + L_6 \cos \theta_6 - L_2 \cos \theta)^2 \qquad (13.56)$$

developing the squared binomial,

$$L_5^2 \cos^2 \theta_5 = L_1^2 + 2L_1 L_6 \cos \theta_6 + L_6^2 \cos^2 \theta_6 - 2(L_1 + L_6 \cos \theta_6)(L_2 \cos \theta) + L_2^2 \cos^2 \theta \qquad (13.57)$$

the following expressions are obtained

$$(L_5 \sin \theta_5)^2 = (L_6 \sin \theta_6 - L_2 \sin \theta)^2 \qquad (13.58)$$

$$L_5^2 \sin^2 \theta_5 = L_6^2 \sin^2 \theta_6 - 2L_2 L_6 \sin \theta \sin \theta_6 + L_2^2 \sin^2 \theta \qquad (13.59)$$

Adding $(L_5 \cos \theta_5)^2$ and $(L_5 \sin \theta_5)^2$, next expression results

$$L_5^2 = L_1^2 + L_6^2 + L_2^2 + 2L_1 L_6 \cos \theta_6 - 2L_1 L_2 \cos \theta - 2L_2 L_6 (\cos \theta \cos \theta_6 + \sin \theta \sin \theta_6) \qquad (13.60)$$

and algebraically rearranging,

$$(2L_1 L_6 - 2L_2 L_6 \cos \theta) \cos \theta_6 - (2L_2 L_6 \sin \theta) \sin \theta_6 + (L_1^2 + L_6^2 + L_2^2 - L_5^2 - 2L_1 L_2 \cos \theta) = 0 \quad (13.61)$$

substituting next trigonometric identities to reduce previous expression,

$$\cos \theta_6 = \frac{1 - \tan^2(\frac{\theta_6}{2})}{1 + \tan^2(\frac{\theta_6}{2})} \qquad (13.62)$$

and

$$\sin \theta_6 = \frac{2 \tan(\frac{\theta_6}{2})}{1 + \tan^2(\frac{\theta_6}{2})} \qquad (13.63)$$

Because the process does not change, an equation like (13.15) is obtained, with factors as:

$$A = 2L_1 L_6 - 2L_2 L_6 \cos \theta \qquad (13.64)$$

$$B = -2L_2 L_6 \sin\theta \qquad (13.65)$$

$$C = L_1^2 + L_6^2 + L_2^2 - L_5^2 - 2L_1 L_2 \cos\theta \qquad (13.66)$$

$$t = \tan\left(\frac{\theta_6}{2}\right) \qquad (13.67)$$

The following process is similar as previously stated from equation (13.15), by arranging as a quadratic general form,

$$t = \frac{-B + \sqrt{B^2 - C^2 + A^2}}{C - A} \qquad (13.68)$$

Let us recall that $t = \tan\left(\frac{\theta_6}{2}\right)$, to solve for θ_6, thus

$$\theta_6 = 2\tan^{-1}(t) \qquad (13.69)$$

Once θ_6 is already solved, then $L_5 \sin\theta_5$ is divided by $L_5 \cos\theta_5$ in order to solve for θ_5,

$$\tan\theta_5 = \left(\frac{L_6 \sin\theta_6 - L_2 \sin\theta}{L_1 + L_6 \cos\theta_6 - L_2 \cos\theta}\right) \qquad (13.70)$$

and subsequently,

$$\theta_5 = \tan^{-1}\left(\frac{L_6 \sin\theta_6 - L_2 \sin\theta}{L_1 + L_6 \cos\theta_6 - L_2 \cos\theta}\right) \qquad (13.71)$$

The following linkage to solve for, is comprised by the rigid links L_6, L_8, L_9, L_{10} arranged as a closed kinematic chain, as depicted in figure 13.1-e). Hence, the following the position equations are stated for the X component

$$L_8 \cos\theta_8 + L_9 \cos\theta_9 = L_6 \cos\theta_6 + L_{10} \cos\theta_{10} \qquad (13.72)$$

and for the Y component,

$$L_8 \sin\theta_8 + L_9 \sin\theta_9 = L_6 \sin\theta_6 + L_{10} \sin\theta_{10} \qquad (13.73)$$

At this stage, it is assumed that the angles θ_6 and θ_8 are already known through a functional form. Thus, in the analysis of actual kinematic chain, the unknown variables are the angles θ_9 and θ_{10}. Therefore, arranging as in the next expressions,

$$L_9 \cos\theta_9 = L_6 \cos\theta_6 + L_{10} \cos\theta_{10} - L_8 \cos\theta_8 \qquad (13.74)$$

and

$$L_9 \sin \theta_9 = L_6 \sin \theta_6 + L_{10} \sin \theta_{10} - L_8 \sin \theta_8 \tag{13.75}$$

They are squared, and algebraically simplified,

$$(L_9 \cos \theta_9)^2 = (L_6 \cos \theta_6 + L_{10} \cos \theta_{10} - L_8 \cos \theta_8)^2 \tag{13.76}$$

$$L_9^2 \cos^2 \theta_9 = L_6^2 \cos^2 \theta_6 + 2L_6 L_{10} \cos \theta_6 \cos \theta_{10}$$
$$+L_{10}^2 \cos^2 \theta_{10} - 2(L_6 \cos \theta_6 + L_{10} \cos \theta_{10})(L_8 \cos \theta_8) + L_8^2 \cos^2 \theta_8 \tag{13.77}$$

hence,

$$(L_9 \sin \theta_9)^2 = (L_6 \sin \theta_6 + L_{10} \sin \theta_{10} - L_8 \sin \theta_8)^2 \tag{13.78}$$

and

$$L_9^2 \sin^2 \theta_9 = L_6^2 \sin^2 \theta_6 + 2L_6 L_{10} \sin \theta_6 \sin \theta_{10} + L_{10}^2 \sin^2 \theta_{10}$$
$$-2(L_6 \sin \theta_6 + L_{10} \sin \theta_{10})(L_8 \sin \theta_8) + L_8^2 \sin^2 \theta_8 \tag{13.79}$$

Adding $(L_9 \cos \theta_9)^2$ and $(L_9 \sin \theta_9)^2$, and algebraically rearranging,

$$(2L_6 L_{10} \cos \theta_6 - 2L_8 L_{10} \cos \theta_8) \cos \theta_{10} + (2L_6 L_{10} \sin \theta_6 - 2L_8 L_{10} \sin \theta_8) \sin \theta_{10} +$$
$$(L_6^2 + L_{10}^2 + L_8^2 - L_9^2 - 2L_6 L_8(\cos \theta_6 \cos \theta_8 + \sin \theta_6 \sin \theta_8) = 0 \tag{13.80}$$

Once again, trigonometric identities are used for further simplification,

$$\cos \theta_{10} = \frac{1 - \tan^2(\frac{\theta_{10}}{2})}{1 + \tan^2(\frac{\theta_{10}}{2})} \tag{13.81}$$

and

$$\sin \theta_{10} = \frac{2\tan(\frac{\theta_{10}}{2})}{1 + \tan^2(\frac{\theta_{10}}{2})} \tag{13.82}$$

We define the following expressions

$$A = 2L_6 L_{10} \cos \theta_6 - 2L_8 L_{10} \cos \theta_8, \tag{13.83}$$

$$B = 2L_6 L_{10} \sin \theta_6 - 2L_8 L_{10} \sin \theta_8, \tag{13.84}$$

$$C = L_6^2 + L_{10}^2 + L_8^2 - L_9^2 - 2L_6 L_8(\cos\theta_6 \cos\theta_8 + \sin\theta_6 \sin\theta_8), \tag{13.85}$$

and

$$t = \tan(\frac{\theta_{10}}{2}) \tag{13.86}$$

hence, solving by using a general quadratic form, the result becomes,

$$t = \frac{-B - \sqrt{B^2 - C^2 + A^2}}{C - A} \tag{13.87}$$

and considering that $t = \tan(\frac{\theta_{10}}{2})$, hence we solve for θ_{10} in the following manner,

$$\theta_{10} = 2\tan^{-1}(t) \tag{13.88}$$

Likewise, with θ_{10} already known, $L_9 \sin\theta_9$ is divided by $L_9 \cos\theta_9$ to obtain the solution for θ_9,

$$\tan\theta_9 = \left(\frac{L_6 \sin\theta_6 + L_{10} \sin\theta_{10} - L_8 \sin\theta_8}{L_6 \cos\theta_6 + L_{10} \cos\theta_{10} - L_8 \cos\theta_8} \right) \tag{13.89}$$

and subsequently,

$$\theta_9 = \tan^{-1}\left(\frac{L_6 \sin\theta_6 + L_{10} \sin\theta_{10} - L_8 \sin\theta_8}{L_6 \cos\theta_6 + L_{10} \cos\theta_{10} - L_8 \cos\theta_8} \right) \tag{13.90}$$

Finally, an analysis for the next linkage comprised of the kinematic chain L_{10}, L_{11}, L_{12} is provided (figure 13.1-f). The following kinematic equations are stated, for the X component,

$$L_{10} \cos\theta_{10} + L_{11} \cos\theta_{11} = L_{12} \cos\theta_{12} \tag{13.91}$$

and for the Y component,

$$L_{10} \sin\theta_{10} + L_{11} \sin\theta_{11} = L_{12} \sin\theta_{12} \tag{13.92}$$

Since the passive angle θ_{10} has already been analytically solved, it is involved as the input angle for the next linkage. So far this stage, only two passive angles θ_{11} and θ_{12} still remain unknown. Thus, arranging as in the following expressions,

$$L_{11} \cos\theta_{11} = L_{12} \cos\theta_{12} - L_{10} \cos\theta_{10} \tag{13.93}$$

and

$$L_{11} \sin \theta_{11} = L_{12} \sin \theta_{12} - L_{10} \sin \theta_{10} \tag{13.94}$$

Then, the expression is squared, and algebraically simplified,

$$(L_{11} \cos \theta_{11})^2 = (L_{12} \cos \theta_{12} - L_{10} \cos \theta_{10})^2 \tag{13.95}$$

then, the squared binomial is expanded,

$$L_{11}^2 \cos^2 \theta_{11} = L_{12}^2 \cos^2 \theta_{12} - 2L_{10}L_{12} \cos \theta_{10} \cos \theta_{12} + L_{10}^2 \cos^2 \theta_{10} \tag{13.96}$$

similarly for the Y component,

$$(L_{11} \sin \theta_{11})^2 = (L_{12} \sin \theta_{12} - L_{10} \sin \theta_{10})^2 \tag{13.97}$$

and

$$L_{11}^2 \sin^2 \theta_{11} = L_{12}^2 \sin^2 \theta_{12} - 2L_{10}L_{12} \sin \theta_{10} \sin \theta_{12} + L_{10}^2 \sin^2 \theta_{10} \tag{13.98}$$

In addition, for algebraic reduction, we add the terms $(L_{11} \cos \theta_{11})^2$ and $(L_{11} \sin \theta_{11})^2$, resulting the next expression,

$$L_{11}^2 = L_{12}^2 - 2L_{10}L_{12}(\cos \theta_{10} \cos \theta_{12} + \sin \theta_{10} \sin \theta_{12}) + L_{10}^2 \tag{13.99}$$

and by algebraically rearranging,

$$- (2L_{10}L_{12} \cos \theta_{10}) \cos \theta_{12} - (2L_{10}L_{12} \sin \theta_{10}) \sin \theta_{12} + (L_{10}^2 - L_{11}^2 + L_{12}^2) = 0 \tag{13.100}$$

Substituting the next trigonometric identities,

$$\cos \theta_{12} = \frac{1 - \tan^2(\frac{\theta_{12}}{2})}{1 + \tan^2(\frac{\theta_{12}}{2})} \tag{13.101}$$

as well as

$$\sin \theta_{12} = \frac{2 \tan(\frac{\theta_{12}}{2})}{1 + \tan^2(\frac{\theta_{12}}{2})} \tag{13.102}$$

Therefore, the following factors are defined,

$$A = -2L_{10}L_{12}\cos\theta_{10},$$
(13.103)

$$B = -2L_{10}L_{12}\sin\theta_{10},$$
(13.104)

and

$$C = L_{10}^2 - L_{11}^2 + L_{12}^2,$$
(13.105)

and

$$t = \tan\left(\frac{\theta_{12}}{2}\right)$$
(13.106)

Solving by using the quadratic general form, it leads to the next solution,

$$t = \frac{-B - \sqrt{B^2 - C^2 + A^2}}{C - A}$$
(13.107)

Considering that $t = \tan(\frac{\theta_{12}}{2})$, and we solve for the passive joint θ_{12}

$$\theta_{12} = 2\tan^{-1}(t)$$
(13.108)

Substituting $a = L_{10}^2 - L_{11}^2 + L_{12}^2$,

$$\theta_{12} = 2\tan^{-1}\left(\frac{(2L_{10}L_{12}\sin\theta_{10}) - \sqrt{(-2L_{10}L_{12}\sin\theta_{10})^2 a^2 + (-2L_{10}L_{12}\cos\theta_{10})^2}}{a + 2L_{10}L_{12}\cos\theta_{10}}\right)$$
(13.109)

Once θ_{12} is known, then $L_{11}\sin\theta_{11}$ is divided by $L_{11}\cos\theta_{11}$ to obtain θ_{11}:

$$\tan(\theta_{11}) = \frac{L_{12}\sin\theta_{12} - L_{10}\sin\theta_{10}}{L_{12}\cos\theta_{12} - L_{10}\cos\theta_{10}}$$
(13.110)

and

$$\theta_{11} = \tan^{-1}\left(\frac{L_{12}\sin\theta_{12} - L_{10}\sin\theta_{10}}{L_{12}\cos\theta_{12} - L_{10}\cos\theta_{10}}\right)$$
(13.111)

13.1.1   Passive joints simulation

Previous analytical solutions were directly coded in C++ programming language, resulting with a very fast computing performance. Likewise, plots were produced with GNUplot. Therefore, by summarising the set of equations modelling the passive joints in terms of the angles of the Jansen mechanism limb, the following expressions are analytic solutions.

$$\theta_1 = 0; \qquad \theta = \theta$$

For the next angles θ_3 and θ_4, let us define $L_{34} = L_1^2 + L_2^2 - L_3^2 + L_4^2$, $\ell_c = 2L_1 L_2 \cos \theta$, $\ell_s = 2L_2 L_4 \sin \theta$, as well as $\ell_z = 2L_2 L_4 \cos \theta$, such that

$$\theta_3 = \tan^{-1} \left(\frac{L_4 \sin \left(2 \tan^{-1} \left(\frac{\ell_s - \sqrt{(-\ell_s)^2 - (L_{34} - \ell_c)^2 + (2L_1 L_4 - \ell_z)^2}}{(L_{34} - \ell_c) - (2L_1 L_4 - 2L_2 L_4 \cos \theta)} \right) \right) - L_2 \sin \theta}{L_1 + L_4 \cos \left(2 \tan^{-1} \left(\frac{(2L_2 L_4 \sin \theta) - \sqrt{-\ell_s^2 - (L_{34} - \ell_c)^2 + (2L_1 L_4 - \ell_z)^2}}{(L_{34} - \ell_c) - (2L_1 L_4 - \ell_z)} \right) \right) - L_2 \cos \theta} \right) \tag{13.112}$$

and

$$\theta_4 = 2 \tan^{-1} \left(\frac{\ell_s - \sqrt{(-\ell_s)^2 - (L_{34} - \ell_c)^2 + (2L_1 L_4 - \ell_z)^2}}{(L_{34} - \ell_c) - (2L_1 L_4 - \ell_z)} \right) \tag{13.113}$$

For the cases of θ_5 and θ_6, we firstly define the expressions $L_{56} = L_1^2 + L_6^2 + L_2^2 - L_5^2 - 2L_1 L_2 \cos \theta$, and $\ell_{56} = 2L_1 L_6 - 2L_2 L_6 \cos \theta$.

$$\theta_5 = \tan^{-1} \left(\frac{L_6 \sin \left(2 \tan^{-1} \left(\frac{2L_2 L_6 \sin \theta + \sqrt{(2L_2 L_6 \sin \theta)^2 - L_{56}^2 + \ell_{56}^2}}{L_{56} - \ell_{56}} \right) \right) - L_2 \sin \theta}{L_1 + L_6 \cos \left(2 \tan^{-1} \left(\frac{2L_2 L_6 \sin \theta + \sqrt{(2L_2 L_6 \sin \theta)^2 - L_{56}^2 + \ell_{56}^2}}{L_{56} - \ell_{56}} \right) \right) - L_2 \cos \theta} \right) \tag{13.114}$$

and

$$\theta_6 = 2 \tan^{-1} \left(\frac{2L_2 L_6 \sin \theta + \sqrt{(-2L_2 L_6 \sin \theta)^2 - L_{56}^2 + \ell_{56}^2}}{L_{56} - \ell_{56}} \right) \tag{13.115}$$

and

$$\theta_7 = \tan^{-1} \left(\frac{L_8 \sin(2 \tan^{-1}(\frac{2L_4 L_8 \sin \theta_4 - \sqrt{(-2L_4 L_8 \sin \theta_4)^2 - (L_8^2 - L_7^2 + L_4^2)^2 + (-2L_4 L_8 \cos \theta_4)^2}}{(L_8^2 - L_7^2 + L_4^2) - (-2L_4 L_8 \cos \theta_4)})) - L_4 \sin \theta_4}{L_8 \cos(2 \tan^{-1}(\frac{2L_4 L_8 \sin \theta_4 - \sqrt{(-2L_4 L_8 \sin \theta_4)^2 - (L_8^2 - L_7^2 + L_4^2)^2 + (-2L_4 L_8 \cos \theta_4)^2}}{(L_8^2 - L_7^2 + L_4^2) - (-2L_4 L_8 \cos \theta_4)})) - L_4 \cos \theta_4} \right)$$

$$\theta_8 = 2 \tan^{-1} \left(\frac{2L_4 L_8 \sin \theta_4 - \sqrt{(-2L_4 L_8 \sin \theta_4)^2 - (L_8^2 - L_7^2 + L_4^2)^2 + (-2L_4 L_8 \cos \theta_4)^2}}{(L_8^2 - L_7^2 + L_4^2) - (-2L_4 L_8 \cos \theta_4)} \right)$$

$$\theta_9 = \tan^{-1}\left(\frac{L_6 \sin\theta_6 + L_{10}\sin\theta_{10} - L_8\sin\theta_8}{L_6\cos\theta_6 + L_{10}\cos\theta_{10} - L_8\cos\theta_8}\right)$$

For the case of θ_{10} let us define the expression $L_d = L_6^2 + L_{10}^2 + L_8^2 - L_9^2 - 2L_6L_8(\cos\theta_6\cos\theta_8 + \sin\theta_6\sin\theta_8)$, $\ell_e = 2L_6L_{10}\cos\theta_6 - 2L_8L_{10}\cos\theta_8$, and $\ell_d = 2L_6L_{10}\sin\theta_6 - 2L_8L_{10}\sin\theta_8$

$$\theta_{10} = 2\tan^{-1}\left(\frac{-\ell_d - \sqrt{\ell_d^2 - L_d^2 + \ell_e^2}}{L_d^2 - \ell_e^2}\right) \tag{13.116}$$

Defining $L_{11} = 2L_{10}L_{12}\sin\theta_{10}$ for θ_{11}

$$\theta_{11} = \tan^{-1}\left(\frac{L_{12}\sin\left(2\tan^{-1}\left(\frac{L_{11}-\sqrt{-L_{11}^2-(L_{10}^2-L_{11}^2+L_{12}^2)^2+(-2L_{10}L_{12}\cos\theta_{10})^2}}{(L_{10}^2-L_{11}^2+L_{12}^2)-(-2L_{10}L_{12}\cos\theta_{10})}\right)\right) - L_{10}\sin\theta_{10}}{L_{12}\cos\left(2\tan^{-1}\left(\frac{L_{11}-\sqrt{-L_{11}^2-(L_{10}^2-L_{11}^2+L_{12}^2)^2+(-2L_{10}L_{12}\cos\theta_{10})^2}}{(L_{10}^2-L_{11}^2+L_{12}^2)-(-2L_{10}L_{12}\cos\theta_{10})}\right)\right) - L_{10}\cos\theta_{10}}\right) \tag{13.117}$$

Likewise, for θ_{12} let us substitute L_{11} too,

$$\theta_{12} = 2\tan^{-1}\left(\frac{L_{11} - \sqrt{-L_{11}^2 - (L_{10}^2 - L_{11}^2 + L_{12}^2)^2 + (-2L_{10}L_{12}\cos\theta_{10})^2}}{(L_{10}^2 - L_{11}^2 + L_{12}^2) - (-2L_{10}L_{12}\cos\theta_{10})}\right) \tag{13.118}$$

Previous summary of passive joints formulae are validated by producing numerical simulations of the entire system. Figure 13.2 shows resulting tracks for each joint in two legs of robot's lateral side. The simulation considered suitable inertial frames transformation, while the two limbs are synchronised. The numerical simulation validates the algebraic approach proposed in this chapter. Each track may be compared with figure 13.1, and with angles θ_0-θ_{12} formulae.

13.2   Robot's global passive movement

This section is mainly focused on the analysis of the mechanical structure end-joint, or contact point with surface. The contact point describes the kinematics of the robot's walking gait, either in terms of planar Cartesian positions, or in terms of velocities. The proposed robotic structure is an octapod, which is depicted in figure 13.3. It consists of four legs per side (lateral), where two actuators drive θ and ϕ as differential control for the two crankshafts. It's important to note that between the first and second pair of legs exist 120° offset in between

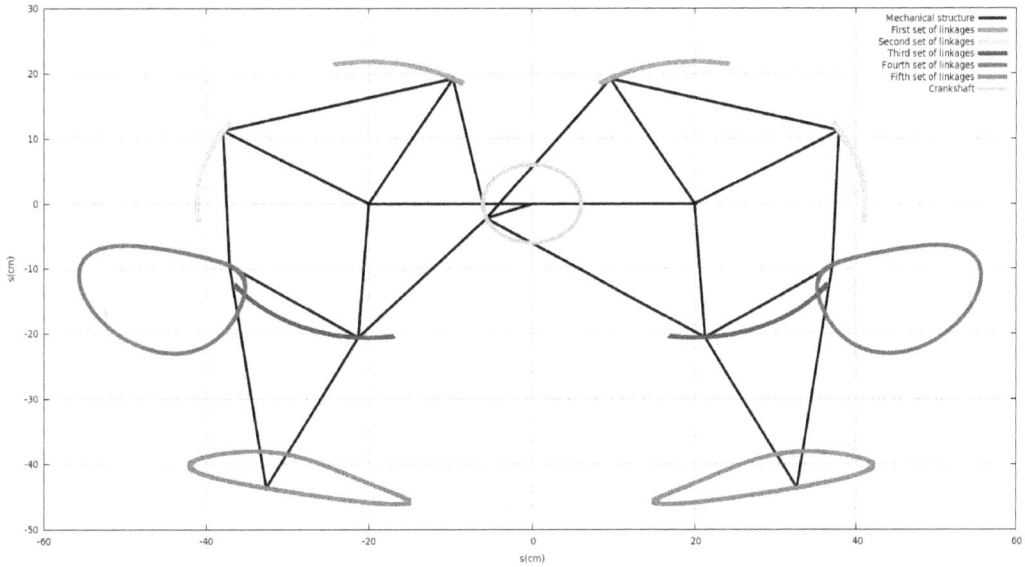

Figure 13.2:   Joints track simulation yielded by the linkage kinematic models θ_0-θ_{12}.

the crankshafts.

This section contains a formal analysis of the contact point position equations as functions of the driven angles θ and ϕ. Then, the study is organised in a section discussion front legs related position kinematics (those using index 1), and a section discussing the rear legs (those using index 2). Variables using index r refer to the right-sided limbs; while those using l refer to the left-sided limbs.

13.2.1   Front legs kinematic position

Lets define and substitute the terms $R_1 = 4L_{10}^2 L_{12}^2 \sin(2\tan^{-1}(D))^2 + 4L_{10}^2 L_{12}^2 \cos(2\tan^{-1}(D))^2$ and $L_s = L_{12}^2 - L_{11}^2 + L_{10}^2$,

$$X_{r_1} = L_{12}\cos\left(2\tan^{-1}\left(\frac{2L_{10}L_{12}\sin(2\tan^{-1}(D)) - \sqrt{R_1 - L_s^2}}{2L_{10}L_{12}\cos(2\tan^{-1}(D)) + L_s}\right)\right) + L_6\cos(A) + L_1 \qquad (13.119)$$

and

$$X_{l_1} = -L_{12}\cos\left(2\tan^{-1}\left(\frac{(2L_{10}L_{12}\sin(2\tan^{-1}(D)) - \sqrt{R_1 - L_s^2}}{2L_{10}L_{12}\cos(2\tan^{-1}(D)) + L_s)}\right)\right) - L_6\cos(A) - L_1 \qquad (13.120)$$

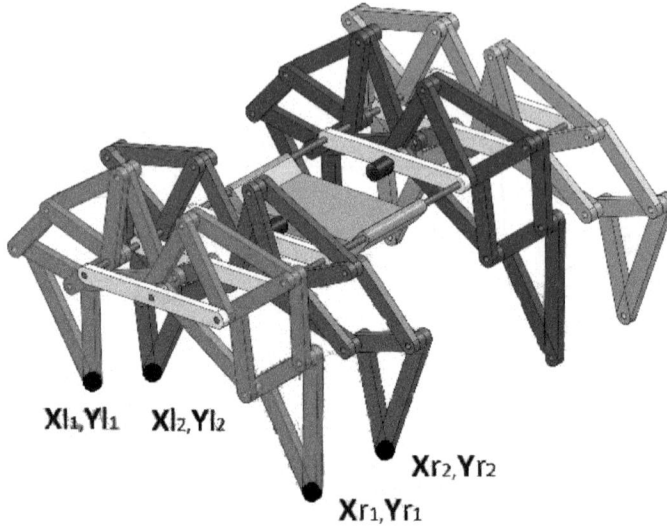

Figure 13.3: Octapod-type walking robot design, with legs' contact point position.

Likewise, $Y_{r_1} = Y_{l_1}$ hence

$$Y_{r_1} = L_{12} \sin \left(2 \tan^{-1} \left(\frac{2L_{10}L_{12} \sin(2\tan^{-1}(D)) - \sqrt{R_1 - L_s^2}}{2L_{10}L_{12} \cos(2\tan^{-1}(D)) + L_s} \right) \right) + L_6 \sin(A) \qquad (13.121)$$

The velocity models are described by their first order derivative,

$$\dot{X}_{r_1} = \frac{dX_{r_1}}{dt} \qquad (13.122)$$

$$\dot{X}_{l_1} = \frac{dX_{l_1}}{dt} \qquad (13.123)$$

and

$$\dot{Y}_{r_1} = \dot{Y}_{l_1} = \frac{dY_{r_1}}{dt} = \frac{dY_{l_1}}{dt} \qquad (13.124)$$

Hence by substituting the following expressions $L_A = L_6^2 - L_5^2 + L_2^2 + L_1^2$, and $L_{As} = 2L_2L_6 \sin(\theta)$,

$$A = 2\tan^{-1} \left(\frac{\sqrt{L_{As}^2 + (2L_1L_6 - 2L_2L_6 \cos(\theta))^2 - (-2L_1L_2 \cos(\theta) + L_A)^2} + L_{As}}{2L_2L_6 \cos(\theta) - 2L_1L_2 \cos(\theta) - 2L_1L_6 + L_A} \right) \qquad (13.125)$$

likewise, by substituting $L_B = L_4^2 - L_3^2 + L_2^2 + L_1^2$, and $L_{Bs} = 2L_2L_4\sin(\theta)$,

$$B = 2\tan^{-1}\left(\frac{L_{Bs} - \sqrt{L_{Bs}^2 + (2L_1L_4 - 2L_2L_4\cos(\theta))^2 - (-2L_1L_2\cos(\theta) + L_B)^2}}{2L_2L_4\cos(\theta) - 2L_1L_2\cos(\theta) - 2L_1L_4 + L_B}\right) \quad (13.126)$$

following with substitutions $L_C = L_8^2 - L_7^2 + L_4^2$

$$C = 2\tan^{-1}\left(\frac{2L_4L_8\sin(B) - \sqrt{4L_4^2L_8^2\sin(B)^2 + 4L_4^2L_8^2\cos(B)^2 - L_B^2}}{2L_4L_8\cos(B) + L_C}\right) \quad (13.127)$$

and defining the term $L_D = L_9^2 + L_8^2 + L_6^2 + L_{10}^2$ the following expression is stated,

$$D = \frac{-\sqrt{d} + 2L_{10}L_8\sin(C) - 2L_{10}L_6\sin(A)}{-2L_6L_8(\sin(A)\sin(C) + \cos(A)\cos(C)) + 2L_{10}L_8\cos(C) - 2L_{10}L_6\cos(A) - L_D} \quad (13.128)$$

Where $d = -(-2L_6L_8(\sin(A)\sin(C) + \cos(A)\cos(C)) - L_D)^2 + (2L_{10}L_6\sin(A) - 2L_{10}L_8\sin(C))^2 + (2L_{10}L_6\cos(A) - 2L_{10}L_8\cos(C))^2$

13.2.2 Rear legs kinematic position

Let us define the expressions $L_{r2} = 2L_{10}L_{12}\sin(2\tan^{-1}(D)$, and $L_{2s} = L_{12}^2 - L_{11}^2 + L_{10}^2$,

$$X_{r_2} = L_{12}\cos\left(\frac{2\tan^{-1}\left(L_{r2} - \sqrt{L_{r2}^2 + 4L_{10}^2L_{12}^2\cos(2\tan^{-1}(D))^2 - L_{2s}^2}\right)}{2L_{10}L_{12}\cos(2\tan^{-1}(D)) + L_{2s}}\right) + L_6\cos(A) + L_1$$

$$(13.129)$$

and

$$X_{1_2} = -L_{12}\cos\left(\frac{2\tan^{-1}\left(L_{r2} - \sqrt{L_{r2}^2 + 4L_{10}^2L_{12}^2\cos(2\tan^{-1}(D))^2 - L_{2s}^2}\right)}{2L_{10}L_{12}\cos(2\tan^{-1}(D)) + L_{2s}}\right) - L_6\cos(A) - L_1$$

$$(13.130)$$

and since $Y_{r_2} = Y_{l_2}$

$$Y_{l,r_2} = L_{12}\sin\left(\frac{2\tan^{-1}\left(L_{r2} - \sqrt{L_{r2}^2 + 4L_{10}^2L_{12}^2\cos(2\tan^{-1}(D))^2 - L_{2s}^2}\right)}{2L_{10}L_{12}\cos(2\tan^{-1}(D)) + L_{2s}}\right) + L_6\sin(A) \quad (13.131)$$

It follows that the Cartesian speeds obtained from firs order derivatives w.r.t. time are stated by

$$\dot{X}_{r_2} = \frac{dX_{r_2}}{dt} \tag{13.132}$$

and

$$\dot{X}_{l_2} = \frac{dX_{l_2}}{dt} \tag{13.133}$$

as well as

$$\dot{Y}_{r_2} = \dot{Y}_{l_2} = \frac{dY_{r_2}}{dt} = \frac{dY_{l_2}}{dt} \tag{13.134}$$

Therefore, by substituting next expressions, $L_{a1} = L_6^2 - L_5^2 + L_2^2 + L_1^2$, $L_{a2} = 2L_2L_6\sin(\theta + \frac{120\pi}{180})$, and $L_{a3} = 2L_2L_6\cos(\theta + \frac{120\pi}{180})$,

$$A = 2\tan^{-1}\left(\frac{\sqrt{L_{a2}^2 + (2L_1L_6 - L_{a3})^2 - (-2L_1L_2\cos(\theta + \frac{120\pi}{180}) + L_{a1})^2} + L_{a2}}{L_{a3} - 2L_1L_2\cos(\theta + \frac{120\pi}{180}) - 2L_1L_6 + L_{a1}}\right) \tag{13.135}$$

Likewise, substituting $L_{b1} = L_4^2 - L_3^2 + L_2^2 + L_1^2$, $L_{b2} = 2L_2L_4\sin(\theta + \frac{120\pi}{180})$, and $L_{b3} = 2L_2L_4\cos(\theta + \frac{120\pi}{180})$,

$$B = 2\tan^{-1}\left(\frac{(L_{b2} - \sqrt{L_{b2}^2 + (2L_1L_4 - L_{a3})^2 - (-2L_1L_2\cos(\theta + \frac{120\pi}{180}) + L_{b1})^2})}{(L_{a3} - 2L_1L_2\cos(\theta + \frac{120\pi}{180}) - 2L_1L_4 + L_{b1})}\right) \tag{13.136}$$

For the case of factor C, it is defined as

$$C = 2\tan^{-1}\left(\frac{2L_4L_8\sin(B) - \sqrt{4L_4^2L_8^2\sin(B)^2 + 4L_4^2L_8^2\cos(B)^2 - (L_8^2 - L_7^2 + L_4^2)^2}}{2L_4L_8\cos(B) + L_8^2 - L_7^2 + L_4^2}\right) \tag{13.137}$$

and for D let us define $L_{d1} = L_9^2 + L_8^2 + L_6^2 + L_{10}^2$

$$D = \frac{-\sqrt[2]{d} + 2L_{10}L_8\sin(C) - 2L_{10}L_6\sin(A)}{-2L_6L_8(\sin(A)\sin(C) + \cos(A)\cos(C)) + 2L_{10}L_8\cos(C) - 2L_{10}L_6\cos(A) - L_{d1}} \tag{13.138}$$

where $d = -(-2L_6L_8(\sin(A)\sin(C) + \cos(A)\cos(C)) - L_{d1})^2 + (2L_{10}L_6\sin(A) - 2L_{10}L_8\sin(C))^2 + (2L_{10}L_6\cos(A) - 2L_{10}L_8\cos(C))^2$.

13.3 Robot analysis with driven angle ϕ

13.3.1 Front limbs

The equations for planar Cartesian positions for the front limbs are stated. Thus, let us define the following terms $R_{r1} = L_{12}^2 - L_{11}^2 + L_{10}^2$, $S_{r1} = 2L_{10}L_{12}\sin(2\tan^{-1}(D))$, and $T_{r1} = 2L_{10}L_{12}\cos(2\tan^{-1}(D))$,

$$X_{r_1} = L_{12}\cos\left(2\tan^{-1}\left(\frac{(S_{r1} - \sqrt{S_{r1}^2 + T_{r1}^2 - R_{r1}^2})}{(T_{r1} + R_{r1})}\right)\right) + L_6\cos(A) + L_1 \qquad (13.139)$$

and,

$$X_{l_1} = -L_{12}\cos(2\tan^{-1}(\frac{(S_{r1} - \sqrt{S_{r1}^2 + 4L_{10}^2 L_{12}^2 \cos(2\tan^{-1}(D))^2 - R_{r1}^2})}{(2L_{10}L_{12}\cos(2\tan^{-1}(D)) + R_{r1})})) - L_6\cos(A) - L_1 \quad (13.140)$$

and since $Y_{r_1} = Y_{l_1}$, then

$$Y_{r_1} = L_{12}\sin\left(2\tan^{-1}\left(\frac{(S_{r1} - \sqrt{R_{r1}^2 + 4L_{10}^2 L_{12}^2 \cos(2\tan^{-1}(D))^2 - (R_{r1})^2)}}{(2L_{10}L_{12}\cos(2\tan^{-1}(D)) + R_{r1})}\right)\right) + L_6\sin(A) \quad (13.141)$$

It follows that the Cartesian speeds are described by the first order derivative w.r.t. time,

$$\dot{X}_{r_2} = \frac{dX_{r_2}}{dt} \qquad (13.142)$$

and

$$\dot{X}_{l_2} = \frac{dX_{l_2}}{dt} \qquad (13.143)$$

and

$$\dot{Y}_{r_2} = \dot{Y}_{l_2} = \frac{dY_{r_2}}{dt} = \frac{dY_{l_2}}{dt} \qquad (13.144)$$

Thus, by defining the following expressions that were used previously. Let us re-define the terms $\ell_a = 2L_1 L_6 - 2L_2 L_6 \cos(\phi)$, and $\tau_A = L_6^2 - L_5^2 + L_2^2 + L_1^2$

$$A = 2\tan^{-1}\left(\frac{\sqrt{4L_2^2 L_6^2 \sin(\phi)^2 + (\ell_a)^2 - (-2L_1 L_2 \cos(\phi) + \tau_A)^2} + 2L_2 L_6 \sin(\phi)}{2L_2 L_6 \cos(\phi) - 2L_1 L_2 \cos(\phi) - 2L_1 L_6 + \tau_A}\right) \qquad (13.145)$$

Likewise, by defining the following expressions that were used previously. Let us re-define the terms $\ell_b = 2L_1L_4 - 2L_2L_4\cos(\phi)$, and $\tau_B = L_4^2 - L_3^2 + L_2^2 + L_1^2$,

$$B = 2\tan^{-1}\left(\frac{2L_2L_4\sin(\phi) - \sqrt{4L_2^2L_4^2\sin(\phi)^2 + \ell_b^2 - (-2L_1L_2\cos(\phi) + \tau_B)^2}}{2L_2L_4\cos(\phi) - 2L_1L_2\cos(\phi) - 2L_1L_4 + \tau_B}\right) \tag{13.146}$$

by defining the following expressions that were used previously. Let us re-define the term $\tau_C = L_8^2 - L_7^2 + L_4^2$,

$$C = 2\tan^{-1}\left(\frac{2L_4L_8\sin(B) - \sqrt{4L_4^2L_8^2\sin(B)^2 + 4L_4^2L_8^2\cos(B)^2 - (\tau_C)^2}}{2L_4L_8\cos(B) + \tau_C}\right) \tag{13.147}$$

and by defining the term $\tau_D = L_9^2 + L_8^2 + L_6^2 + L_{10}^2$,

$$D = \frac{-\sqrt{d} + 2L_{10}L_8\sin(C) - 2L_{10}L_6\sin(A)}{-2L_6L_8(\sin(A)\sin(C) + \cos(A)\cos(C)) + 2L_{10}L_8\cos(C) - 2L_{10}L_6\cos(A) - \tau_D} \tag{13.148}$$

where $d = -(-2L_6L_8(\sin(A)\sin(C) + \cos(A)\cos(C)) - \tau_D)^2 + (2L_{10}L_6\sin(A) - 2L_{10}L_8\sin(C))^2 + (2L_{10}L_6\cos(A) - 2L_{10}L_8\cos(C))^2$.

13.3.2 Rear limbs

The equations for planar Cartesian positions for the front limbs are stated. Thus, let us use the terms previously defined R_{r1}, S_{r1} and T_{r1},

$$X_{r_2} = L_{12}\cos\left(2\tan^{-1}\left(\frac{S_{r1} - \sqrt{S_{r1}^2 + T_{r1}^2 - R_{r1}^2}}{T_{r1} + R_{r1}}\right)\right) + L_6\cos(A) + L_1 \tag{13.149}$$

Likewise, following with the same substitution terms R_{r1}, S_{r1}, and T_{r1},

$$X_{1_2} = -L_{12}\cos\left(2\tan^{-1}\left(\frac{(S_{r1} - \sqrt{S_{r1}^2 + T_{r1}^2 - R_{r1}^2})}{(T_{r1} + R_{r1})}\right)\right) - L_6\cos(A) - L_1 \tag{13.150}$$

and

$$Y_{r_2} = Y_{l_2} = L_{12}\sin\left(2\tan^{-1}\left(\frac{S_{r1} - \sqrt{S_{r1}^2 + T_{r1}^2 - (R_{r1})^2}}{T_{r1} + R_{r1}}\right)\right) + L_6\sin(A) \tag{13.151}$$

In addition, let us define the Cartesian speed of the rear limbs' contact point,

$$\dot{X}_{r_2} = \frac{dX_{r_2}}{dt}$$

and

$$\dot{X}_{l_2} = \frac{dX_{l_2}}{dt}$$

as well as,

$$\dot{Y}_{r_2} = \dot{Y}_{l_2} = \frac{dY_{r_2}}{dt} = \frac{dY_{l_2}}{dt}$$

By substituting the factors used in previous equations, the expressions for A, B, C, and D are defined. In addition, let us redefine $\ell_1 = 2L_2L_6\cos(\phi + \frac{120\pi}{180})$, $\ell_2 = 2L_1L_2\cos(\phi + \frac{120\pi}{180})$, $\ell_3 = 2L_2L_6\sin(\phi + \frac{120\pi}{180})$, and $L_A = L_5^2 + L_2^2 + L_1^2$

$$A = 2\tan^{-1}\left(\frac{\sqrt[2]{\ell_3^2 + (2L_1L_6 - \ell_1)^2 - (-\ell_2 + L_6^2 - L_A)^2} + \ell_3}{(\ell_1 - \ell_2 + L_6^2 - 2L_1L_6 - L_A)}\right) \tag{13.152}$$

and defining $L_{41} = L_4^2 - L_3^2 + L_2^2 + L_1^2$, and $\ell_4 = 2L_2L_4\sin(\phi + \frac{120\pi}{180})$,

$$B = 2\tan^{-1}\left(\frac{\ell_4 - \sqrt{\ell_4^2 + (2L_1L_4 - 2L_2L_4\cos(\phi + \frac{120\pi}{180}))^2 - (-\ell_2 + L_{41})^2}}{2L_2L_4\cos(\phi + \frac{120\pi}{180}) - \ell_2 - 2L_1L_4 + L_{41}}\right) \tag{13.153}$$

Similarly,

$$C = 2\tan^{-1}\left(\frac{2L_4L_8\sin(B) - \sqrt{4L_4^2L_8^2\sin(B)^2 + 4L_4^2L_8^2\cos(B)^2 - (L_8^2 - L_7^2 + L_4^2)^2}}{2L_4L_8\cos(B) + L_8^2 - L_7^2 + L_4^2}\right) \tag{13.154}$$

Finally, by defining the term $L_{19} = L_9^2 + L_8^2 + L_6^2 + L_{10}^2$

$$D = \frac{-\sqrt{d} + 2L_{10}L_8\sin(C) - 2L_{10}L_6\sin(A)}{-2L_6L_8(\sin(A)\sin(C) + \cos(A)\cos(C)) + 2L_{10}L_8\cos(C) - 2L_{10}L_6\cos(A) - L_{19}} \tag{13.155}$$

where $d = -(-2L_6L_8(\sin(A)\sin(C) + \cos(A)\cos(C)) - L_{19})^2 + (2L_{10}L_6\sin(A) - 2L_{10}L_8\sin(C))^2 + (2L_{10}L_6\cos(A) - 2L_{10}L_8\cos(C))^2$.

13.4 Robot's motion control

On studying how the mechanisms walking evolve, results of critical importance to provide the robot with the ability of self-balancing. Synchronised contact of the eight legs with ground may grant suitable stability for the global robot's manoeuvrability. Figure 13.4 depicts a top view of the octapod, as well as hash-tag symbols plus a number indicating the synchronization order of the limbs during walking. The step's displacements magnitude were set arbitrary for both: for the purpose of simulation, and for an hoe-made experimental prototype robot. In addition, this order of walking is a preamble for any type of controller design.

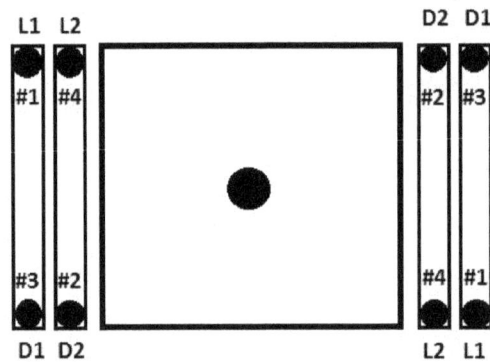

Figure 13.4:   Octapod's top view. The hashtag symbols plus a number indicate the synchronising order for the limbs during walking.

Such as depicted in figure 13.4 (XZ-plane), an eight-leg walking pattern is indicated. Where the big black dot located at the robot's centre determines the robotic structure's centroid, while the smaller black dots represent each limb's contact point (i.e. $L1$, $L2$, $D1$, and $D2$). Further, the gait sequence follows the next order:

1. The front left-sided leg (L_1) steps first.

2. Second, the rear right-sided leg (D_2) steps.

3. Third, the front right-sided leg (D_1) steps next.

4. Finally the rear left-sided leg (L_2) moves.

5. Repeats again from 1.

From previous walking sequence, and kinematic analysis, figure 13.5 illustrates a side view (XY plane) of the gait patterns, and the walking behaviour for the set of three limbs during a normal walking task. Two front-side limbs (black color for the right-sided, and red color for the left-sided limbs), and one rear-side limb (blue color). The walking pattern yielded by the two actuators θ and ϕ is cyclic and records the same type of gait as any other limb does, although with constant angular offsets.

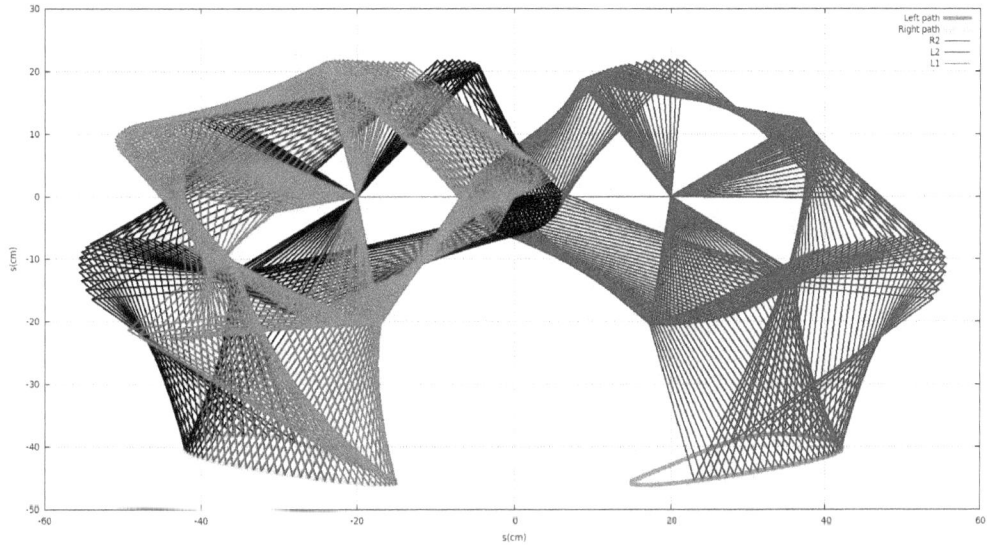

Figure 13.5: Walking patterns simulation of three limbs.

In addition, the robot's global displacement along the X-axis is contributed by the eight limbs moving in coordination. Thus, the robot's displacement was characterised by setting arbitrary link sizes preserving same rate even when lengths are changed. The robot displacement is numerically expressed by the following limbs' contribution, where such linear displacement will depend on the limbs touching the ground in turn. At least two limbs synchronised that simultaneously are producing the same displacement.

$$\Delta_x \approx \begin{cases} 24.2cm, & \Delta_{R_1} \\ 19cm, & \Delta_{R_2} \\ 24.2cm, & \Delta_{L_1} \\ 19cm, & \Delta_{L_2} \end{cases} \tag{13.156}$$

The robot's trajectory control is one of the main issues of this chapter. In terms of the robot's global kinematic control, the instantaneous input vector $\mathbf{u}_t \in \mathbb{R}^2$ is defined. Its components are the instantaneous velocity v_t, and the instantaneous angular velocity ω_t,

$$\mathbf{u}_t = \begin{pmatrix} v_t \\ \omega_t \end{pmatrix}$$

On formulating v_t, the robot's absolute velocity represents an averaged displacement of the eight legs along the longitudinal X-axis. The eight legs averaged displacements arise from their first order derivatives w.r.t. time Δt. Therefore, a formulation that approaches v_t is defined by

Postulate 13.4.1 (Octapod's instantaneous absolute velocity). *The absolute velocity is an average speed value of the eight limbs' contact point.*

$$v_t = \frac{1}{8} \left(\dot{X}_{\theta R_1} + \dot{X}_{\theta R_2} + \dot{X}_{\theta L_1} + \dot{X}_{\theta L_2} + \dot{X}_{\phi R_1} + \dot{X}_{\phi R_2} + \dot{X}_{\phi L_1} + \dot{X}_{\phi L_2} \right) \qquad (13.157)$$

The indexes θ and ϕ denote the lateral side of the actuators positions Besides the robot's lateral control is related to the instantaneous angular velocity. Therefore, when the difference of sides speed yields an angular velocity that gradually changes the robot's direction. Thus, any change in lateral velocities will produce a speed component projected along the transversal axis, also known the normal component that impulses the robots to spin. Therefore, the definition of ω_t is provided next,

$$\omega_t = \frac{\hat{v}_t cos(\alpha)}{r} \qquad (13.158)$$

Where parameters and variables are depicted in figure 13.6. The differential velocity basically is the speed difference between both robot's sides as defined by the next expressions,

$$\hat{v}_t = \Sigma_R V_R - \Sigma_L V_L \qquad (13.159)$$

Such previous expression is known as the instantaneous differential velocity, and is equivalent to the sum of all limbs' velocity. The right-sides velocities are considered positives (counter-

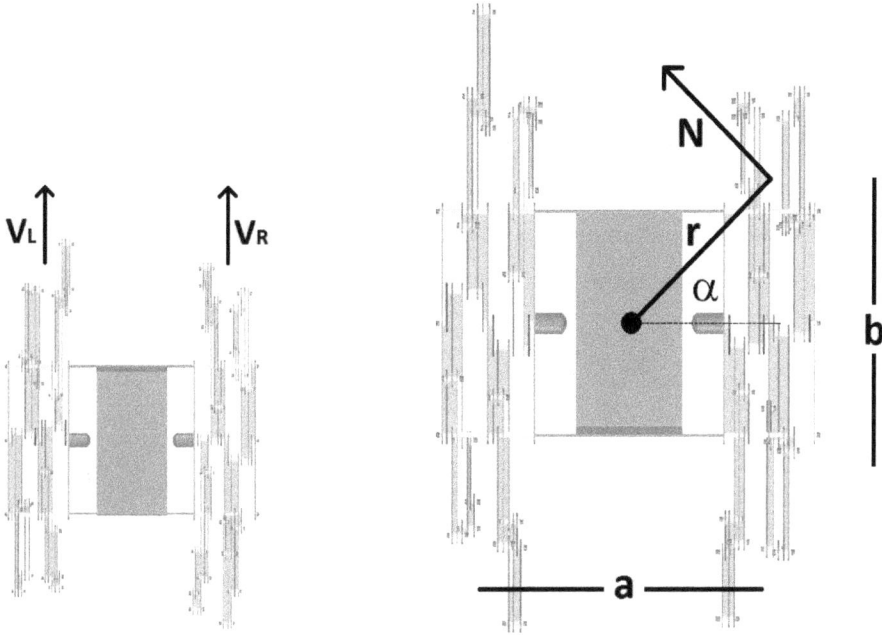

Figure 13.6: Robot's top view with kinematic configuration.

clockwise), while the left-sided velocities are assumed negative as they turn clockwise direction.

$$\cos(\alpha) = \frac{a}{2r} \tag{13.160}$$

where the functional form of r is stated by

$$r = \sqrt[2]{\left(\frac{a}{2}\right)^2 + \left(\frac{b}{2}\right)^2} \tag{13.161}$$

Being r the average distance of contact points with the ground of each one of the legs w.r.t. the robot's centroid. Likewise, a and b are constants and represent the lateral and longitudinal distances respectively among the limbs' contact point. Finally,

$$\omega_t = \frac{(v_R - v_L)(\frac{a}{2r})}{\sqrt{\frac{a^2+b^2}{4}}} \tag{13.162}$$

thus, algebraically arranging,

$$\omega_t = \frac{\frac{(v_R - v_L)a}{2r}}{\frac{\sqrt{a^2+b^2}}{2}} \tag{13.163}$$

reorganizing division terms,

$$\omega_t = \frac{(v_R - v_L)(a)(2)}{2r\sqrt[2]{a^2 + b^2}}$$

(13.164)

by reducing terms,

Postulate 13.4.2 (Octapod's instantaneous yaw velocity model). *The octapod's yaw speed is postulated in terms of its kinematic structure, and difference between lateral velocities.*

$$\omega_t = \frac{a(v_R - v_L)}{r\sqrt[2]{a^2 + b^2}}$$

(13.165)

The instantaneous ω_t physical units are given in rad/s. When there exists a difference in the subtraction of lateral velocities, a numeric change for ω_t exists and $\omega_t \neq 0$, which is consequently produced by any change in θ and/or ϕ. In addition, when $v_t = v_{t-1}$, then there will not be any difference in velocities $v_t = \Sigma V_R - \Sigma V_L = 0$, and $\omega_t = 0$. For any instant $\hat{v}_t$ the projection upon the transversal axis is what determines the angular velocity. Mathematical analysis of this structure led to obtain a control vector of the octapod structure based on Theo Jansen's mechanism. The study of angles in all linkages allowed to gain better understanding on how the system behaves and evolves solely with two driven angles.

Bibliography

[1] Ingram A.J., *Numerical kinematic and kinetic analysis of a new class of twelve bar linkage for walking machines*, Master thesis University of Johannesburg, Faculty of Engineering, Dec 2004.

[2] Moldovan F., Dolga V., Cinios O., Pop C., *CAD design and analytical model of a twelve bar walking mechanism*, U.P.B. Sci. Bull., Series D, vol.73(2), pp.35–48, 2011, issn: 1454-2358.

[3] F. Asano, M. Yamakita, N. Kamamichi, and L. Zhi-Wei, *A novel gait generation for biped walking robots based on mechanical energy constraint*, IEEE Transactions Robotics and Automation, vol. 20(3), pp. 565 - 573, 2004, doi: dx.doi.org10.1109TRA.2004.824685.

[4] Sun-Wook K., Sung Hyun H., Dong Hun K., *Analysis of a crab robot based on Jansen mechanism*, 11th Intl. Conf. on Control, Automation and Systems, Cyeonggi-do, Korea, pp. 858–860, 2011.

[5] Ghassaei A., *The design and optimization of a crank-based leg mechanism*, Master Thesis, Pomona College, Department of Physics and Astronomy, USA, 2011.

[6] McGeer T., *Passive dynamic walking*, The International Journal of Robotics Research.vol. 9(2), pp. 62–82, 1990, issn:0278-3649, doi:10.1177027836499000900206

[7] Nansai S., Mohan R.E.,Tan N., Rojas N., Iwase M., *Dynamic Modeling and Nonlinear Position Control of a Quadruped Robot with Theo Jansen Linkage Mechanisms and a Single Actuator*, Journal of Robotics, Vol. 2015, pp.-15, 2015, doi: 10.11552015315673.

Chapter 14

RECONFIGURABLE GAIT PATTERNS OF A KLANN-BASED ROBOT

Jaichandar K. Sheba[1,3], Rajesh E. Mohan[2], Edgar A. Martínez García[3], Le Tan-Phuc[1]

[1]*Singapore Polytechnic, Singapore.*
[2]*Singapore University of Technology and Design, Singapore.*
[3]*Laboratorio de Robótica, Institute of Engineering and Technology, Universidad Autónoma de Ciudad Juárez, Mexico.*

Legged robots are well suited to walk over abrupt terrains and they are effective in using isolated footholds which optimize support and traction. Design based on one degree-of-freedom planar linkages can be energy efficient however their locomotion is limited by the range of gaits produced. In this chapter, novel reconfigurable mechanism based upon Klann linkage to generate wide range of gait cycles has been investigated, opening new possibilities for innovative applications. A robot that has a fixed structure of movement mechanism faces issues related to constrained set of gaits that it can produce. Numerous research efforts have been dedicated to this end involving varied design strategies. However, such an approach results in increasing difficulty in control algorithm when more and more units are connected with each other. Another design approach in the literature achieves gait variations via parametric changes of leg structure ‾ . Our aim in this research is to design a robot using Klann mechanism which can change its linkage configurations to adapt their gait according to changes of surrounding environment. In this chapter, Klann based reconfigurable design and implementation is presented, where a robot changes its structure morphology by changing its components and sub-assemblies parameters to adapt to multi-terrain and multi-task. The leg linkage length ratio of the robot platform has been modified to create a large number of gait patterns which can be selectively used by the robots to explore, test and apply to get real feedback. The proposed reconfigurable mechanism approach can be extended from two symmetry legged assembly to four or more legged assembly. This design supports multiple legged robots with adaptive changes when one leg fails or to use a leg as a tool to perform functions other than locomotion. This design can be applied to both homogeneous and heterogeneous legged platform.

14.1   Klann mechanism position analysis

The Klann linkage ("Klann leg" can be used as term in this chapter) named after its inventor is a one degree of freedom planar mechanism which is formed by six bars connected with each other by revolute joints . This linkage was designed to emulate a smooth walking motion with only one actuator. The limitation of this mechanism is that it can produce only one walking gait for one specific design of linkages. It can be observed that the foot trajectory of the standard Klann leg is similar to a specific animal walking gait ' . The main challenge in identifying foot trajectory for further reconfigurable design is to create efficient approaches to solve the position analysis problem of the Klann leg. A common method is to solve a system of non linear equation with the number of equation equal to number of unknown variable . However, the elimination process will give out a large number of solutions caused by trigonometric parts and tangent half-angle problem. Using bilateration method solution can be found for this type of problem .

The bilateration problem consists in finding feasible locations of a point P_c, given its distances to two other points, say P_a and P_b, whose locations are known. Then, according to figure 14.1-b), the solution to this problem can be expressed in matrix form as:

$$p_{a,c} = Z_{a,b,c}\, p_{a,b} \qquad (14.1)$$

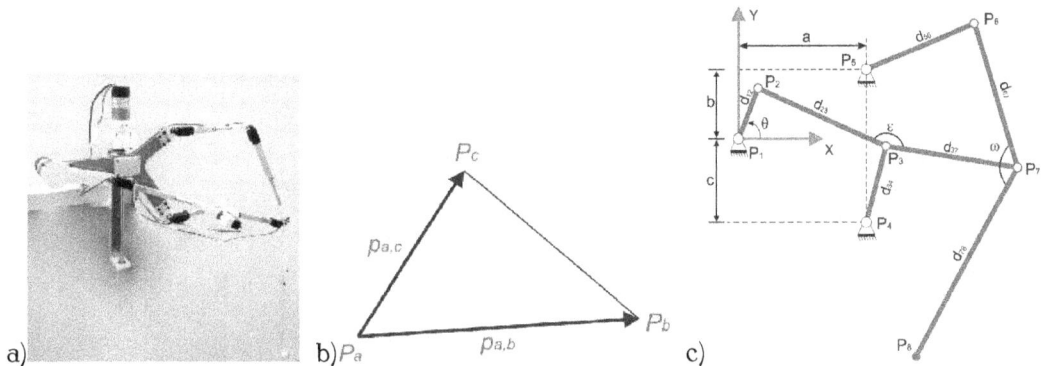

Figure 14.1:   a) Photo of an experimental Klann-based prototype limb. b) Geometric foundation of the bilateration problem. c) Klann linkage kinematics.

Where $p_{a,b} = \overrightarrow{P_a P_b}$ and,

$$Z_{a,b,c} = \frac{1}{2s_{a,b}} \begin{pmatrix} s_{a,b} + s_{a,c} - s_{b,c} & -4A_{a,b,c} \\ 4A_{a,b,c} & s_{a,b} + s_{a,c} - s_{b,c} \end{pmatrix} \tag{14.2}$$

is called a bilateration matrix, with $s_{a,b} = d_{a,b}^2 = \|P_{a,b}\|^2$, the squared distance between P_a and P_b, and

$$A_{a,b,c} = \pm \frac{1}{4}\sqrt{(s_{a,b} + s_{a,c} + s_{b,c})^2 - 2(s_{a,b}^2 + s_{a,c}^2 + s_{b,c}^2)} \tag{14.3}$$

The oriented area of $\triangle P_a P_b P_c$ which is defined as positive if P_c is to the left of vector $\overrightarrow{p_{a,b}}$, and vice versa. The interested reader can refer to the work for a derivation of equation. By using bilateration matrices, the position analysis problem of linkages such as Klann leg is greatly simplified. Next, we apply the bilateration method on Klann leg for solving the position analysis problem of the end point of the leg.

Figure 14.2 shows a Klann leg with 5 links, they are $\overline{P_1 P_2}$, $\overline{P_3 P_4}$, $\overline{P_5 P_6}$, $\overline{P_2 P_3 P_7}$, $\overline{P_6 P_7 P_8}$. This one-degree-of-freedom planar linkage consists of the frame ($\triangle P_1 P_4 P_5$), one crank ($\overline{P_1 P_2}$), two grounded rockers (segments $\overline{P_3 P_4}$, $\overline{P_5 P_6}$), and two couplers ($\triangle P_2 P_3 P_7$, $\triangle P_6 P_7 P_8$) all connected by revolute joints. These links and frame's dimensions including links length $d_{1,2}$, $d_{2,3}$, $d_{3,4}$, $d_{5,6}$, $d_{3,7}$, $d_{6,7}$ and $d_{7,8}$, fixed angle ε and ω of two couplers are all known with an angle θ for the input link. The Cartesian coordinate plane O_{xy} with original point was placed on joint P_1 together with axis directions as shown. The position analysis problem for Klann leg is then calculating all possible Cartesian locations of end point P_8.

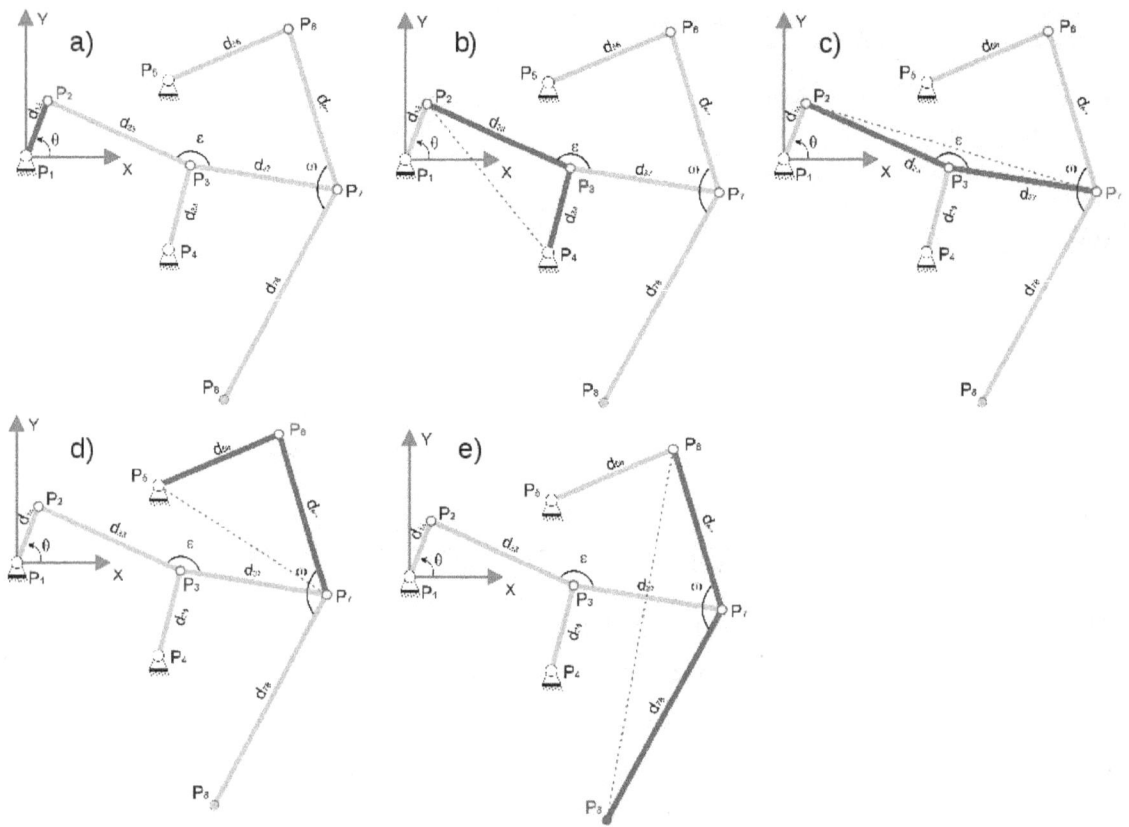

Figure 14.2:   Bilateration method applied to red-highlighted chain of links.

In this case, squared distances and bilateration matrix are used to compute the corresponding location of end point P_8 based on angle θ. First, let us locate the position of P_2 based on the input angle θ and the origin P_1 through a simple cosine equation as following:

$$\overrightarrow{P_1P_2} = d_{1,2} \begin{pmatrix} \cos\theta \\ \sin\theta \end{pmatrix} \tag{14.4}$$

Figure 14.2-a) shows the corresponding link for the calculation of equation (2). Figure 14.2-b) shows the next step of the process, the position of P_3 is located by computing $p_{2,3}$ from θ and the position of previous calculated joints P_2 and P_4 using equation (1).

$$p_{2,3} = Z_{2,3,4}p_{2,4} \tag{14.5}$$

it follows that,

$$\overrightarrow{P_2P_3} = \frac{1}{2s_{2,4}} \begin{pmatrix} s_{2,4} + s_{2,3} - s_{4,3} & -4A_{2,4,3} \\ 4A_{2,4,3} & s_{2,4} + s_{2,3} - s_{4,3} \end{pmatrix} \overrightarrow{P_2P_4} \tag{14.6}$$

with

$$\overrightarrow{P_2P_4} = \overrightarrow{P_4} - \overrightarrow{P_2} = \begin{pmatrix} x_4 - d_{1,2}\cos\theta \\ y_4 - d_{1,2}\sin\theta \end{pmatrix} \tag{14.7}$$

and

$$A_{2,4,3} = \frac{1}{4}\sqrt{(s_{2,4} + s_{2,3} + s_{4,3})^2 - 2(s_{2,4}{}^2 + s_{2,3}{}^2 + s_{4,3}{}^2)} \tag{14.8}$$

P_3 is to the left of vector $\overrightarrow{P_2P_4}$ in this case; and,

$$s_{2,4} = \left\| \overrightarrow{P_2P_4} \right\|^2 = (x_4 - d_{1,2}\cos\theta)^2 + (y_4 - d_{1,2}\sin\theta)^2 \tag{14.9}$$

After getting $\overrightarrow{P_2P_3}$, we can calculate $\overrightarrow{P_3} = \overrightarrow{P_2P_3} + \overrightarrow{P}_2$. Next, we locate P_7 based on P_2 and P_3 as shown in figure 14.2-c):

$$P_{2,7} = Z_{2,3,7}p_{2,3} \tag{14.10}$$

therefore,

$$\overrightarrow{P_2P_7} = \frac{1}{2s_{2,3}} \begin{pmatrix} s_{2,3} + s_{2,7} - s_{3,7} & -4A_{2,3,7} \\ 4A_{2,3,7} & s_{2,3} + s_{2,7} - s_{3,7} \end{pmatrix} \overrightarrow{P_2P_3} \tag{14.11}$$

with

$$s_{2,7} = s_{3,2} + s_{3,7} - 2d_{3,2}d_{3,7}\cos(\varepsilon), \tag{14.12}$$

and

$$A_{2,3,7} = \frac{1}{4}\sqrt[2]{(s_{2,3} + s_{2,7} + s_{3,7})^2 - 2(s_{2,3}^2 + s_{2,7}^2 + s_{3,7}^2)} \tag{14.13}$$

given that P_7 is to the left of vector $\overrightarrow{P_2P_3}$ in this case. thus,

$$\overrightarrow{P_7} = \overrightarrow{P_2P_7} + \overrightarrow{P_2} \tag{14.14}$$

Then, from known position of P_7 and P_5, P_6 is located (figure 14.2-d),

$$p_{5,6} = Z_{5,7,6}p_{5,7} \tag{14.15}$$

hence,

$$\overrightarrow{P_5P_6} = \frac{1}{2s_{5,7}}\begin{pmatrix} s_{5,7} + s_{5,6} - s_{7,6} & -4A_{5,7,6} \\ 4A_{5,7,6} & s_{5,7} + s_{5,6} - s_{5,6} \end{pmatrix}\overrightarrow{P_5P_7} \tag{14.16}$$

with $s_{5,7} = \|\overrightarrow{P_5P_7}\|^2$,

$$A_{5,7,6} = \frac{1}{4}\sqrt[2]{(s_{5,7} + s_{5,6} + s_{7,6})^2 - 2(s_{5,7}^2 + s_{5,6}^2 + s_{7,6}^2)} \tag{14.17}$$

where P_6 is to the left of vector $\overrightarrow{P_5P_7}$ in this case.

$$\overrightarrow{P_6} = \overrightarrow{P_5P_6} + P_5 \tag{14.18}$$

and finally, we locate the end point P_8 of the linkage based on previous calculated joints P_6 and P_7. From figure 14.2-e, we have:

$$p_{6,8} = Z_{6,7,8}p_{6,7} \tag{14.19}$$

Therefore,

$$\overrightarrow{P_6P_8} = \frac{1}{2s_{6,7}}\begin{pmatrix} s_{6,7} + s_{6,8} - s_{7,8} & -4A_{6,7,8} \\ 4A_{6,7,8} & s_{6,7} + s_{6,8} - s_{7,8} \end{pmatrix}\overrightarrow{P_6P_7} \tag{14.20}$$

with

$$s_{6,8} = s_{7,6} + s_{7,8} - 2d_{7,6}d_{7,8}\cos(\omega) \tag{14.21}$$

and

$$A_{6,7,8} = -\frac{1}{4}\sqrt[2]{(s_{6,7} + s_{6,8} + s_{7,8})^2 - 2(s_{6,7}^2 + s_{6,8}^2 + s_{7,8}^2)} \qquad (14.22)$$

where P_8 is to the right of vector $\overrightarrow{P_6P_7}$ in this case. Then

$$\overrightarrow{P_8} = \overrightarrow{P_6P_8} + \overrightarrow{P_6} \qquad (14.23)$$

From equations (18), (20), position of points P_7, P_6, is located respectively. Equation (14.23) defines the position of point P_8, the foot of Klann leg, which depends on the set of link dimensions (L) including fixed angle ω and ε, input angle (θ), the location of P_1, P_4, P_5 and the oriented areas $A_{2,4,3}$, $A_{2,3,7}$, $A_{5,7,6}$, and $A_{6,7,8}$. To this end, instead of using independent loop-closure equations with joint angles, for a specific set of link dimensions (S) and input angle θ of Klann leg, a unique position of point end point P_8 is located using bilateration method mentioned above.

14.2   Foot traces generation beyond Klann linkage

Our aim is to generate useful gait pattern based on novel reconfigurable Klann mechanism yet maintaining the efficiency and simplicity of the actuation. As this research is an initial step for design of a complete reconfigurable platform using Klann linkage, we chose one leg for the analysis presented in this work. We compute position of every point using a series of equations presented in previous section. By connecting all calculated points, we are able to trace the gait pattern. With a different set of link dimensions, we can acquire a distinct foot trajectory for each set. Hence, if the Klann leg can change the link dimensions itself, the linkage will generate a numerous new and different coupler curves. With this basic principle in mind, our objective in this study is to identify whether by performing small variations in the lengths of the links of a standard Klann leg, novel foot trajectories of interest for a walking platform can be obtained. Hence, a simple exploratory method is conducted in which we change the standard dimensions of every links of the Klann leg within a limit of except the crank $d_{1,2}$, the leg $d_{7,8}$ and the frame P_1, P_4, P_5. Then, all the foot resulting coupler curves are computed using the above method for each change of link dimensions. To be classified as reconfigurable at least one of the following features should vary a) the effective number of links and/or joints, b) the kinematic type i.e.,

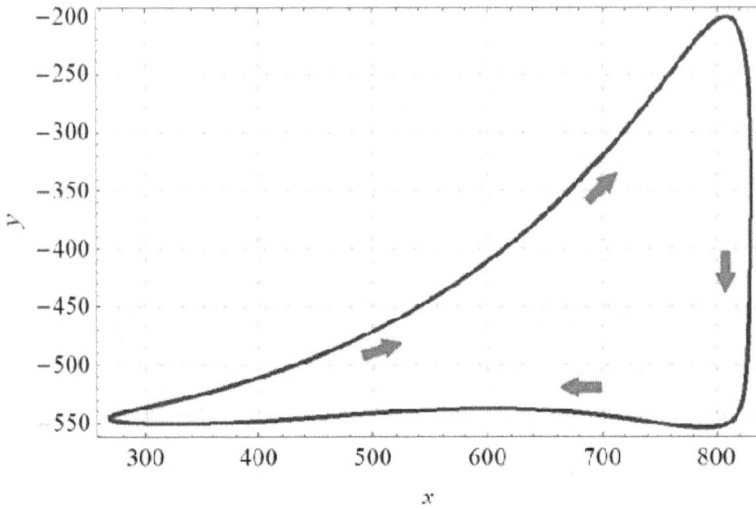

Figure 14.3: Traced foot trajectory of standard Klann leg with trajectory direction.

the contact constraint of some joints, c) the adjacency and incidence of links and joints, and d) the relative arrangement between joints. By reallocating the joint position new and different coupler curves can be generated. By changing the link lengths, five distinctive gait patterns that extend the original Klann linkage has been presented in table 14.1 achieved through four links, five links and six links transformations.

Digitigrade Locomotion   The standard foot trajectory of a Klann leg is similar to a kind of spider's locomotion with a long stride. On the other hand, digitigrades walk on their digits or toes. Due to the short floor touch of the foot of digitigrades, these kinds of animal perform less friction and use less energy than others. This interest makes digitigrade locomotion of great interest for the development of walking platforms.

Jam avoidance (Walking on soft, sticky terrain)   When walking on soft or sticky terrains such as semi-wet mud, walkers could easily get jam because of the soil conditions. Usually, a change in walker's gait pattern has to be made where the stride become shorter and touch-down as well as lift-up leg angle is increased nearly to overcome such situation.

Step climbing   One of the advantages of a standard Klann leg is that it has a relative high foot step at 350 (units) with highest point at −200 (units), which can handle with uneven terrain or non-significant obstacles such as gravel or small stone. However, the difficulty is increased when the leg faces with higher step level (above -200 units) or obstacles higher than its limit (350 units).

Hammering motion   Beside those walking-related foot patterns that we have presented above, in the process of analysing foot trajectories for reconfigurable design, we have found this potential pattern whose shape is similar to a hammering motion. The hammering action provides a repeated short, rapid impact to an object with high force on a small area.

Digging motion   For those kinds of legged animal such as dog or cat, legs are used not only for walking or running but also for others functions, pawing, clawing or digging, etc.

As shown in figure 14.1-a) a fully function reconfigurable Klann leg has been designed with two DC servos and five linear servos, suitable for performing the transformation previously discussed. Transformations are achieved by changing link length variable. To facilitate up down movement during reconfigurable process the base link is designed with ball screw with a ball screw and a slider joint. The base link is designed to allow the up and down movement of the reconfigurable Klan leg during transformation process. The base system and extendable links are the principal components of the proposed design of reconfigurable klann leg.

14.3   Experimental result

The first experiment consisted in comparing the simulated and experimental leg trajectories for the different gait patterns . The linear servos represent link dimensions of the reconfigurable Klann leg, and were set according to the number of active links, together with five cycles of input crank. A percent error of less than 25% is obtained for all cases. The origin of such errors is principally due to the non-conformity of link lengths and the presence of joint clearances in the prototype. Both error sources are almost inherent to the fabrication of mechanical designs and are the typical elements that affect the performance of linkages and mechanisms. The highest error in the digging motion pattern, 24.9% in height, is caused by the dramatically change in moment generated by the mechanism. It can be easily shown by simulation that the farthest center of mass point, respect to the base link, is achieved in such pattern, as a consequence, the joint clearance (backlash effect) affects more this pattern than

Patterns	4 links	5 links	6 links
Digitigrade locomotion			
Jam avoidance			
Step climbing			

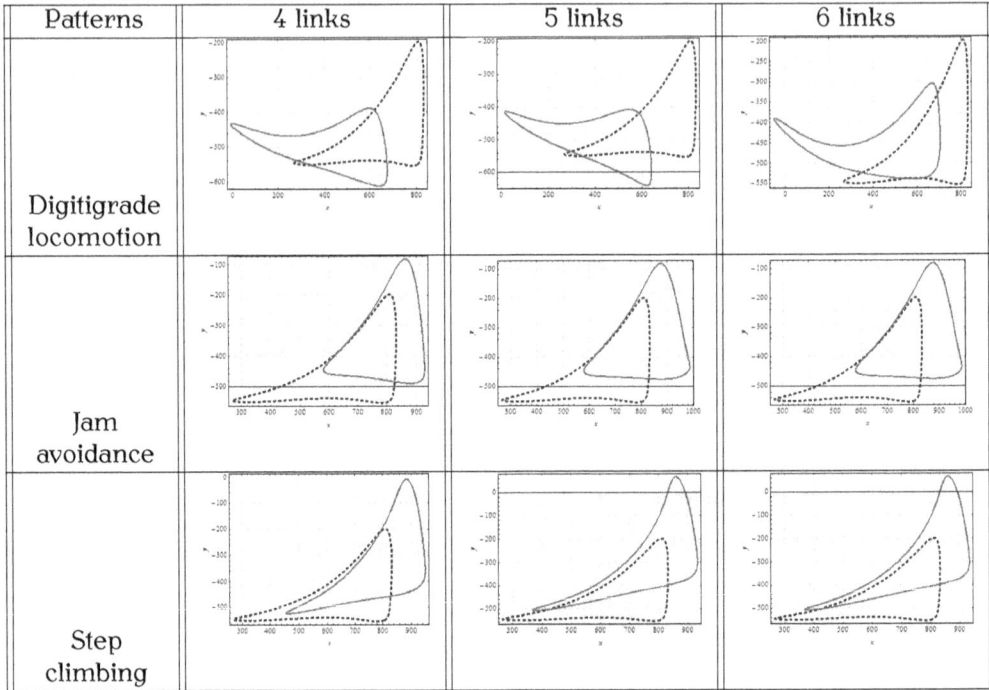

Table 14.1: Foot trajectory patterns for reconfiguration application.

the others. The second type of experiment consisted in verifying the transformation process between patterns. To this end, the transformation from jam avoidance locomotion to digging motion was tested. The results verify that the transformation is carried out without undesired floor contacts.

All experimental results of leg trajectories are obtained by getting data of five cycles of input joint and calculating their median value with seven implemented actuators. This design which has the ability to transform its links to produce five different gait patterns is still considered as a one degree of freedom linkage when operating in normal cycle. A simple but effective method to solve the position analysis problem for Klann linkage based on bilateration matrix has been introduced together with leg transformation technique. Five potential gait patterns of reconfigurable design have been classified and evaluated beside a transformation method to swap among those gaits. These typical gait patterns have shown that this simple but original Klann linkage can produce foot trajectories not only used for walking purposes but also behaved as a tool with other functions. A real prototype of Klann leg is built based on the preliminary design to test output trajectories.

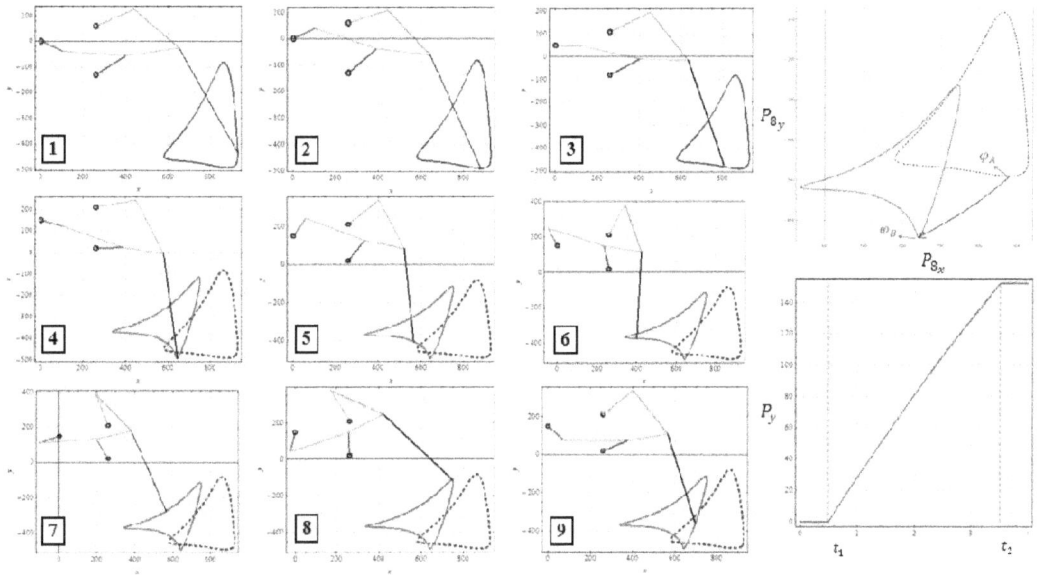

Figure 14.4: Simulation of the leg transformation from jam avoidance to digging motion through several steps.

Bibliography

[1] Tsujita K., Kobayashi T., Inoura T., Masuda T., *Gait transition by tuning muscle tones using pneumatic actuators in quadruped locomotion*, IEEE/RSJ Intl. Conf. on Intelligent Robots and Systems, France, pp. 2453–2458, 2008.

[2] Sheikh f.I., Pfeifer R., *Adaptive locomotion on varying ground conditions via a reconfigurable leg length hopper*, Adaptive Mobile Robotics, pp. 527–535, 2012.

[3] Alamdari A., Hắrin R., Krovi V. N., *Quantitative kinematic performance comparison of reconfigurable leg-wheeled vehicles*, Proceedings of the 16th Intl. Conf. on Climbing and Walking Robots, CLAWAR 2013.

[4] Klann J.C., *Walking device*, Patent No. 6260862, USA, 2001.

[5] Fukuoka Y., Kimura H., Cohen A.H., *Adaptive dynamic walking of a quadruped robot on irregular terrain based on biological concepts*, The Intl. Journal of Robotics Research, vol. 22, pp. 187–202, 2003.

[6] Gilho B.V., Bezerra J.M., *Mechatronic Design of a Chair for Disabled with Locomotion by Legs*, ABCM Symposium Series in Mechatronics vol.5, pp. 1142–1149, 2012.

[7] Rojas N. Thomas F., *On Closed Form Solutions to the Position Analysis of Baranov Trusses*, Mechanism and Machine Theory 50, pp.179–196, 2012.

[8] Chen X., Li-Quan W., Xiu-fen Y., Wang G., Hai-long W., *Prototype development and gait planning of biologically inspired multi-legged crablike robot*, Mechatronics vol. 23, pp.429–444, 2013.

Chapter 15

Analysis and modelling of a Hoekens-Jansen biped

Julio Reyes Muñoz and Edgar Alonso Martínez García

Laboratorio de Robótica, Institute of Engineering and Technology
Universidad Autónoma de Ciudad Juárez, Mexico.

In this chapter a detailed kinematic analysis of walking trajectory for an experimental biped robot comprised of hybrid limbs Hoekens-Jansen is presented. Mathematical and numerical modelling of the Hoekens and Jansen mechanisms are discussed separately. Subsequently, a novel hybrid mechanical limb Hoekens-Jansen is described. The both mechanisms are discussed in terms of their planar motion ′ . In addition, the Jansen mechanism is a mechanical linkage that was taken as a foundation for the development of the proposed mechanical design. The Jansen mechanism has one active independent control variable, similarly to the Hoekens linkage. However, the Jansen type poses more passive joints that the Hoekens. Therefore, the whole limb's movement is transmitted by only one rotatory actuator, and the hybrid limb's theoretical trajectory geometry is validated with its experimental trajectory motion.

15.1   Kinematic analysis of the Hoekens mechanism

The Hoekens mechanism is one of the mechanical linkages used for the basis of the proposed limb design. The Hoekens mechanism (figure 15.1) is built with two Cartesian variables of the workspace that describe the geometry of motion (two degrees of freedom) of the final link l_4's xy components. The Hoekens mechanism is comprised of a closed 3-link kinematic chain $l_0 + l_3$ (rotation radius and link 3), l_1(chassis), and link l_2. In addition, one independent active variable controls the rotation motion of l_0, which is the motion that inputs the first link l_3. Likewise, the end point of the link l_4 moves along a non linear trajectory, which is defined according to the numerical geometric parameters of the mechanism. The motion track of l_4 is developed regardless of the actuators speed. Because of the the geometrical relation between the links of the mechanism, the output trajectory is traditionally described by a semi-elliptic curve that is closed by a straight line ′ .

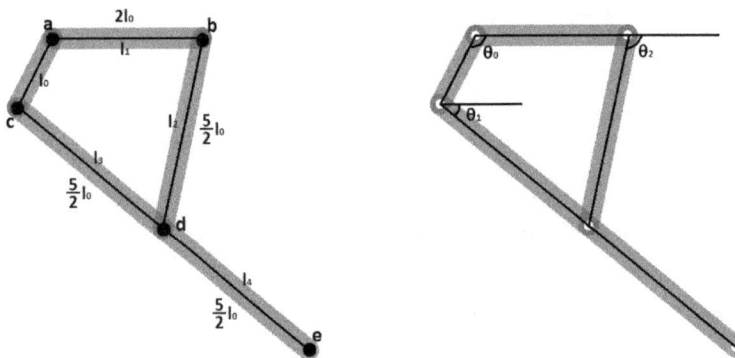

Figure 15.1: Hoekens mechanism's rigid links(left). Angles of interest for premier analysis (right).

Figure 15.1 shows all Hoekens mechanism's points of interest. Where a is the control variable or rotatory actuation. b is a fixed passive joint with no translation over time. c and d are passive joints. e is the point describing the output trajectory. In addition, the metric relation for each link's length is depicted and provided next, where the link l_0 is the starting length parameter to establish the relation for the rest of the links:

$$l_1 = 2l_0; \qquad l_2 = \frac{5}{2}l_0; \qquad l_3 = \frac{5}{2}l_0; \qquad l_4 = \frac{5}{2}l_0$$

The active angular joint θ_0 is the only angle measured over time, while the other angles are inferred and described through the mechanism's kinematics model.

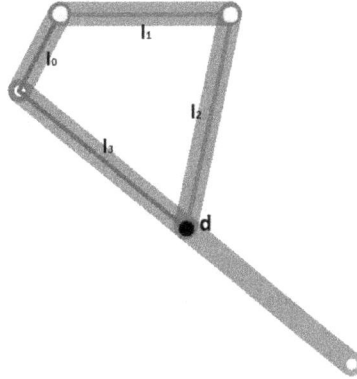

Figure 15.2: Kinematic chains to arrive to point d.

According to figure 15.2, in order to infer until joint d, the following model is established:

Axiom 15.1.1 (Closed kinematic chain). *Point d position in planar coordinates is formulated by*

$$l_0 \cos \theta_0 + l_3 \cos \theta_1 = l_1 + l_2 \cos \theta_2 \tag{15.1}$$

and

$$l_0 \sin \theta_0 + l_3 \sin \theta_1 = l_2 \sin \theta_2 \tag{15.2}$$

Hence, having a system of two equations θ_1 and θ_2 are unknown. Thus, dropping off the following terms in each equation,

$$l_2 \cos \theta_2 = l_0 \cos \theta_0 + l_3 \cos \theta_1 - l_1 \tag{15.3}$$

and

$$l_2 \sin \theta_2 = l_0 \sin \theta_0 + l_3 \sin \theta_1 \tag{15.4}$$

Both equations are squared, and by substituting the trigonometric identities for sine and cosine (see chapter 1.1),

$$l_2^2 \cos^2 \theta_2 = l_1^2 + l_0^2 \cos^2 \theta_0 + l_3^2 \cos^2 \theta_1 + 2l_0 l_3 \cos \theta_0 \cos \theta_1 - 2l_0 l_1 \cos \theta_0 - 2l_1 l_3 \cos \theta_1$$

and

$$l_2^2 \sin^2 \theta_2 = l_0^2 \sin^2 \theta_0 + l_3^2 \sin^2 \theta_1 + 2l_0 l_3 \sin \theta_0 \sin \theta_1$$

Thus, by simplifying equations,

$$l_2^2 = l_1^2 + l_3^2 + l_0^2 + 2l_0 l_3 \cos \theta_0 \cos \theta_1 - 2l_0 l_1 \cos \theta_0 - 2l_1 l_3 \cos \theta_1 + 2l_0 l_3 \sin \theta_0 \sin \theta_1$$

By rearranging equations,

$$(2l_0 l_3 \cos \theta_0 - 2l_1 l_3) \cos \theta_1 + (2l_0 l_3 \sin \theta_0) \sin \theta_1 + (l_1^2 + l_3^2 + l_0^2 - l_2^2 - 2l_0 l_1 \cos \theta_0) = 0 \qquad (15.5)$$

By using the following trigonometric identities:

$$\cos \theta = \frac{1 - tan^2(\frac{\theta}{2})}{1 + tan^2(\frac{\theta}{2})}$$

as well as,

$$\sin \theta = \frac{2 \tan(\frac{\theta}{2})}{1 + tan^2(\frac{\theta}{2})}$$

Equation (15.5) is simplified and rewritten to provide the following expression

Postulate 15.1.2 (Quadratic general equation). *The quadratic general equation as a function of* $\tan(\theta_1/2)$ *is postulated by*

$$A \left(\frac{1 - t^2}{1 + t^2} \right) + B \left(\frac{2t}{1 + t^2} \right) + C = 0$$

Where, for simplicity the next notation is used,

$$t = tan \left(\frac{\theta_1}{2} \right) \qquad (15.6)$$

Thus,

$$A = 2l_0 l_3 \cos \theta_0 - 2l_1 l_3$$

and

$$B = 2l_0 l_3 \sin \theta_0$$

and

$$C = l_1^2 + l_3^2 + l_0^2 - l_2^2 - 2l_0 l_1 \cos \theta_0$$

therefore, by rewriting the next equation

$$(C - A)t^2 + (2B)t + (A + C) = 0 \tag{15.7}$$

and by solving the quadratic equation by using the general form,

$$t = \frac{-B - \sqrt{A^2 + B^2 - C^2}}{C - A}$$

The solution for θ_1 from equation (15.6) is obtained,

Corollary 15.1.3 (Passive joint θ_1). *The equation that models the passive joint θ_1,*

$$\theta_1 = 2 \arctan \left(\frac{-B - \sqrt{A^2 + B^2 - C^2}}{C - A} \right) \tag{15.8}$$

Likewise, a solution for θ_2 is provided when dividing equation (15.4) by equation (15.3),

$$\tan(\theta_2) = \frac{l_0 \sin \theta_0 + l_3 \sin \theta_1}{l_0 \cos \theta_0 + l_3 \cos \theta_1 - l_1}$$

Thus, by solving for θ_2, the next corollary has been proved,

Corollary 15.1.4 (Passive joint θ_2). *The equation that models the passive joint θ_2,*

$$\theta_2 = \arctan \left(\frac{l_0 \sin \theta_0 + l_3 \sin \theta_1}{l_0 \cos \theta_0 + l_3 \cos \theta_1 - l_1} \right) \tag{15.9}$$

Therefore, a kinematic analysis of the kinematic closed chain will be obtained until the final point e. The vector of position (15.10) is stated by

$$\mathbf{p}_e = \begin{pmatrix} x_e \\ y_e \end{pmatrix} = \begin{pmatrix} l_0 \cos(\theta_0) + (l_3 + l_4) \cos(\theta_1) \\ l_0 \sin(\theta_0) + (l_3 + l_4) \sin(\theta_1) \end{pmatrix} \tag{15.10}$$

The Cartesian trajectories of each point in the mechanism are depicted in figure 15.3. The position vector a starts with Cartesian components

$$\mathbf{p}_a = \begin{pmatrix} x_a \\ y_a \end{pmatrix} = \begin{pmatrix} 0 \\ 0 \end{pmatrix}$$

likewise, the position vector of point b:

$$\mathbf{p}_b = \begin{pmatrix} x_b \\ y_b \end{pmatrix} = \begin{pmatrix} l_1 \\ 0 \end{pmatrix}$$

the position vector of point c

$$\mathbf{p}_c = \begin{pmatrix} x_c \\ y_c \end{pmatrix} = \begin{pmatrix} l_0 \cos(\theta_0) \\ l_0 \sin(\theta_0) \end{pmatrix}$$

and the position vector of point d

$$\mathbf{p}_d = \begin{pmatrix} x_d \\ y_d \end{pmatrix} = \begin{pmatrix} l_0 \cos(\theta_0) + l_3 \cos(\theta_1) \\ l_0 \sin(\theta_0) + l_3 \sin(\theta_1) \end{pmatrix}$$

Eventually, in order to plot the trajectory of point e, the position model was previously provided by the expression (15.10). It is worth noting that the trajectories plotted in figure 15.3 correspond to a complete cycle of the link l_0 denoted by the control angle θ_0, ranging from 0 to 2π radians. The lengths for each link were established according to the following parameters (given in mm): $l_0 = 50$; $l_1 = 100$; $l_2 = 125$; $l_3 = 125$; and $l_4 = 125$.

The advantage on using this mechanism is its simplicity, because only four links are needed to describe a semi-elliptic trajectories. Furthermore, despite this mechanical simplicity, it is possible to modify the output trajectory by solely changing the Cartesian position of the fixed

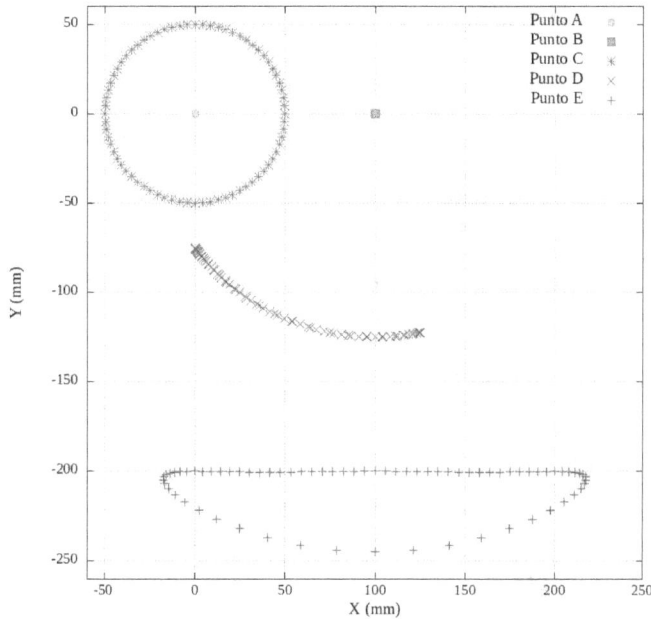

Figure 15.3: Trajectories described by each point of the Hoekens mechanism.

passive joints, and/or by numerically changing the links lengths. As a matter of fact, the low number of links makes a light mechanism, which is suitable to manage as required the ending section of the proposed design.

15.2 Kinematic analysis of the Jansen mechanism

The Jansen mechanism is another of the mechanical linkages taken as foundation for the development of the proposed limb design. Similarly as the Hoekens mechanism, the Jansen mechanism is actuated by deploying one rotatory joint. Nevertheless, the Jansen mechanism posses even more links (eleven), and passive joints than the Hoekens mechanism. Using similar notation as previous sections, the Jansen actuation is transmitted by only one rotatory control variable the inputs link l_0.

Figure 15.4 shows the Jansen mechanism with all its joints and links. a is the driven joint. b and i are fixed passive joints at fixed positions. c, d, e, f, and g are the passive joints. Finally, h is the point describing the output trajectory. Likewise, the mechanism has a special metric ratio of the links w.r.t. l_0 (driven by the actuated joint): $l_1 = 0.52l_0$, $l_2 = 2.53l_0$, $l_3 = 3.33l_0$,

$l_4 = 4.13l_0$, $l_5 = 2.76l_0$, $l_6 = 3.72l_0$, $l_7 = 2.67l_0$, $l_8 = 2.62l_0$, $l_9 = 2.63l_0$, $l_{10} = 2.45l_0$, $l_{11} = 3.27l_0$, $l_{12} = 4.38l_0$.

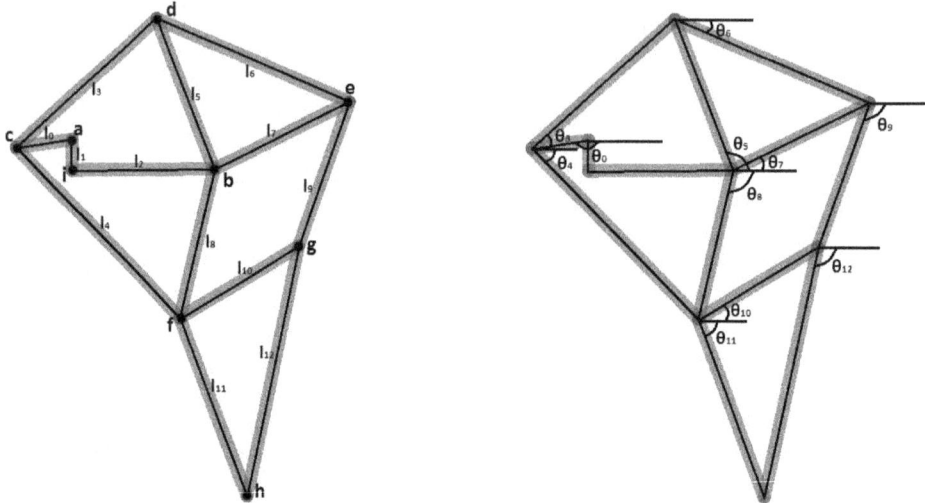

Figure 15.4:   The Jansen mechanism's links (left). Angles of each link w.r.t. the horizontal (right).

In order to obtain the kinematic model of the Jansen linkage, the whole set of angular values of the passive joints are required to be expressed with a mathematical functional form $\cdot$. The kinematic analysis is very similar to one previously described for the Hoekens mechanism. The mathematical expression is stated w.r.t. the driven angle θ_0, and the links lengths associated. Thus, by developing a similar method as the previous section, the equation (15.7) was established and solved by using the next quadratic formula,

Postulate 15.2.1 (Quadratic solution for $\tan(\theta_i/2)$). *For the set of closed kinematic chains, the general quadratic solution for passive angles is*

$$t_i = \frac{-B_i \pm \sqrt{A_i^2 + B_i^2 - C_i^2}}{C_i - A_i} \qquad (15.11)$$

Where A, B, and C are expressions, which depend on the known driven angle and its links lengths. The index i identifies a closed set of kinematic chains. Thus, in order to obtain an analytical solution for the unknown angles θ_3 and θ_5, the kinematic chains are shown in figure 15.5. The following system of equations is obtained:

Axiom 15.2.2 (Kinematic chain b). *The kinematic chain model to solve position of point b,*

$$l_5 \cos \theta_5 + l_2 = l_0 \cos \theta_0 + l_3 \cos \theta_3 \tag{15.12}$$

and

$$l_5 \sin \theta_5 - l_1 = l_0 \sin \theta_0 + l_3 \sin \theta_3 \tag{15.13}$$

And, by firstly isolating a term of each equation containing one of the unknown variables,

$$l_3 \cos \theta_3 = l_5 \cos \theta_5 - l_0 \cos \theta_0 + l_2 \tag{15.14}$$

likewise,

$$l_3 \sin \theta_3 = l_5 \sin \theta_5 - l_0 \sin \theta_0 - l_1 \tag{15.15}$$

The expression are subsequently squared

$$l_3^2 \cos^2 \theta_3 - l_5^2 \cos^2 \theta_5 + l_0^2 \cos^2 \theta_0 + l_2^2 - 2l_0l_5 \cos \theta_0 \cos \theta_5 - 2l_0l_2 \cos \theta_0 + 2l_2l_5 \cos \theta_5 \tag{15.16}$$

as well as

$$l_3^2 \sin^2 \theta_3 = l_5^2 \sin^2 \theta_5 + l_0^2 \sin^2 \theta_0 + l_1^2 - 2l_0l_5 \sin \theta_0 \sin \theta_5 + 2l_0l_1 \sin \theta_0 - 2l_1l_5 \sin \theta_5 \tag{15.17}$$

and algebraically simplifying the equations

$$l_3^2 = l_0^2 + l_1^2 + l_2^2 + l_5^2 - 2l_0l_5c_{\theta_0}c_{\theta_5} - 2l_0l_2c_{\theta_0} + 2l_2l_5c_{\theta_5} - 2l_0l_5s_{\theta_0}s_{\theta_5} + 2l_0l_1s_{\theta_0} - 2l_1l_5s_{\theta_5} \tag{15.18}$$

and

$$(2l_2l_5 - 2l_0l_5c_{\theta_0})c_{\theta_5} + (-2l_1l_5 - 2l_0l_5s_{\theta_0})s_{\theta_5} + (l_0^2 + l_1^2 + l_2^2 + l_5^2 - l_3^2 - 2l_0l_2c_{\theta_0} + 2l_0l_1s_{\theta_0}) = 0 \tag{15.19}$$

and by obtaining the expressions A, B and C for this pair of closed kinematic chains:

$$A_1 = 2l_2l_5 - 2l_0l_5 \cos \theta_0 \tag{15.20}$$

$$B_1 = -2l_1 l_5 - 2l_0 l_5 \sin \theta_0 \tag{15.21}$$

and

$$C_1 = l_0^2 + l_1^2 + l_2^2 + l_5^2 - l_3^2 - 2l_0 l_2 \cos \theta_0 + 2l_0 l_1 \sin \theta_0 \tag{15.22}$$

hereafter, from equation (15.11), it provides an analytical solution

$$t_1 = \frac{-B_1 - \sqrt{A_1^2 + B_1^2 - C_1^2}}{C_1 - A_1}$$

and considering that

$$t_1 = \tan\left(\frac{\theta_5}{2}\right)$$

Thus, the solution for θ_5:

$$\theta_5 = 2\tan^{-1}\left(\frac{-B_1 - \sqrt{A_1^2 + B_1^2 - C_1^2}}{C_1 - A_1}\right) \tag{15.23}$$

In addition, dividing equation (15.15) by expression (15.14):

$$\tan \theta_3 = \frac{l_5 \sin \theta_5 - l_0 \sin \theta_0 - l_1}{l_5 \cos \theta_5 - l_0 \cos \theta_0 + l_2}$$

hence, the solution for θ_3:

$$\theta_3 = \tan^{-1}\left(\frac{l_5 \sin \theta_5 - l_0 \sin \theta_0 - l_1}{l_5 \cos \theta_5 - l_0 \cos \theta_0 + l_2}\right) \tag{15.24}$$

In order to obtain the next pair of angles, the kinematic chains shown in figure 15.5 were algebraically analysed. Thus, the kinematic path from the point b to the point e is stated

$$l_5 \cos \theta_5 + l_6 \cos \theta_6 = l_7 \cos \theta_7 \tag{15.25}$$

as well as

$$l_5 \sin \theta_5 + l_6 \sin \theta_6 = l_7 \sin \theta_7 \tag{15.26}$$

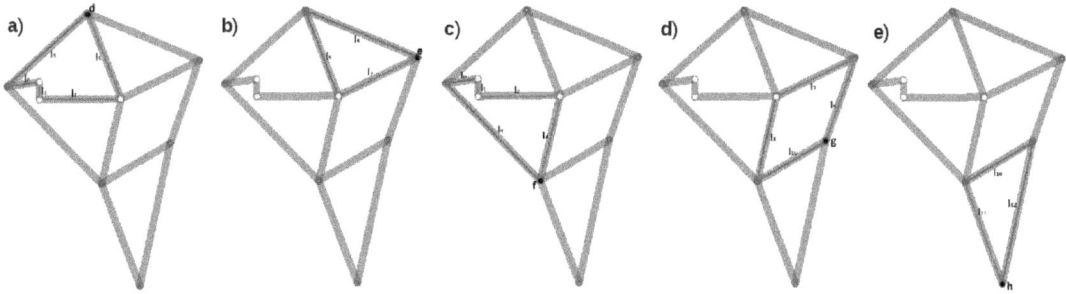

Figure 15.5:   a) Kinematic chain of links until joint d. b) Kinematic chain of links until joint g. c) Kinematic chain of links until joint f. d) Kinematic chain of links until joint e. e) Kinematic chain of links until joint h.

and the resulting equations are subsequently squared,

$$l_7^2 \cos^2 \theta_7 = l_5^2 \cos^2 \theta_5 + l_6^2 \cos^2 \theta_6 + 2l_5l_6 \cos \theta_5 \cos \theta_6 \qquad (15.27)$$

and

$$l_7^2 \sin^2 \theta_7 = l_5^2 \sin^2 \theta_5 + l_6^2 \sin^2 \theta_6 + 2l_5l_6 \sin \theta_5 \sin \theta_6 \qquad (15.28)$$

by algebraically simplifying the expressions,

$$l_7^2 = l_5^2 + l_6^2 + 2l_5l_6 \cos \theta_5 \cos \theta_6 + 2l_5l_6 \sin \theta_5 \sin \theta_6 \qquad (15.29)$$

and

$$(2l_5l_6 \cos \theta_5) \cos \theta_6 + (2l_5l_6 \sin \theta_5) \sin \theta_6 + (l_5^2 + l_6^2 - l_7^2) = 0 \qquad (15.30)$$

after simplifying previous expressions, A, B and C are obtained and the related pair of closed kinematic chains are solved

$$A_2 = 2l_5l_6 \cos \theta_5$$

$$B_2 = 2l_5l_6 \sin \theta_5$$

and

$$C_2 = l_5^2 + l_6^2 - l_7^2$$

Hence, by using equation (15.11) and solving for θ_6

$$\theta_6 = 2\tan^{-1}\left(\frac{-B_2 + \sqrt{A_2^2 + B_2^2 - C_2^2}}{C_2 - A_2}\right) \tag{15.31}$$

Likewise, solving for θ_7 equation (15.26) is divided by equation (15.25):

$$\theta_7 = \tan^{-1}\left(\frac{l_5\sin\theta_5 + l_6\sin\theta_6}{l_5\cos\theta_5 + l_6\cos\theta_6}\right) \tag{15.32}$$

In order to obtain the unknown angles θ_4 and θ_8, the kinematic path from point a until point f is modelled by

$$l_8\cos\theta_8 + l_2 = l_0\cos\theta_0 + l_4\cos\theta_4 \tag{15.33}$$

and

$$l_8\sin\theta_8 - l_1 = l_0\sin\theta_0 + l_4\sin\theta_4 \tag{15.34}$$

Thus, by isolating the terms containing the variables of interest

$$l_4\cos\theta_4 = l_8\cos\theta_8 - l_0\cos\theta_0 + l_2 \tag{15.35}$$

as well as

$$l_4\sin\theta_4 = l_8\sin\theta_8 - l_0\sin\theta_0 - l_1 \tag{15.36}$$

In addition, equations are squared to algebraically re-arrange,

$$l_4^2\cos^2\theta_4 = l_8^2\cos^2\theta_8 + l_0^2\cos^2\theta_0 + l_2^2 - 2l_0l_8\cos\theta_0\cos\theta_8 - 2l_0l_2\cos\theta_0 + 2l_2l_8\cos\theta_8$$

and

$$l_4^2\sin^2\theta_4 = l_8^2\sin^2\theta_8 + l_0^2\sin^2\theta_0 + l_1^2 - 2l_0l_8\sin\theta_0\sin\theta_8 + 2l_0l_1\sin\theta_0 - 2l_1l_8\sin\theta_8$$

by defining $\ell = l_0^2 + l_1^2 + l_2^2 + l_8^2$, and algebraically simplifying equations:

$$l_4^2 = \ell - 2l_0l_8c_{\theta_0}c_{\theta_8} - 2l_0l_2c_{\theta_0} + 2l_2l_8c_{\theta_8} - 2l_0l_8s_{\theta_0}s_{\theta_8} + 2l_0l_1s_{\theta_0} - 2l_1l_8s_{\theta_8} \tag{15.37}$$

and

$$(2l_2 l_8 - 2l_0 l_8 c_{\theta_0})c_{\theta_8} + (-2l_1 l_8 - 2l_0 l_8 s_{\theta_0})s_{\theta_8} + (\ell - l_4^2 - 2l_0 l_2 c_{\theta_0} + 2l_0 l_1 s_{\theta_0}) = 0 \qquad (15.38)$$

The solution for expressions A, B and C are obtained for the actual pair of closed kinematic chains:

$$A_3 = 2l_2 l_8 - 2l_0 l_8 \cos \theta_0$$

$$B_3 = -2l_1 l_8 - 2l_0 l_8 \sin \theta_0$$

and

$$C_3 = l_0^2 + l_1^2 + l_2^2 + l_8^2 - l_4^2 - 2l_0 l_2 \cos \theta_0 + 2l_0 l_1 \sin \theta_0$$

likewise, by using equation (15.11) to solve θ_8

$$\theta_8 = 2\tan^{-1}\left(\frac{-B_3 + \sqrt{A_3^2 + B_3^2 - C_3^2}}{C_3 - A_3}\right) \qquad (15.39)$$

In addition, equation (15.36) is divided by (15.35) to solve for θ_4

$$\theta_4 = \tan^{-1}\left(\frac{l_8 \sin \theta_8 - l_0 \sin \theta_0 - l_1}{l_8 \cos \theta_8 - l_0 \cos \theta_0 + l_2}\right) \qquad (15.40)$$

It follows that in order to solve for the unknown angles θ_9 and θ_{10}, the kinematic path from point b to point g is stated next:

$$l_7 \cos \theta_7 + l_9 \cos \theta_9 = l_8 \cos \theta_8 + l_{10} \cos \theta_{10} \qquad (15.41)$$

and

$$l_7 \sin \theta_7 + l_9 \sin \theta_9 = l_8 \sin \theta_8 + l_{10} \sin \theta_{10} \qquad (15.42)$$

Isolating the terms containing variables of interest in both expression,

$$l_{10} \cos \theta_{10} = l_7 \cos \theta_7 + l_9 \cos \theta_9 - l_8 \cos \theta_8 \qquad (15.43)$$

and

$$l_{10} \sin \theta_{10} = l_7 \sin \theta_7 + l_9 \sin \theta_9 - l_8 \sin \theta_8 \qquad (15.44)$$

Squaring both sides of the expression to subsequently arrange algebraically,

$$l_{10}^2 c_{\theta_{10}}^2 = l_7^2 c_{\theta_7}^2 + l_8^2 c_{\theta_8}^2 + l_9^2 c_{\theta_9}^2 + 2l_7 l_9 c_{\theta_7} c_{\theta_9} - 2l_7 l_8 c_{\theta_7} c_{\theta_8} - 2l_8 l_9 c_{\theta_8} c_{\theta_9} \tag{15.45}$$

and

$$l_{10}^2 s_{\theta_{10}}^2 = l_7^2 s_{\theta_7}^2 + l_8^2 s_{\theta_8}^2 + l_9^2 s_{\theta_9}^2 + 2l_7 l_9 s_{\theta_7} s_{\theta_9} - 2l_7 l_8 s_{\theta_7} s_{\theta_8} - 2l_8 l_9 s_{\theta_8} s_{\theta_9} \tag{15.46}$$

The term $\ell = l_7^2 + l_8^2 + l_9^2$ is defined, and by algebraically simplifying

$$l_{10}^2 = \ell + 2l_7 l_9 c_{\theta_7} c_{\theta_9} - 2l_7 l_8 c_{\theta_7} c_{\theta_8} - 2l_8 l_9 c_{\theta_8} c\theta_9 + 2l_7 l_9 s_{\theta_7} s_{\theta_9} - 2l_7 l_8 s\theta_7 s_{\theta_8} - 2l_8 l_9 s_{\theta_8} s_{\theta_9} \tag{15.47}$$

and

$$(2l_7 l_9 c_{\theta_7} - 2l_8 l_9 c_{\theta_8})c_{\theta_9} + (2l_7 l_9 s_{\theta_7} - 2l_8 l_9 s_{\theta_8})s_{\theta_9} + (\ell - l_{10}^2 - 2l_7 l_8 c_{\theta_7} c_{\theta_8} - 2l_7 l_8 s_{\theta_7} s_{\theta_8}) = 0 \tag{15.48}$$

hence, solving for expressions A, B and C of the actual closed kinematic chains,

$$A_4 = 2l_7 l_9 \cos\theta_7 - 2l_8 l_9 \cos\theta_8 \tag{15.49}$$

$$B_4 = 2l_7 l_9 \sin\theta_7 - 2l_8 l_9 \sin\theta_8 \tag{15.50}$$

and

$$C_4 = l_7^2 + l_8^2 + l_9^2 - l_{10}^2 - 2l_7 l_8 \cos\theta_7 \cos\theta_8 - 2l_7 l_8 \sin\theta_7 \sin\theta_8 \tag{15.51}$$

solving θ_9 through equation (15.11)

$$\theta_9 = 2\tan^{-1}\left(\frac{-B_4 + \sqrt{A_4^2 + B_4^2 - C_4^2}}{C_4 - A_4}\right) \tag{15.52}$$

Similarly, equation (15.44) is divided by expression (15.43) to solve for θ_{10}

$$\theta_{10} = \tan^{-1}\left(\frac{l_7 \sin\theta_7 + l_9 \sin\theta_9 - l_8 \sin\theta_8}{l_7 \cos\theta_7 + l_9 \cos\theta_9 - l_8 \cos\theta_8}\right) \tag{15.53}$$

Similarly, in order to obtain the unknown angles θ_{11} and θ_{12}, the kinematic path from point f to point h is developed next,

$$l_{10} \cos \theta_{10} + l_{12} \cos \theta_{12} = l_{11} \cos \theta_{11} \tag{15.54}$$

and

$$l_{10} \sin \theta_{10} + l_{12} \sin \theta_{12} = l_{11} \sin \theta_{11} \tag{15.55}$$

further, both sides of equations are squared and algebraically arranged,

$$l_{11}^2 \cos^2 \theta_{11} = l_{10}^2 \cos^2 \theta_{10} + l_{12}^2 \cos^2 \theta_{12} + 2l_{10}l_{12} \cos \theta_{10} \cos \theta_{12} \tag{15.56}$$

and

$$l_{11}^2 \sin^2 \theta_{11} = l_{10}^2 \sin^2 \theta_{10} + l_{12}^2 \sin^2 \theta_{12} + 2l_{10}l_{12} \sin \theta_{10} \sin \theta_{12} \tag{15.57}$$

by simplifying the expressions

$$l_{11}^2 = l_{10}^2 + l_{12}^2 + 2l_{10}l_{12} \cos \theta_{10} \cos \theta_{12} + 2l_{10}l_{12} \sin \theta_{10} \sin \theta_{12} \tag{15.58}$$

and

$$(2l_{10}l_{12} \cos \theta_{10}) \cos \theta_{12} + (2l_{10}l_{12} \sin \theta_{10}) \sin \theta_{12} + (l_{10}^2 + l_{12}^2 - l_{11}^2) = 0 \tag{15.59}$$

solving expressions A, B and C:

$$A_5 = 2l_{10}l_{12} \cos \theta_{10}$$

$$B_5 = 2l_{10}l_{12} \sin \theta_{10}$$

and

$$C_5 = l_{10}^2 + l_{12}^2 - l_{11}^2$$

Eventually, expression (15.11) is used to solve θ_{12}

Corollary 15.2.3 (Solution of passive joint θ_{12}). *the model solution for the passive joint θ_{12}*

$$\theta_{12} = 2 \tan^{-1} \left(\frac{-B_5 + \sqrt{A_5^2 + B_5^2 - C_5^2}}{C_5 - A_5} \right) \tag{15.60}$$

Likewise, equation (15.55) is divided by expression (15.54) to solve for θ_{11},

Corollary 15.2.4 (Solution of passive joint θ_{11}). *the model solution for the passive joint θ_{11} is described by*

$$\theta_{11} = \tan^{-1}\left(\frac{l_{10}\sin\theta_{10} + l_{12}\sin\theta_{12}}{l_{10}\cos\theta_{10} + l_{12}\cos\theta_{12}}\right) \tag{15.61}$$

Hereafter, the analytical solutions for each Jansen mechanism's passive joints have been found, hence it is possible to state the position vector model that corresponds to the mechanism's contact point h and the ground. Following a kinematic chains starting from point a (origin of the mechanism) to the point h through the links l_0, l_4 and l_{11}, the vector of position (15.62) is obtained.

Theorem 15.2.5 (Jansen limb's contact point position). *The Jansen limb's contact point position in planar coordinates is stated by the next vector:*

$$\mathbf{p}_h = \begin{pmatrix} x_h \\ y_h \end{pmatrix} = \begin{pmatrix} l_0\cos(\theta_0) + l_4\cos(\theta_4) + l_{11}\cos(\theta_{11}) \\ l_0\sin(\theta_0) + l_4\sin(\theta_4) + l_{11}\sin(\theta_{11}) \end{pmatrix} \tag{15.62}$$

Validation of the solution model is provided through numerical simulations depicted in figure 15.6. Each mechanism's link produced a kinematic trajectory from corresponding position vectors in the mechanism that will be summarised next. Thus, the position vector for point a is stated,

$$\mathbf{p}_a = \begin{pmatrix} x_a \\ y_a \end{pmatrix} = \begin{pmatrix} 0 \\ 0 \end{pmatrix} \tag{15.63}$$

likewise, the position vector of point b,

$$\mathbf{p}_b = \begin{pmatrix} x_b \\ y_b \end{pmatrix} = \begin{pmatrix} l_2 \\ -l_1 \end{pmatrix} \tag{15.64}$$

the position vector of point c:

$$\mathbf{p}_c = \begin{pmatrix} x_c \\ y_c \end{pmatrix} = \begin{pmatrix} l_0 \cos(\theta_0) \\ l_0 \sin(\theta_0) \end{pmatrix} \tag{15.65}$$

the position vector of point d:

$$\mathbf{p}_d = \begin{pmatrix} x_d \\ y_d \end{pmatrix} = \begin{pmatrix} l_0 \cos(\theta_0) + l_3 \cos(\theta_3) \\ l_0 \sin(\theta_0) + l_3 \sin(\theta_3) \end{pmatrix} \tag{15.66}$$

the position vector of point e:

$$\mathbf{p}_e = \begin{pmatrix} x_e \\ y_e \end{pmatrix} = \begin{pmatrix} l_2 + l_7 \cos \theta_7 \\ -l_1 + l_7 \sin \theta_7 \end{pmatrix} \tag{15.67}$$

the position vector of point f:

$$\mathbf{p}_f = \begin{pmatrix} x_f \\ y_f \end{pmatrix} = \begin{pmatrix} l_2 + l_8 \cos \theta_8 \\ -l_1 + l_8 \sin \theta_8 \end{pmatrix} \tag{15.68}$$

Finally, the position vector model for the point g is defined by

$$\mathbf{p}_g = \begin{pmatrix} x_g \\ y_g \end{pmatrix} = \begin{pmatrix} l_2 + l_8 \cos \theta_8 + l_{10} \cos \theta_{10} \\ -l_1 + l_{10} \sin \theta_{10} \end{pmatrix} \tag{15.69}$$

The trajectory of the point h is governed by the position vector (15.62), and plotted in figure 15.6 (Punto H). All trajectories behave as a result of a complete turn of the driven angle θ_0 from 0 to 2π radians. For this numerical simulations the parameters of each link were the following (in mm): $l_0 = 50$, $l_1 = 126.65$, $l_2 = 26$, $l_3 = 166.67$, $l_4 = 206.35$, $l_5 = 138.30$, $l_6 = 186$, $l_7 = 133.71$, $l_8 = 131$, $l_9 = 131.33$, $l_{10} = 122.33$, $l_{11} = 163.41$, $l_{12} = 219.06$.

The advantage on using this type of planar mechanism in the proposed limbs design is because its high stiffness, and it is able to describe gait trajectories suitable for walking machines. Furthermore, the forces involved during a gait are not transmitted from the contact point to the actuator, rather the stiffness of the ensemble of links only allows to transmit movement from the actuator towards the contact point.

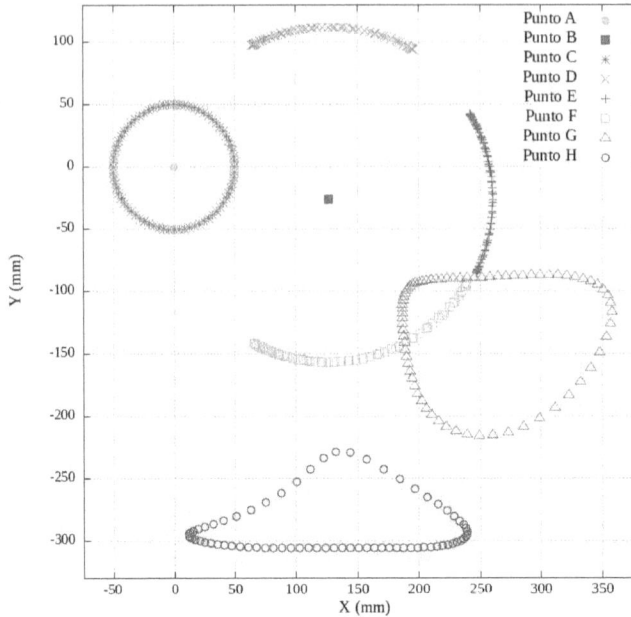

Figure 15.6:   Trajectories described by each point of the Jansen mechanism.

Thus, it makes a non-back drivable system that would function analogously to biological organisms.

15.3   Kinematic analysis of a hybrid limb

In this section we detail the kinematics of a hybrid mechanical design that mixes the Jansen with the Hoekens mechanism to work as a single limb. The main purpose is to control the trajectory of the Jansen mechanism in combination with the Hoekens motion to generate although complex, but a stable and flexible output trajectory. The proposed extremity design provides some advantages since both mechanisms abilities are exploited $^-$.   The Jansen mechanism provided a good stiffness to the design, making it non-back drivable. At the same time, the incorporation of the Hoekens linkage supplied the flexibility for achieving a more complex output trajectory. The hybrid mechanism has fourteen links with only an actuated joint θ_0. Figure 15.3 shows the proposed mechanism with all its joints of interest.

Where a is the actuated joint; b and i are passive joints fixed to the chassis. c, d, e, f, g, h, j, and k are passive joints analysed in previous sections of this chapter. l is a point describing

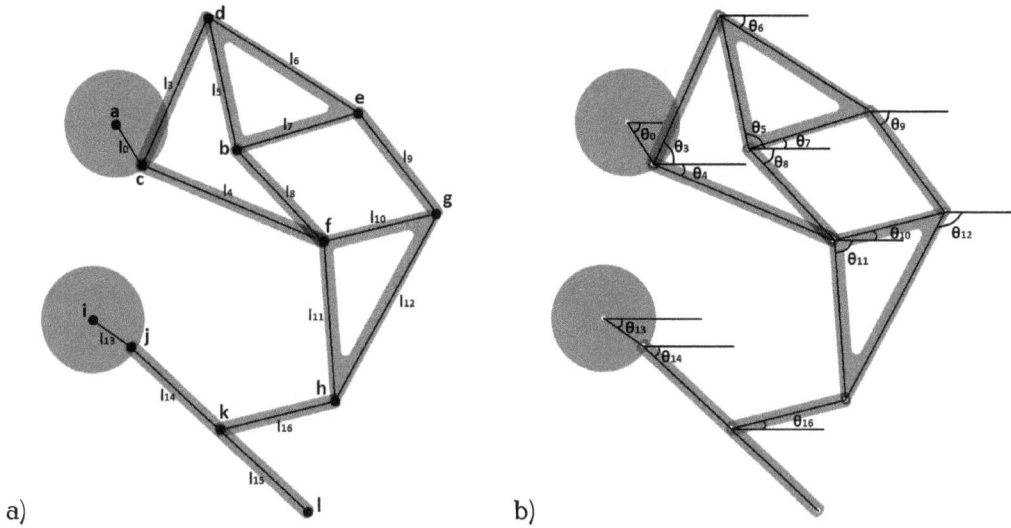

Figure 15.7:   a) Points of interest and links of the proposed limb. b) Angles of each link w.r.t. the horizontal.

the output trajectory or the limb's contact point. The links' length ratio is keep the same as previously explained in other sections (Hoekens and Jansen). Likewise, the relationship between the Hockens and the Jansen mechanism was preserved by 1 : 1, its ratio is $l_0 = l_{13}$. This analysis carried out in this section for the hybrid version is very similar to analysis discussed previously. However, minimal modifications have been carried out according to the new design features.

Firstly, to obtain the hybrid limb design kinematic model, the passive joint angles must analytically be solved, in accordance to the links chain as depicted in figure 15.8. Each passive angle directly depends on the actuator's position, and the links length. The final trajectory will depend on the position of point i w.r.t. the axis of rotation of the link l_{13} in close relation to the point a, which is the axis of rotation of l_0. Likewise, it will depends on the angle β (not depicted in figure 15.8), which is the angular phase difference of the link l_0 w.r.t. the actual angle of l_{13} mechanically connected by a crossed chain. The first part of the analysis focuses on the Jansen mechanism. Although, the kinematic analysis is developed similar to previous sections, but now a virtual link l_a and angle θ_a is included to preserve the equations homogeneity. The virtual link equals the polar representation of a vector describing the position of the point b w.r.t. a. In order to solve for the unknown angles θ_3 and θ_5, the kinematic chains involved are depicted in figure 15.8 from where our analysis is developed. The first path taken goes from the point a to the point d. Thus, in order to obtain the unknown angles, a system of two equations previously stated must be solved. for this case, there are six closed link trajectories, with angles solution obtained through equation (15.11).

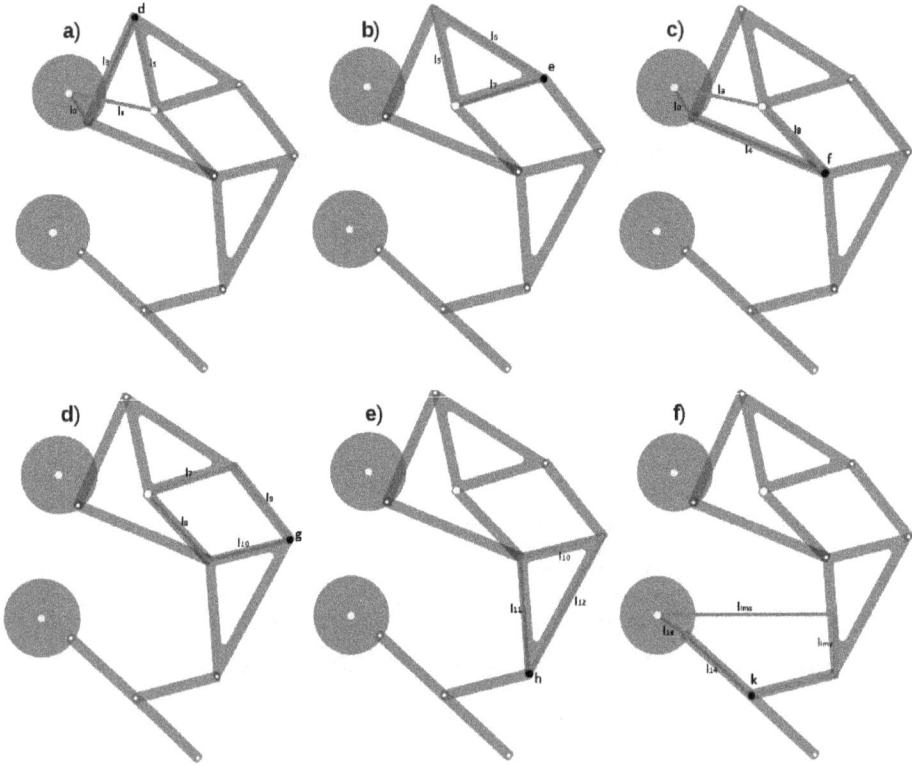

Figure 15.8:   Kinematic chains to arrive to point d .

Thus, let us start our analysis by stating the following axiom:

Axiom 15.3.1 (Hybrid limb initial kinematic chain). *The initial set of kinematic equations for the hybrid limb starting from the actuated angle is stated by*

$$l_a \cos \theta_a + l_5 \cos \theta_5 = l_0 \cos \theta_0 + l_3 \cos \theta_3$$

and

$$l_a \sin \theta_a + l_5 \sin \theta_5 = l_0 \sin \theta_0 + l_3 \sin \theta_3$$

It follows, that by isolating a term of each equation containing one of the variables of interest,

$$l_3 \cos \theta_3 = l_a \cos \theta_a + l_5 \cos \theta_5 - l_0 \cos \theta_0 \qquad (15.70)$$

and

$$l_3 \sin \theta_3 = l_a \sin \theta_a + l_5 \sin \theta_5 - l_0 \sin \theta_0 \qquad (15.71)$$

Similarly, again the resulting equations are now squared for a subsequent algebraic arrangement,

$$l_3^2 c_{\theta_3}^2 = l_a^2 c_{\theta_a}^2 + l_5^2 c_{\theta_5}^2 + l_0^2 c_{\theta_0}^2 + 2 l_a l_5 c_{\theta_a} c_{\theta_5} - 2 l_0 l_a c_{\theta_0} c_{\theta_a} - 2 l_0 l_5 c_{\theta_0} c_{\theta_5} \qquad (15.72)$$

and

$$l_3^2 s_{\theta_3}^2 = l_a^2 s_{\theta_a}^2 + l_5^2 s_{\theta_5}^2 + l_0^2 s_{\theta_0}^2 + 2 l_a l_5 s_{\theta_a} s_{\theta_5} - 2 l_0 l_a s_{\theta_0} s_{\theta_a} - 2 l_0 l_5 s_{\theta_0} s_{\theta_5} \qquad (15.73)$$

Thus, defining $\ell_h = l_a^2 + l_5^2 + l_0^2$ by simplifying the two expression

$$l_3^2 = \ell_h + 2 l_a l_5 c_{\theta_a} c_{\theta_5} - 2 l_0 l_a c_{\theta_0} c_{\theta_a} - 2 l_0 l_5 c_{\theta_0} c_{\theta_5} + 2 l_a l_5 s_{\theta_a} s_{\theta_5} - 2 l_0 l_a s_{\theta_0} s_{\theta_a} - 2 l_0 l_5 s_{\theta_0} s_{\theta_5} \qquad (15.74)$$

similarly,

$$(2 l_a l_5 c_{\theta_a} - 2 l_0 l_5 c_{\theta_0}) c_{\theta_5} + (2 l_a l_5 s_{\theta_a} - 2 l_0 l_5 s_{\theta_0}) s_{\theta_5} + (\ell_h - l_3^2 - 2 l_0 l_a c_{\theta_0} c_{\theta_a} - 2 l_0 l_a s_{\theta_0} s_{\theta_a}) = 0 \qquad (15.75)$$

Obtaining the solution for expressions A, B and C,

$$A_1 = 2 l_a l_5 \cos \theta_a - 2 l_0 l_5 \cos \theta_0 \qquad (15.76)$$

$$B_1 = 2 l_a l_5 \sin \theta_a - 2 l_0 l_5 \sin \theta_0 \qquad (15.77)$$

and

$$C_1 = l_a^2 + l_0^2 + l_5^2 - l_3^2 - 2 l_0 l_a \cos \theta_0 \cos \theta_a - 2 l_0 l_a \sin \theta_0 \sin \theta_a \qquad (15.78)$$

Hence, by using equation (15.11) to solve for θ_5,

$$\theta_5 = 2 \tan^{-1} \left(\frac{-B_1 - \sqrt{A_1^2 + B_1^2 - C_1^2}}{C_1 - A_1} \right) \qquad (15.79)$$

and equation (15.71) is divided by equation (15.70) to solve for θ_3,

$$\theta_3 = \tan^{-1}\left(\frac{l_a \sin\theta_a + l_5 \sin\theta_5 - l_0 \sin\theta_0}{l_a \cos\theta_a + l_5 \cos\theta_5 - l_0 \cos\theta_0}\right) \tag{15.80}$$

It follows that in order to obtain the unknown angles θ_6 and θ_7, the kinematic chains of the path between the point b and the point e is described by

$$l_5 \cos\theta_5 + l_6 \cos\theta_6 = l_7 \cos\theta_7 \tag{15.81}$$

and

$$l_5 \sin\theta_5 + l_6 \sin\theta_6 = l_7 \sin\theta_7 \tag{15.82}$$

Thus, squaring in both sides of the equations

$$l_7^2 \cos^2\theta_7 = l_5^2 \cos^2\theta_5 + l_6^2 \cos^2\theta_6 + 2l_5 l_6 \cos\theta_5 \cos\theta_6 \tag{15.83}$$

and

$$l_7^2 \sin^2\theta_7 = l_5^2 \sin^2\theta_5 + l_6^2 \sin^2\theta_6 + 2l_5 l_6 \sin\theta_5 \sin\theta_6. \tag{15.84}$$

Algebraically simplifying the following is obtained:

$$l_7^2 = l_5^2 + l_6^2 + 2l_5 l_6 \cos\theta_5 \cos\theta_6 + 2l_5 l_6 \sin\theta_5 \sin\theta_6 \tag{15.85}$$

and

$$(2l_5 l_6 \cos\theta_5)\cos\theta_6 + (2l_5 l_6 \sin\theta_5)\sin\theta_6 + (l_5^2 + l_6^2 - l_7^2) = 0 \tag{15.86}$$

Now stating the solution of expressions A, B and C

$$A_2 = 2l_5 l_6 \cos\theta_5$$

$$B_2 = 2l_5 l_6 \sin\theta_5 \tag{15.87}$$

and

$$C_2 = l_5^2 + l_6^2 - l_7^2 \tag{15.88}$$

Likewise, by using equation (15.11) to solve for θ_6

$$\theta_6 = 2 \tan^{-1} \left(\frac{-B_2 - \sqrt{A_2^2 + B_2^2 - C_2^2}}{C_2 - A_2} \right) \tag{15.89}$$

and equation (15.82) is divided by equation (15.81) to solve for θ_7,

$$\theta_7 = \tan^{-1} \left(\frac{l_5 \sin \theta_5 + l_6 \sin \theta_6}{l_5 \cos \theta_5 + l_6 \cos \theta_6} \right) \tag{15.90}$$

Following to solve for the next kinematic chain of links, to obtain the unknown angles θ_4 and θ_8, the kinematic chains from point a to point f are stated next.

$$l_a \cos \theta_a + l_8 \cos \theta_8 = l_0 \cos \theta_0 + l_4 \cos \theta_4 \tag{15.91}$$

and

$$l_a \sin \theta_a + l_8 \sin \theta_8 = l_0 \sin \theta_0 + l_4 \sin \theta_4 \tag{15.92}$$

Dropping off the terms containing the unknown variables of interest in both equations

$$l_4 \cos \theta_4 = l_a \cos \theta_a + l_8 \cos \theta_8 - l_0 \cos \theta_0 \tag{15.93}$$

and

$$l_4 \sin \theta_4 = l_a \sin \theta_a + l_8 \sin \theta_8 - l_0 \sin \theta_0 \tag{15.94}$$

It follows to square both sides of the equations

$$l_4^2 \cos^2 \theta_4 = l_a^2 \cos^2 \theta_a + l_8^2 \cos^2 \theta_8 + l_0^2 \cos^2 \theta_0 + 2 l_a l_8 \cos \theta_a \cos \theta_8 - 2 l_0 l_a \cos \theta_0 \cos \theta_a - 2 l_0 l_8 \cos \theta_0 \cos \theta_8$$

and

$$l_4^2 \sin^2 \theta_4 = l_a^2 \sin^2 \theta_a + l_8^2 \sin^2 \theta_8 + l_0^2 \sin^2 \theta_0 + 2 l_a l_8 \sin \theta_a \sin \theta_8 - 2 l_0 l_a \sin \theta_0 \sin \theta_a - 2 l_0 l_8 \sin \theta_0 \sin \theta_8$$

By defining $\ell_f = l_a^2 + l_8^2 + l_0^2$ Algebraically simplifying,

$$l_4^2 = \ell_f + 2 l_a l_8 c_{\theta_a} c_{\theta_8} - 2 l_0 l_a c_{\theta_0} c_{\theta_a} - 2 l_0 l_8 c_{\theta_0} c_{\theta_8} + 2 l_a l_8 s_{\theta_a} s_{\theta_8} - 2 l_0 l_a s_{\theta_0} s_{\theta_a} - 2 l_0 l_8 s_{\theta_0} s_{\theta_8} \tag{15.95}$$

and

$$(2l_a l_8 c_{\theta_a} - 2l_0 l_8 c_{\theta_0})c_{\theta_8} + (2l_a l_8 s_{\theta_a} - 2l_0 l_8 s_{\theta_0})s_{\theta_8} + (\ell_f - l_4^2 - 2l_0 l_a c_{\theta_0} c_{\theta_a} - 2l_0 l_a s_{\theta_0} s_{\theta_a}) = 0 \quad (15.96)$$

Obtaining a solution for expressions A, B and C,

$$A_3 = 2l_a l_8 \cos\theta_a - 2l_0 l_8 \cos\theta_0$$

$$B_3 = 2l_a l_8 \sin\theta_a - 2l_0 l_8 \sin\theta_0$$

and

$$C_3 = l_a^2 + l_0^2 + l_8^2 - l_4^2 - 2l_0 l_a \cos\theta_0 \cos\theta_a - 2l_0 l_a \sin\theta_0 \sin\theta_a$$

Through equation (15.11) we solve for θ_8,

$$\theta_8 = 2\tan^{-1}\left(\frac{-B_3 - \sqrt{A_3^2 + B_3^2 - C_3^2}}{C_3 - A_3}\right)$$

As well as equation (15.94) divided by equation (15.93) for solving θ_4,

$$\theta_4 = \tan^{-1}\left(\frac{l_a \sin\theta_a + l_8 \sin\theta_8 - l_0 \sin\theta_0}{l_a \cos\theta_a + l_8 \cos\theta_8 - l_0 \cos\theta_0}\right)$$

Similarly, it follows that to obtain the next unknown angles θ_9 and θ_{10}, the kinematic chains form the path from the point b to the point g modelled by

$$l_7 \cos\theta_7 + l_9 \cos\theta_9 = l_8 \cos\theta_8 + l_{10} \cos\theta_{10}$$

and

$$l_7 \sin\theta_7 + l_9 \sin\theta_9 = l_8 \sin\theta_8 + l_{10} \sin\theta_{10}$$

Isolating the terms having the variables of interest in both equations,

$$l_{10} \cos\theta_{10} = l_7 \cos\theta_7 + l_9 \cos\theta_9 - l_8 \cos\theta_8 \quad (15.97)$$

and

$$l_{10} \sin \theta_{10} = l_7 \sin \theta_7 + l_9 \sin \theta_9 - l_8 \sin \theta_8 \tag{15.98}$$

and squaring in both sides of equations

$$l_{10}^2 \cos^2 \theta_{10} = l_7^2 \cos^2 \theta_7 + l_9^2 \cos^2 \theta_9 + l_8^2 \cos^2 \theta_8 + 2l_7 l_9 \cos \theta_7 \cos \theta_9 - 2l_7 l_8 \cos \theta_7 \cos \theta_8 - 2l_8 l_9 \cos \theta_8 \cos \theta_9$$

and

$$l_{10}^2 \sin^2 \theta_{10} = l_7^2 \sin^2 \theta_7 + l_9^2 \sin^2 \theta_9 + l_8^2 \sin^2 \theta_8 + 2l_7 l_9 \sin \theta_7 \sin \theta_9 - 2l_7 l_8 \sin \theta_7 \sin \theta_8 - 2l_8 l_9 \sin \theta_8 \sin \theta_9$$

Algebraically simplifying with definition $\ell_g = l_7^2 + l_9^2 + l_8^2$

$$l_{10}^2 = \ell_g + 2l_7 l_9 c_{\theta_7} c_{\theta_9} - 2l_7 l_8 c_{\theta_7} c_{\theta_8} - 2l_8 l_9 c_{\theta_8} c_{\theta_9} + 2l_7 l_9 s_{\theta_7} s_{\theta_9} - 2l_7 l_8 s_{\theta_7} s_{\theta_8} - 2l_8 l_9 s_{\theta_8} s_{\theta_9} \tag{15.99}$$

and

$$(2l_7 l_9 c_{\theta_7} - 2l_8 l_9 c_{\theta_8})c_{\theta_9} + (2l_7 l_9 s_{\theta_7} - 2l_8 l_9 s_{\theta_8})s_{\theta_9} + (\ell_g - l_{10}^2 - 2l_7 l_8 c_{\theta_7} c_{\theta_8} - 2l_7 l_8 s_{\theta_7} s_{\theta_8}) = 0 \tag{15.100}$$

Thus, obtaining the solution of expressions A, B and C,

$$A_4 = 2l_7 l_9 \cos \theta_7 - 2l_8 l_9 \cos \theta_8 \tag{15.101}$$

$$B_4 = 2l_7 l_9 \sin \theta_7 - 2l_8 l_9 \sin \theta_8 \tag{15.102}$$

and

$$C_4 = l_7^2 + l_9^2 + l_8^2 - l_{10}^2 - 2l_7 l_8 \cos \theta_7 \cos \theta_8 - 2l_7 l_8 \sin \theta_7 \sin \theta_8 \tag{15.103}$$

Using expression (15.11) to solve θ_9,

$$\theta_9 = 2 \tan^{-1} \left(\frac{-B_4 - \sqrt{A_4^2 + B_4^2 - C_4^2}}{C_4 - A_4} \right) \tag{15.104}$$

and equation (15.98) is divided by equation (15.97) to solve θ_{10},

$$\theta_{10} = \tan^{-1}\left(\frac{l_7 \sin \theta_7 + l_9 \sin \theta_9 - l_8 \sin \theta_8}{l_7 \cos \theta_7 + l_9 \cos \theta_9 - l_8 \cos \theta_8}\right) \tag{15.105}$$

In order to obtain an analytical solution for the unknown angles θ_{11} and θ_{12}, the kinematic path from the point f to the point h is developed next,

$$l_{10} \cos \theta_{10} + l_{12} \cos \theta_{12} = l_{11} \cos \theta_{11} \tag{15.106}$$

and

$$l_{10} \sin \theta_{10} + l_{12} \sin \theta_{12} = l_{11} \sin \theta_{11} \tag{15.107}$$

squaring in both sides of the resulting equations,

$$l_{11}^2 \cos^2 \theta_{11} = l_{10}^2 \cos^2 \theta_{10} + l_{12}^2 \cos^2 \theta_{12} + 2l_{10}l_{12} \cos \theta_{10} \cos \theta_{12} \tag{15.108}$$

and

$$l_{11}^2 \sin^2 \theta_{11} = l_{10}^2 \sin^2 \theta_{10} + l_{12}^2 \sin^2 \theta_{12} + 2l_{10}l_{12} \sin \theta_{10} \sin \theta_{12} \tag{15.109}$$

simplifying expressions,

$$l_{11}^2 = l_{10}^2 + l_{12}^2 + 2l_{10}l_{12} \cos \theta_{10} \cos \theta_{12} + 2l_{10}l_{12} \sin \theta_{10} \sin \theta_{12} \tag{15.110}$$

and

$$(2l_{10}l_{12} \cos \theta_{10}) \cos \theta_{12} + (2l_{10}l_{12} \sin \theta_{10}) \sin \theta_{12} + (l_{10}^2 + l_{12}^2 - l_{11}^2) = 0$$

hence the solutions for expressions A, B and C are obtained,

$$A_5 = 2l_{10}l_{12} \cos \theta_{10}$$

$$B_5 = 2l_{10}l_{12} \sin \theta_{10}$$

and

$$C_5 = l_{10}^2 + l_{12}^2 - l_{11}^2$$

By using equation (15.11), θ_{12} is solved,

$$\theta_{12} = 2\tan^{-1}\left(\frac{-B_5 - \sqrt{A_5^2 + B_5^2 - C_5^2}}{C_5 - A_5}\right)$$

likewise, with an algebraic division of expressions (15.107) and (15.106) θ_{11} is solved,

$$\theta_{11} = \tan^{-1}\left(\frac{l_{10}\sin\theta_{10} + l_{12}\sin\theta_{12}}{l_{10}\cos\theta_{10} + l_{12}\cos\theta_{12}}\right)$$

In order to find a functional form to solve for the angle θ_{13}, which describes the angular motion on point i, the input angle θ_0 was involved with motion transmission of 1:1 relation, with inverse sense of rotation and angular offset β

$$\theta_{13} = \left(-\frac{l_0}{l_{13}}\theta_0 - \beta\right) \tag{15.111}$$

From the provided formulation, it follows to include the Hoekens mechanism model, where for the hybrid case the point h is actually not fixed, depending only on the output trajectory of the Jansen linkage. There is assumed an imaginary link l_{im} represented by a vector position on the point h w.r.t. the point i. such vector components are

$$l_{imX} = h_x - i_x \tag{15.112}$$

and

$$l_{imY} = h_y - i_y \tag{15.113}$$

In addition, the connection between the two mechanisms is in principle by obtaining a solution for the unknown angles θ_{14} and θ_{16}. The kinematic path from the point i to the point k is defined by

$$l_{imX} + l_{16}\cos\theta_{16} = l_{13}\cos\theta_{13} + l_{14}\cos\theta_{14} \tag{15.114}$$

and

$$l_{imY} + l_{16}\sin\theta_{16} = l_{13}\sin\theta_{13} + l_{14}\sin\theta_{14} \tag{15.115}$$

Similarly as in previous methodology by isolating the terms having the unknown variables of interest at this stage

$$l_{16} \cos \theta_{16} = l_{13} \cos \theta_{13} + l_{14} \cos \theta_{14} - l_{imX} \tag{15.116}$$

and

$$l_{16} \sin \theta_{16} = l_{13} \sin \theta_{13} + l_{14} \sin \theta_{14} - l_{imY} \tag{15.117}$$

The resulting equations are squared in both sides of the expressions

$$l_{16}^2 \cos^2 \theta_{16} = l_{13}^2 \cos^2 \theta_{13} + l_{14}^2 \cos^2 \theta_{14} + l_{imX}^2 + 2l_{13}l_{14} \cos \theta_{13} \cos \theta_{14} - 2l_{13}l_{imX} \cos \theta_{13} - 2l_{14}l_{imX} \cos \theta_{14}$$

and

$$l_{16}^2 \sin^2 \theta_{16} = l_{13}^2 \sin^2 \theta_{13} + l_{14}^2 \sin^2 \theta_{14} + l_{imY}^2 + 2l_{13}l_{14} \sin \theta_{13} \sin \theta_{14} - 2l_{13}l_{imY} \sin \theta_{13} - 2l_{14}l_{imY} \sin \theta_{14}$$

simplifying algebraically the expressions substituting $\ell_m = l_{13}^2 + l_{14}^2 + l_{imX}^2 + l_{imY}^2$

$$l_{16}^2 = \ell_m + 2l_{13}l_{14}c_{\theta_{13}}c_{\theta_{14}} - 2l_{14}l_{imX}c_{\theta_{14}} + 2l_{13}l_{14}s_{\theta_{13}}s_{\theta_{14}} - 2l_{14}l_{imY}s_{\theta_{14}} - 2l_{13}l_{imX}c_{\theta_{13}} - 2l_{13}l_{imY}s_{\theta_{13}} \tag{15.118}$$

$$(2l_{13}l_{14}c_{\theta_{13}} - 2l_{14}l_{imX})c_{\theta_{14}} + (2l_{13}l_{14}s_{\theta_{13}} - 2l_{14}l_{imY})s_{\theta_{14}} + (\ell_m - l_{16}^2 - 2l_{13}l_{imX}c_{\theta_{13}} - 2l_{13}l_{imY}s_{\theta_{13}}) = 0 \tag{15.119}$$

Hence, it follows to obtain the expressions A, B and C,

$$A_6 = 2l_{13}l_{14} \cos \theta_{13} - 2l_{14}l_{imX} \tag{15.120}$$

$$B_6 = 2l_{13}l_{14} \sin \theta_{13} - 2l_{14}l_{imY} \tag{15.121}$$

and

$$C_6 = l_{13}^2 + l_{14}^2 + l_{imX}^2 + l_{imY}^2 - l_{16}^2 - 2l_{13}l_{imX} \cos \theta_{13} - 2l_{13}l_{imY} \sin \theta_{13} \tag{15.122}$$

By using equation (15.11) to solve θ_{14}, The passive joint angle θ_{14} is modelled by

$$\theta_{14} = 2 \tan^{-1} \left(\frac{-B_6 - \sqrt{A_6^2 + B_6^2 - C_6^2}}{C_6 - A_6} \right) \tag{15.123}$$

Likewise, (15.117) is divided by (15.116) to solve θ_{16},

Corollary 15.3.2. *The passive joint angle θ_{16} is modelled by*

$$\theta_{16} = \tan^{-1}\left(\frac{l_{13}\sin\theta_{13} + l_{14}\sin\theta_{14} - l_{imY}}{l_{13}\cos\theta_{13} + l_{14}\cos\theta_{14} - l_{imX}}\right) \qquad (15.124)$$

To obtain the position vector of the contact point with the ground l, the kinematic chain formed by the links l_0, l_4, l_{11}, l_{16} and l_{15} was used to state (15.125).

Theorem 15.3.3 (Limb's contact point position). *The limb's contact point kinematic position is defined by the next vector:*

$$\mathbf{p}_l = \begin{pmatrix} x_l \\ y_l \end{pmatrix} = \begin{pmatrix} l_0\cos\theta_0 + l_4\cos\theta_4 + l_{11}\cos\theta_{11} + l_{16}\cos\theta_{16} + l_{15}\cos\theta_{15} \\ l_0\sin\theta_0 + l_4\sin\theta_4 + l_{11}\sin\theta_{11} + l_{16}\sin\theta_{16} + l_{15}\sin\theta_{15} \end{pmatrix} \qquad (15.125)$$

To validate the position model, numerous numerical simulations were produced as the ones depicted in figure 15.9. The trajectories plotted basically complete a revolution by the driven joint θ_0 from 0 to 2π radians, with links length in mm: $l_0 = 50$, $l_a = 129.29$, $l_3 = 166.67$, $l_4 = 206.35$, $l_5 = 138.30$, $l_6 = 186$, $l_7 = 133.71$, $l_8 = 131$, $l_9 = 131.33$, $l_{10} = 122.33$, $l_{11} = 163.41$, $l_{12} = 219.06$, $l_{13} = 50$, $l_{14} = 125$, $l_{15} = 125$, $l_{16} = 125$, $l_{17} = -25$, $l_{18} = -175$. Simulation results are produced by the summarised set of links' vector models. The position vector of the point a is provided next,

$$\mathbf{p}_a = \begin{pmatrix} x_a \\ y_a \end{pmatrix} = \begin{pmatrix} 0 \\ 0 \end{pmatrix} \qquad (15.126)$$

The position vector of the point b,

$$\mathbf{p}_b = \begin{pmatrix} x_b \\ y_b \end{pmatrix} = \begin{pmatrix} l_a\cos\theta_a \\ l_a\sin\theta_a \end{pmatrix} \qquad (15.127)$$

The position vector of the point c,

$$\mathbf{p}_c = \begin{pmatrix} x_c \\ y_c \end{pmatrix} = \begin{pmatrix} l_0\cos\theta_0 \\ l_0\sin\theta_0 \end{pmatrix} \qquad (15.128)$$

The position vector of the point d,

$$\mathbf{p}_d = \begin{pmatrix} x_d \\ y_d \end{pmatrix} = \begin{pmatrix} l_0 \cos \theta_0 + l_3 \cos \theta_3 \\ l_0 \sin \theta_0 + l_3 \sin \theta_3 \end{pmatrix} \tag{15.129}$$

The position vector of the point e,

$$\mathbf{p}_e = \begin{pmatrix} x_e \\ y_e \end{pmatrix} = \begin{pmatrix} l_0 \cos \theta_0 + l_3 \cos \theta_3 + l_6 \cos \theta_6 \\ l_0 \sin \theta_0 + l_3 \sin \theta_3 + l_6 \sin \theta_6 \end{pmatrix} \tag{15.130}$$

The position vector of the point f,

$$\mathbf{p}_f = \begin{pmatrix} x_f \\ y_f \end{pmatrix} = \begin{pmatrix} l_0 \cos \theta_0 + l_4 \cos \theta_4 \\ l_0 \sin \theta_0 + l_4 \sin \theta_4 \end{pmatrix} \tag{15.131}$$

The position vector of the point g,

$$\mathbf{p}_g = \begin{pmatrix} x_g \\ y_g \end{pmatrix} = \begin{pmatrix} l_0 \cos \theta_0 + l_4 \cos \theta_4 + l_{10} \cos \theta_{10} \\ l_0 \sin \theta_0 + l_4 \sin \theta_4 + l_{10} \sin \theta_{10} \end{pmatrix} \tag{15.132}$$

the position vector of the point h,

$$\mathbf{p}_h = \begin{pmatrix} x_h \\ y_h \end{pmatrix} = \begin{pmatrix} l_0 \cos \theta_0 + l_4 \cos \theta_4 + l_{10} \cos \theta_{10} + l_{12} \cos \theta_{12} \\ l_0 \sin \theta_0 + l_4 \sin \theta_4 + l_{10} \sin \theta_{10} + l_{12} \sin \theta_{12} \end{pmatrix} \tag{15.133}$$

The position vector of the point i:

$$\mathbf{p}_i = \begin{pmatrix} x_i \\ y_i \end{pmatrix} = \begin{pmatrix} -l_{17} \\ -l_{18} \end{pmatrix} \tag{15.134}$$

The position vector of the point j,

$$\mathbf{p}_j = \begin{pmatrix} x_j \\ y_j \end{pmatrix} = \begin{pmatrix} l_{13} \cos \theta_{13} - l_{17} \\ l_{13} \sin \theta_{13} - l_{18} \end{pmatrix} \tag{15.135}$$

The position vector of the point k,

$$\mathbf{p}_k = \begin{pmatrix} x_k \\ y_k \end{pmatrix} = \begin{pmatrix} l_{13} \cos \theta_{13} + l_{14} \cos \theta_{14} - l_{17} \\ l_{13} \sin \theta_{13} + l_{14} \sin \theta_{14} - l_{18} \end{pmatrix} \qquad (15.136)$$

For plotting the trajectory of the point l, the position vector (15.125) was numerically valued.

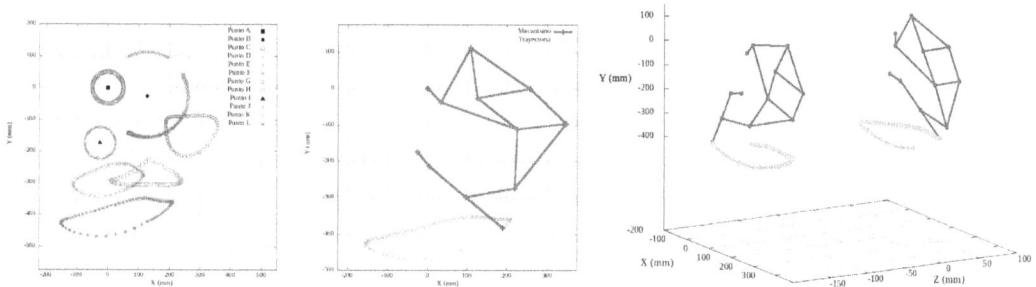

Figure 15.9: Trajectories described by each point of the hybrid limb (left). Walking gait simulation (centre). 3D biped's limbs simulation (right).

Alternative to gait validation by numerical simulations (figure 15.9), numerous experimental measurements of the gait trajectory tracking were carried out (figure 15.10-a)). A home-made experimental prototype robot was built in our Robotics Lab shown in figure 15.10-c), and its real gait shape was compared with numerical models obtaining accurate and satisfactory results. Each experiment consisted of capturing the set of images while the limbs walked. By deploying only one actuator per limb, the mechanism movement generation alternated movements between the two parallel limbs imitating a bipedal gait. An artificial visual landmark was laterally placed on the Limb's contact point l as a manner to track it by an external computer vision system, as shown by figure 15.10-a). Therefore, by using the real kinematic parameters of the walker robot, the corresponding numerical gait simulation (theoretical) was matched with the mechanism plot in Cartesian space (observation), finding a minimal error due to several factors: manufacturing inaccuracies, gravitational and inertial effects.

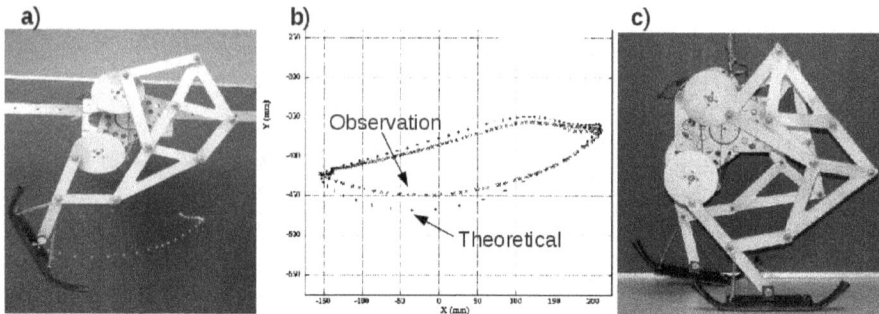

Figure 15.10:   a) Experimental robot's gait tracking by a computer vision system. b) Comparative plot between a theoretical gait and the empirical observation. c) A photo of the experimental biped prototype.

Comparative results between the mathematical theoretical model, and the empirical observation through the tracking vision system are depicted in figure 15.10-b). Therefore, it is concluded that the experimental robot prototype emulating bipedal gaits, acted closely as predicted by the kinematic model. Both trajectories had similar kinematic behaviour, and such results validated the theoretical analysis provided in this chapter. In addition, in order to achieve under-actuated walking gaits, there exist the need to deploy actuators with optimal torque. Likewise, actuators velocity control systems are required to accomplish synchronization between the two parallel extremities, achieving gait coordination as expected according to the kinematic formulation.

Bibliography

[1] Giesbrecht D., Wu C.Q., Sepehri N., *Design and Optimization of an Eight-bar Legged Walking Mechanism Imitating a Kinetic Sculpture*, Transactions of the Canadian Society for Mechanical Engineering, vol. 36(4), pp. 343–355, 2012.

[2] Asano F., Zhi-Wei L., *Underactuated virtual passive dynamic walking with an upper body*, IEEE Intl. Conf. on Robotics and Automation, CA, USA, pp. 2441–2446, 2008.

[3] L. Guangri, H. Qiang, X. Qian, L. Guodong, L. Jing, and L. Min, *Design of a small biped mechanism with 7 DOFs legs and double spherical hip joint*, 8th World Congress on Intelligent Control and Automation (WCICA), Jinan, China, pp. 264–269, 2010.

[4] Marghitu D.B., Crocker M.J., *Analytical elements of mechanisms*, Cambridge University Press, 2001, isbn 0-521-62383-9.

[5] Kim S., Hyun H.S., Hun K. D., *Analysis of a Crab Robot Based on Jansen Mechanism*, 11th Intl. Conf. on Control, Automation and Systems, Gyeonggi-do, Korea, pp. 858–860, 2011.

[6] Norton P.L., *Design of Machinery: An introduction to the synthesis and analysis of mechanisms and machines*, 4th Ed. 2007, isbn 978-0073290980.

[7] H. Qingjiu, T. Hase, and K. Ono, *Passive/active unified dynamic walking for biped locomotion*, IEEE Intl. Conf. on Robotics and Biomimetics, Sanya, China, pp. 964–971, 2007.

[8] R.C. Luo, H.P. Yi, C. Chwan-Hsen, R.C. Jia, and Y.L. Cheng, *Design and implementation of humanoid biped walking robot mechanism towards natural walking*, IEEE Intl. Conf. on Robotics and Biomimetics, Phuket, Thailand, pp. 1165–1170, 2011.

[9] Collins S.H., Ruina A., *A Bipedal Walking Robot with Efficient and Human-Like Gait*, IEEE Intl. Conf. on Robotics and Automation, Barcelona, Spain, pp.1983–1988, 2005.

[10] D. Foerg, H. Ulbirch, and A. Seyfarth, *Study of a Bipedal Robot with Elastic Elements*, 6th German Conference on Robotics (ROBOTIK), Munich, Germany, pp. 1–7, 2010.

[11] D. Tlalolini, C. Chevallereau, and Y. Aoustin, *Optimal reference walking with rotation of the stance feet in single support for a 3D biped*, IEEE/RSJ Intl. Conf. on Intelligent Robots and Systems, Nice, France,pp. 1091–1096, 2008.

[12] Ting-Ying W., Yeh T., *Optimal design and implementation of an energy-efficient semi-active biped*, IEEE Intl. Conf. on Robotics and Automation, CA, USA, pp. 1252–1257, 2008.

Part V

Robot Modelling for 3D Navigation

Chapter 16

MODELLING A HEXAPOD-TYPE

AMPHIBIOUS ROBOT

Angel A. Maldonado Ramírez[2], Edgar A. Martínez García[1] and L. Abril Torres Méndez[2]

[1]*Laboratorio de Robótica, Institute of Engineering and Technology, Universidad Autónoma de Ciudad Juárez, Mexico.*

[2]*Robotics and Advanced Manufacturing Group, CINVESTAV Campus Saltillo, Coahuila, Mexico.*

Mathematical modelling of physical systems is one of the major foundations to develop applied control systems, and process design. The mathematical model of a system is derived either from physical laws or experimental data. In this chapter, the main focus is on modelling a complete amphibious robotic system comprised of six propulsive paddles. This analysis starts from the experimental modelling of the actuators (DC-motors), and rotatory sensors (encoders), until reaching an Euler-Lagrange formulation of the full robotic platform. Experimental results are presented to show the application of the model for controlling the actuators. Systems modelling is an important process in all fields of science and engineering. It provides a deeper understanding of the behaviour of any system. Particularly, when specifically it is desired to control such a system. Thus, by taking into account its full mathematical dynamic model, more sophisticated is the system design, and hence more robust results are obtained. Nevertheless, obtaining an analytical solution of a model is not always an easy task specially when there exist numerous unknown perturbation phenomena that are difficult to define. The mathematical model of a system can be derived either from physical laws as analytical solutions, or experimental data through numerical solutions. In some works, a mathematical model is obtained and then through experimentation the values of certain parameters are calculated ' . Another approach for modelling a system is the Euler-Lagrange formulation, which is based on the energy of the system as a function of the generalized coordinates. As long as, we know how to relate the generalized coordinates with the energy of the system, we can use the Lagrangian approach. This approach is widely used in the robotics field, some example of applications has been reported ' . In the last part of this work, we will focus on the modelling of the entire underwater robotic platform by using an Euler-Lagrange formulation.

16.1 Actuator's experimental-theoretical model

In the first part of this chapter, the actuators angular velocity of an amphibious robotic system is modelled by using a measuring experimental approach to subsequently state the control of the paddles angular speed. This approach approximates a full model that scopes the present uncertainties within the actuator's empirical model. The empirical model is obtained by deploying simple rotatory encoder devices. The encoders are sensors that measure the angular position of an actuator device (i.e. a DC-motor). The angular position is obtained by the number of pulses, and the relationship between the pulses and the angular position is given by

$$\phi_i = \frac{2\pi}{r_{enc}} n_i,$$ (16.1)

where r_{enc} is the resolution of the encoder and n_i is the instantaneous number of pulses. In order to infer the angular speed, we obtain the numerical derivative (see section 1.7) of the angular position with respect to time, as provided by the following expression:

$$\hat{\dot{\phi}}_i = \frac{\frac{2\pi n_i}{r_{enc}} - \frac{2\pi n_{i-1}}{r_{enc}}}{t_i - t_{i-1}} = \frac{2\pi}{r_{enc}} \left(\frac{n_i - n_{i-1}}{t_i - t_{i-1}} \right).$$ (16.2)

By providing the numerical derivative w.r.t. time of the angular velocity, the angular acceleration is obtained by

$$\hat{\ddot{\phi}}_i = \frac{\dot{\phi}_i - \dot{\phi}_{i-1}}{t_i - t_{i-1}} = \frac{\frac{2\pi}{r_{enc}} \frac{n_i - n_{i-1}}{t_i - t_{i-1}} - \frac{2\pi}{r_{enc}} \frac{n_{i-1} - n_{i-2}}{t_{i-1} - t_{i-2}}}{t_i - t_{i-1}},$$

Assuming constant sampling times $\Delta t = t_i - t_{i-1}$, the next equation is also obtained,

$$c_i = \frac{2\pi}{r_{enc}} \left(\frac{\frac{n_i - n_{i-1}}{\Delta t} - \frac{n_{i-1} - n_{i-2}}{\Delta t}}{\Delta t} \right) = \frac{2\pi}{r_{enc}} \left(\frac{n_i - 2n_{i-1} + n_{i-2}}{\Delta t} \right). \tag{16.3}$$

Therefore, the following postulate is hereafter the actuator's observation, which feedbacks the rate of change of the angular position w.r.t. time.

Postulate 16.1.1 (angle observation model). *The actuator's averaged value of k measurements $\dot{\hat{\phi}}$ is the derivative of the angular position w.r.t. to t.*

$$\dot{\phi}(d) = \frac{1}{k} \sum_{i=1}^{k} \dot{\hat{\phi}}_i = \sum_{i=1}^{k} \frac{2\pi}{r_{enc}} \left(\frac{n_i - n_{i-1}}{t_i - t_{i-1}} \right) \tag{16.4}$$

As a part to control the actuators, it is important to establish the mathematical model of the actuator's variables, which are of interest as the control input and output of the robot system. For instance, it is fundamental to find the mathematical relationship between the digital control, and the motor's tangential and/or angular speeds. Since the models for the actuator and for the inherent perturbations are unknown, the empirical model is obtained experimentally.

In our particular case, we are able model the relation between a digital control command and the speed of the motor. All the possible control commands were applied to two motors and their average speed response were registered (16.4). The motor type, encoders, and mechanical structure are identical. The collected data are shown by figure 16.1.

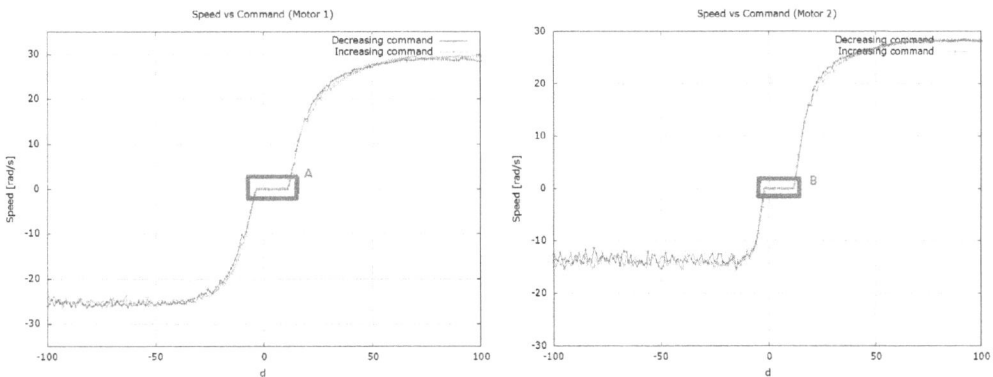

Figure 16.1: Angular speed as a function of a digital control (two physical motors).

Although both motors were under the same conditions, the sets of collected data were different. Also, it is seen in figure 16.1 that there is a region where no motors' response was performed regardless the digital control commands (i.e. regions A and B). These regions lack of angular speeds due to the shaft frictions and mechanical loads applied to the motors (also frictions due to shaft misalignments). We obtained approximated theoretical models for each motor by using a non-linear regression applied to the empirical model. A residual error (r_i) between the empirical data and a theoretical model ($\dot{\phi}_*$) was defined. A third degree polynomial as a theoretical model was proposed to fit well the speed behaviour.

$$r_i = \dot{\phi}_i - \dot{\phi}_*,$$

$$r_i = \dot{\phi}_i - (a_0 + a_1 d_i + a_2 d_i^2 + a_3 d_i^3), \tag{16.5}$$

where d_i is the digital control command. Then, we define the sum of the quadratic errors (see section 1.6.3),

$$S = \sum_{i=1}^{n} r_i^2 = \sum_{i=1}^{n} [\dot{\phi}_i - (a_0 + a_1 d_i + a_2 d_i^2 + a_3 d_i^3)]^2. \tag{16.6}$$

Furthermore, to define a separate model for each coefficient of interest, the partial derivatives of S w.r.t. each parameter a_0, a_1, a_2 and a_3 are developed to set an approximation $S \approx 0$.

$$\frac{\partial S}{\partial a_0} = 2 \sum_{i=1}^{n} \left(\dot{\phi}_i - a_0 - a_1 d_i - a_2 d_i^2 - a_3 d_i^3 \right) (c - 1),$$

$$\frac{\partial S}{\partial a_1} = 2 \sum_{i=1}^{n} \left(\dot{\phi}_i - a_0 - a_1 d_i - a_2 d_i^2 - a_3 d_i^3 \right) (-d_i),$$

$$\frac{\partial S}{\partial a_2} = 2 \sum_{i=1}^{n} \left(\dot{\phi}_i - a_0 - a_1 d_i - a_2 d_i^2 - a_3 d_i^3 \right) (-d_i^2), \tag{16.7}$$

$$\frac{\partial S}{\partial a_3} = 2 \sum_{i=1}^{n} \left(\dot{\phi}_i - a_0 - a_1 d_i - a_2 d_i^2 - a_3 d_i^3 \right) (-d_i^3),$$

Setting each derivative to zero,

$$0 = 2\sum_{i=1}^{n}(\dot{\phi}_i - a_0 - a_1 d_i - a_2 d_i^2 - a_3 d_i^3)(-1),$$

$$0 = 2\sum_{i=1}^{n}(\dot{\phi}_i - a_0 - a_1 d_i - a_2 d_i^2 - a_3 d_i^3)(-d_i), \tag{16.8}$$

$$0 = 2\sum_{i=1}^{n}(\dot{\phi}_i - a_0 - a_1 d_i - a_2 d_i^2 - a_3 d_i^3)(-d_i^2),$$

$$0 = 2\sum_{i=1}^{n}(\dot{\phi}_i - a_0 - a_1 d_i - a_2 d_i^2 - a_3 d_i^3)(-d_i^3),$$

then, by performing the products and rearranging the terms, the following is obtained

$$\sum_{i=1}^{n}\dot{\phi}_i = \sum_{i=1}^{n}a_0 + \sum_{i=1}^{n}a_1 d_i + \sum_{i=1}^{n}a_2 d_i^2 + \sum_{i=1}^{n}a_3 d_i^3,$$

$$\sum_{i=1}^{n}\dot{\phi}_i d_i = \sum_{i=1}^{n}a_0 d_i + \sum_{i=1}^{n}a_1 d_i^2 + \sum_{i=1}^{n}a_2 d_i^3 + \sum_{i=1}^{n}a_3 d_i^4, \tag{16.9}$$

$$\sum_{i=1}^{n}\dot{\phi}_i d_i^2 = \sum_{i=1}^{n}a_0 d_i^2 + \sum_{i=1}^{n}a_1 d_i^3 + \sum_{i=1}^{n}a_2 d_i^4 + \sum_{i=1}^{n}a_3 d_i^5,$$

$$\sum_{i=1}^{n}\dot{\phi}_i d_i^3 = \sum_{i=1}^{n}a_0 d_i^3 + \sum_{i=1}^{n}a_1 d_i^4 + \sum_{i=1}^{n}a_2 d_i^5 + \sum_{i=1}^{n}a_3 d_i^6.$$

Thus, arranging the equations (16.9) in the matrix form,

$$\begin{pmatrix} \sum_{i=1}^{n}\dot{\phi}_i \\ \sum_{i=1}^{n}\dot{\phi}_i d_i \\ \sum_{i=1}^{n}\dot{\phi}_i d_i^2 \\ \sum_{i=1}^{n}\dot{\phi}_i d_i^3 \end{pmatrix} = \begin{pmatrix} n & \sum_{i=1}^{n}d_i & \sum_{i=1}^{n}d_i^2 & \sum_{i=1}^{n}d_i^3 \\ \sum_{i=1}^{n}d_i & \sum_{i=1}^{n}d_i^2 & \sum_{i=1}^{n}d_i^3 & \sum_{i=1}^{n}d_i^4 \\ \sum_{i=1}^{n}d_i^2 & \sum_{i=1}^{n}d_i^3 & \sum_{i=1}^{n}d_i^4 & \sum_{i=1}^{n}d_i^5 \\ \sum_{i=1}^{n}d_i^3 & \sum_{i=1}^{n}d_i^4 & \sum_{i=1}^{n}d_i^5 & \sum_{i=1}^{n}d_i^6 \end{pmatrix} \cdot \begin{pmatrix} a_0 \\ a_1 \\ a_2 \\ a_3 \end{pmatrix} \tag{16.10}$$

Hence, the compact form of (16.10) may be expressed as a linear form,

$$\mathbf{x} = \mathbf{A} \cdot \lambda. \tag{16.11}$$

Since, $\mathbf{A}$ is a positive-definite matrix, then it always has an inverse. Thus, by solving for λ (see section 1.2.3),

$$\mathbf{x} = \mathbf{A} \cdot \lambda,$$

$$\lambda = \mathbf{A}^{-1} \cdot \mathbf{x}. \tag{16.12}$$

Thus, it follows the next proposition of the actuators theoretical model,

Proposition 16.1.2 (actuator theoretical model). *The parameters a_0, a_1, a_2, and a_3 fits the real angular velocity behaviour through the theoretical model $\dot{\phi}_i(d_i)$ as a polynomial function of the digital control d_i,*

$$\dot{\phi}_i = a_0 + a_1 d_i + a_2 d_i^2 + a_3 d_i^3. \tag{16.13}$$

By applying the above process to the empirical data, and obtaining the next values for the parameters,

	a_0	a_1	a_2	a_3
$\dot{\phi}_1$ (-)	-14.4229	1.91041	-0.0275454	0.000129814
$\dot{\phi}_1$ (+)	-0.758746	1.36245	0.0227922	0.000117595
$\dot{\phi}_2$ (+)	-13.4632	1.90241	-0.0283169	0.000136413
$\dot{\phi}_2$ (-)	-5.88016	0.568018	0.0111437	0.0000645076

Table 16.1: Actuators third degree polynomial coefficients found.

In applied control it is required to know the inverse solution of (16.13) in order to calculate d, given that $\dot{\phi}(d)$ is known, then to find one real roots of direct form,

$$0 = a_0 - \dot{\phi}(d) + a_1 d + a_2 d^2 + a_3 d^3 = \mathbf{p}(d), \tag{16.14}$$

Nevertheless, in order to find a close-form solution to this problem, it became a complex algebraic process. Instead, an iterative numerical method was used to solve it, the Newton-Raphson method (see section 1.4). Therefore, as a first mathematical step, a first order derivative poly-

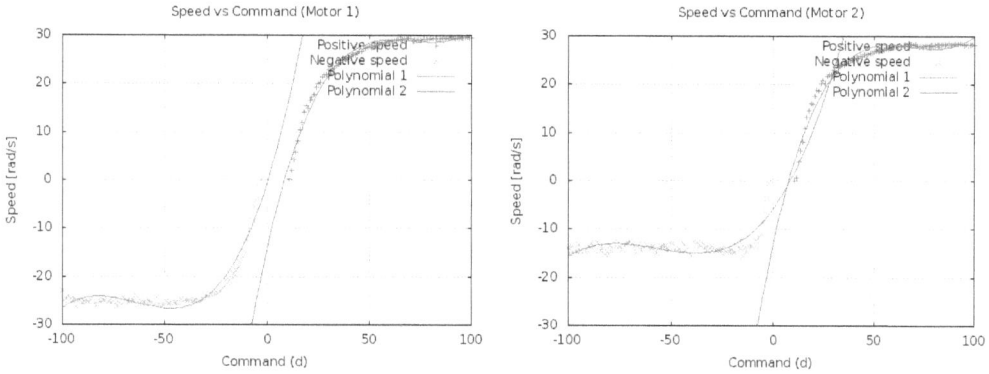

Figure 16.2: Two motors' empirical velocities approximated with third degree polynomials.

nomial is approached through the Maclaurin and Taylor series of $\mathbf{p}(d)$ (see section 1.3)

$$\mathbf{p}(d) = \frac{\mathbf{p}(d_0)}{0!} + \frac{\mathbf{p}'(d_0)(d - d_0)}{1!} + \frac{\mathbf{p}'(d_0)(d - d_0)^2}{2!} + ... \qquad (16.15)$$

Then, $\mathbf{p}(d) = 0$, and let us ignore the high order terms to obtain

$$0 = \mathbf{p}(d_0) + \mathbf{p}'(d_0)(d - d_0). \qquad (16.16)$$

Solving for d, we have

$$d = d_0 - \frac{\mathbf{p}(d_0)}{\mathbf{p}'(d_0)}. \qquad (16.17)$$

An approximated solution for (16.14) is obtained. By doing this process iteratively, we can find the desired d. Expressing this process with equation (16.18).

Corollary 16.1.3. *(actuator's inverse solution) The inverse solution $d(\dot{\phi})$ is stated from previous proposition about the theoretical model $\dot{\phi}(d)$.*

$$d_{k+1} = d_k - \frac{a_0 - \dot{\phi}^d + a_1 d_k + a_2 d_k^2 + a_3 d_k^3}{a_1 + 2a_2 d_k + 3a_3 d_k^2}\Big|_1^{\epsilon < \varepsilon}, \qquad .(16.18)$$

where $\epsilon = \mathbf{p}(d_k)$ and ε is a small value that defines the precision of the result. We will use (16.18) iteratively until $\epsilon < \varepsilon$.

The model is used to directly control the actuators desired speed $\dot{\phi}(d)$, and by using the Newton-Raphson method it is obtained the digital command d to control the motor.

16.2 2nd-order paddle control

The actuator's physical model is critical to propose a control law for approaching the angular oscillations of robot's paddles. Based on different tests and analysis, a 2nd-order angular position resulted reliable and feasible for implementation.

16.2.1 Proportional velocity control

First, we propose a control law for the angular speed of the motor.

$$\dot{\phi}_i = \dot{\phi}_{i-1} + \alpha(\dot{\phi}_{ref} - \hat{\dot{\phi}}_i), \tag{16.19}$$

where $\dot{\phi}_i, \dot{\phi}_{i-1}$ are the values for controlling speed in the current and previous instant. $\dot{\phi}_i$ is used in (16.18) to obtain the digital control input. The second term of the sum is the difference between the desired speed $\dot{\phi}_{ref}$ and the observed speed $\hat{\dot{\phi}}_i$, all this weighted by a factor α. By substituting equation (16.2) in equation (16.19) the following recursive approach is postulated

Postulate 16.2.1 (recursive $\dot{\phi}$ control). *Having a reference model $\dot{\phi}_{ref}$, the angular speed $\dot{\phi}$ is recursively controlled by feedback of its proportional observation error.*

$$\dot{\phi}_i = \dot{\phi}_{i-1} + \alpha \left(\dot{\phi}_{ref} - \frac{2\pi}{r_{enc}} \left[\frac{n_i - n_{i-1}}{t_i - t_{i-1}} \right] \right) \tag{16.20}$$

with

$$\alpha = \begin{cases} \alpha^*, |\dot{\phi}_{ref} - \hat{\dot{\phi}}_i| \geq 1 \\ 1 - \alpha^*, |\dot{\phi}_{ref} - \hat{\dot{\phi}}_i| < 1 \end{cases}, \tag{16.21}$$

and

$$\alpha^* \in (0, 1)$$

Experimental tests were carried out by using expression (16.20) to control the paddle angular velocity. Figure 16.3 shows the results obtained. It can be seen that the constant reference velocity is reached with considerable reliability.

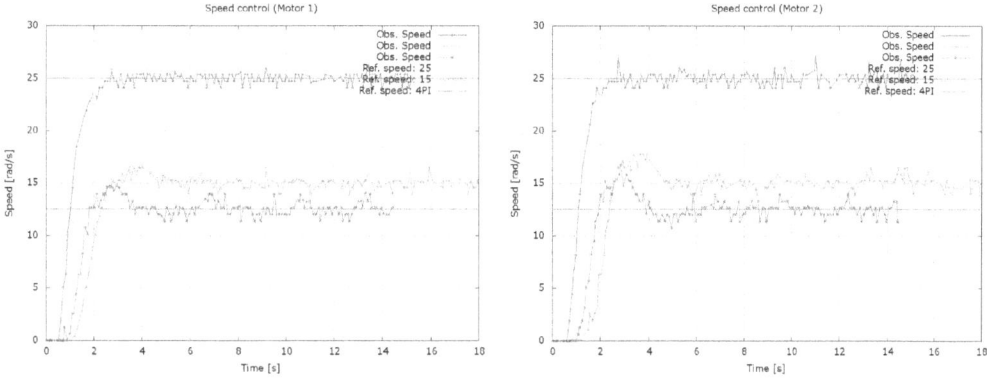

Figure 16.3:   Actuators controlled angular velocity with constant reference models.

Additionally, other experiments were performed adjusting for different values of α. Such results are depicted in figure 16.4. It can be seen that the smaller the value of α the faster the motor reaches the desired speed.

16.2.2   Acceleration control

When slow or unstable controlled speeds are yielded, control of the acceleration may outperform the results instead. And an acceleration control may be preferred in such cases. We propose the next control law:

$$\ddot{\phi}_i = \ddot{\phi}_{i-1} + \kappa(\ddot{\phi}_{ref} - \hat{\ddot{\phi}}_i), \tag{16.22}$$

with

$$\kappa = \begin{cases} \kappa^*, |\ddot{\phi}_{ref} - \hat{\ddot{\phi}}_i| \geq 1 \\ 1 - \kappa^*, |\ddot{\phi}_{ref} - \hat{\ddot{\phi}}_i| < 1 \end{cases} \tag{16.23}$$

and

$$\kappa^* \in (0, 1),$$

where $\ddot{\phi}_i, \ddot{\phi}_{i-1}$ are the values for controlling acceleration, in the current and previous instant. $\ddot{\phi}_{ref}$ is the desired acceleration and $\hat{\ddot{\phi}}_i$ the observed acceleration, all of them weighted by a factor κ. By substituting equation (16.3) in equation (16.22), the following expresson is proposed,

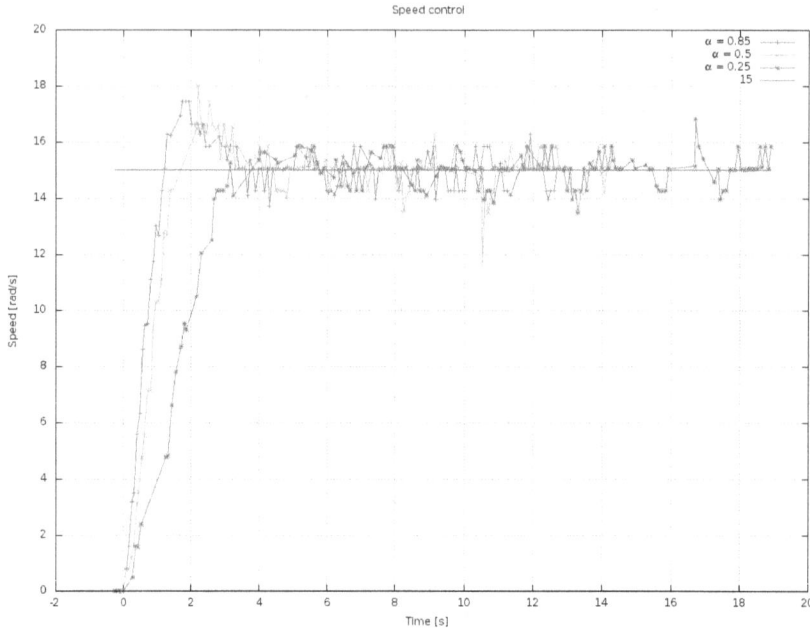

Figure 16.4: Actuator's controlled angular velocity behaviours with different convergence factors α.

Proposition 16.2.2. (controlled $\ddot{\phi}$) *The actuator angular acceleration is recursively controlled by feedback of its error observation of the angular rate of change.*

$$\ddot{\phi}_i = \ddot{\phi}_{i-1} + \kappa \left[\ddot{\phi}_{ref} - \frac{2\pi}{r_{enc}} \left(\frac{n_i - 2n_{i-1} + n_{i-2}}{\Delta t} \right) \right], \tag{16.24}$$

From equation (16.24) in previous proposition, it follows to establish an inverse solution for the digital control variable. To achieve that, we use the trapezoidal rule of the numerical integration methods (see section 1.7) to integrate the acceleration into speed. The obtained speed can be converted into digital control command.

$$\dot{\phi}_i = \int_0^{t_i} \ddot{\phi}_i dt \approx \dot{\phi}_0 + \sum_{j=1}^{i} \frac{(t_j - t_{j-1})(\ddot{\phi}_j + \ddot{\phi}_{j-1})}{2}. \tag{16.25}$$

A recursive expression is derived to perform numerical integration

$$\dot{\phi}_i = \dot{\phi}_0 + \sum_{j=1}^{i} \frac{(t_j - t_{j-1})(\ddot{\phi}_j + \ddot{\phi}_{j-1})}{2}, \qquad (16.26)$$

then, factorizing the term i from the sum,

$$\dot{\phi}_i = \dot{\phi}_0 + \sum_{j=1}^{i-1} \left[\frac{(t_j - t_{j-1})(\ddot{\phi}_j + \ddot{\phi}_{j-1})}{2} \right] + \frac{(t_i - t_{i-1})(\ddot{\phi}_i + \ddot{\phi}_{i-1})}{2}, \qquad (16.27)$$

and by considering $\dot{\phi}_{i-1} = \dot{\phi}_0 + \sum_{j=1}^{i-1} \left[\frac{(t_j - t_{j-1})(\ddot{\phi}_j + \ddot{\phi}_{j-1})}{2} \right]$,

$$\dot{\phi}_i = \dot{\phi}_{i-1} + \frac{(t_i - t_{i-1})(\ddot{\phi}_i + \ddot{\phi}_{i-1})}{2}. \qquad (16.28)$$

Substituting $\dot{\phi}_i$ by $\dot{\phi}_d$ in (16.28), and then in (16.18), we obtain the equation that provide the digital command, given a known acceleration $\ddot{\phi}_i$.

Corollary 16.2.3 (inverse of $\ddot{\phi}$). *The digital control d solution is recursively provided as a function of observations $\dot{\phi}$ and $\ddot{\phi}$.*

$$d_{k+1} = d_k - \frac{a_0 - \dot{\phi}_{i-1} - \frac{(t_i - t_{i-1})(\ddot{\phi}_i + \ddot{\phi}_{i-1})}{2} + a_1 d_k + a_2 d_k^2 + a_3 d_k^3}{a_1 + 2a_2 d_k + 3a_3 d_k^2} \Big|_1^{\epsilon < \varepsilon} \qquad (16.29)$$

The experimental controlled acceleration results obtained are depicted in figure 16.5. The desired acceleration values for this experiment was set to $\ddot{\phi}_{ref} = 0.1$. Being the motor's speed a straight line with slope $m = 0.1$, which was confirmed by setting another straight line with same slope 0.1, but different translation parameter. It is important to highlight that the motor was intentionally subjected to perturbations during the experiment. It was concluded that the acceleration control compensated quite well under the influence of existing perturbations.

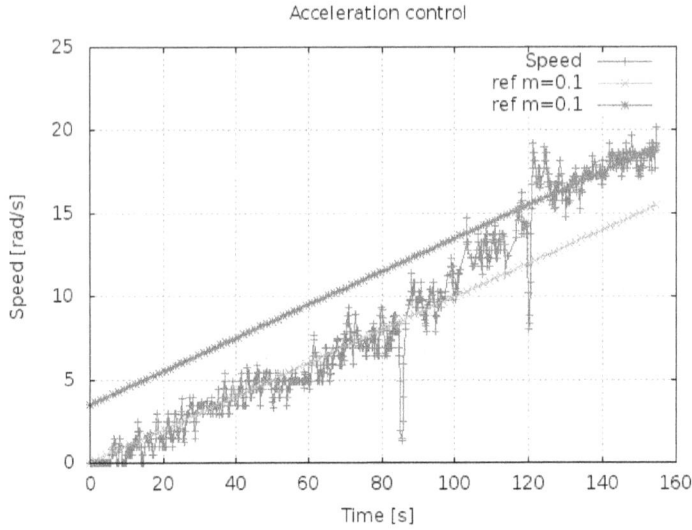

Figure 16.5: Actuator's behaviour with reference model $a(t) = a_0 + a_1 t$, and two different values for a_0.

16.2.3 Acceleration model reference

Plamondon described a function that relates the oscillation of a paddle submerged in water and the force generated by it. The geometry and shape of the paddle is shown in figure 16.6.

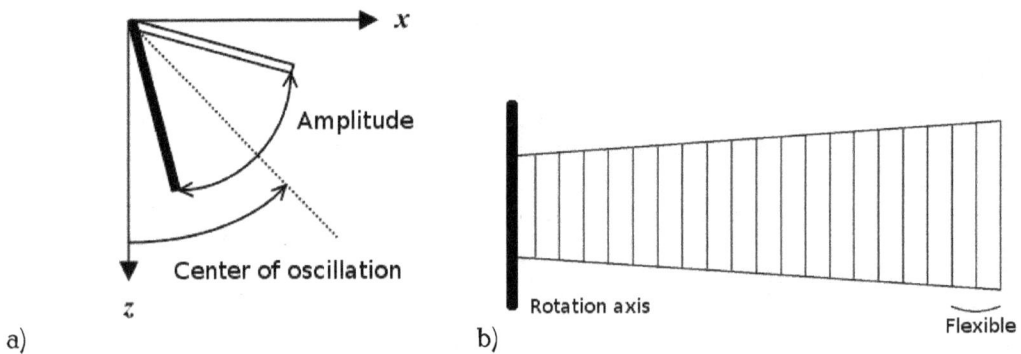

Figure 16.6: a) Centre of oscillation of one paddle (front/back side). b) Paddle's shape (top side).

The movement of the paddle is provided by the following definition:

Definition 16.2.4 (reference oscillation model). The paddle oscillation movement is established as a reference movement by the sinusoidal function:

$$\phi(t) = \frac{A}{2} \text{sen} \left(\frac{2\pi}{P} t + \delta \right) + \lambda \tag{16.30}$$

where A is the oscillation amplitude; P is the oscillation period; δ is the phase, and λ is the centre of oscillation. The relationship between the paddle movement and the force generated is described by the following postulation.

Postulate 16.2.5 (paddle propulsion force). *The experimental propulsion force as a function of the paddle's oscillatory motion is produced according to the next model,*

$$f_p = 0.1963 \frac{(w_1 + 2w_2)l^2}{3} \rho \frac{A}{P} - 0.1554 \tag{16.31}$$

where w_1 and w_2 are the dimensions of the paddle with $w_1 < w_2$. l is the length (all the dimensions in meters) and ρ the density of the water in $\frac{kg}{m^3}$. The direction of the force is determined by the angle λ. We calculate the second derivative w.r.t. time of the equation (16.30) to obtain the acceleration which will be used as the reference value in the equation (16.24).

$$\ddot{\phi}_{ref}(t) = -\frac{2A\pi^2}{P^2} \text{sen} \left(\frac{2\pi}{P} t + \delta \right). \tag{16.32}$$

Therefore, by substituting (16.32) in (16.24), the following corollary is stated,

Corollary 16.2.6. *[model reference based $\ddot{\phi}$] The paddle angular velocity control is given as a recursive tracking function of the non linear reference model.*

$$\ddot{\phi}_i = \ddot{\phi}_{i-1} + \kappa \left[-\frac{2A\pi^2}{P^2} \text{sen} \left(\frac{2\pi}{P} t_i + \delta \right) - \frac{2\pi}{r_{enc}} \left(\frac{n_i - 2n_{i-1} + n_{i-2}}{\Delta t} \right) \right] \tag{16.33}$$

An underwater robot with oscillating paddles was emulated by deploying two motors and controlling their acceleration through equation (16.33). The results obtained whit varying amplitude of oscillations are shown in figure 16.7.

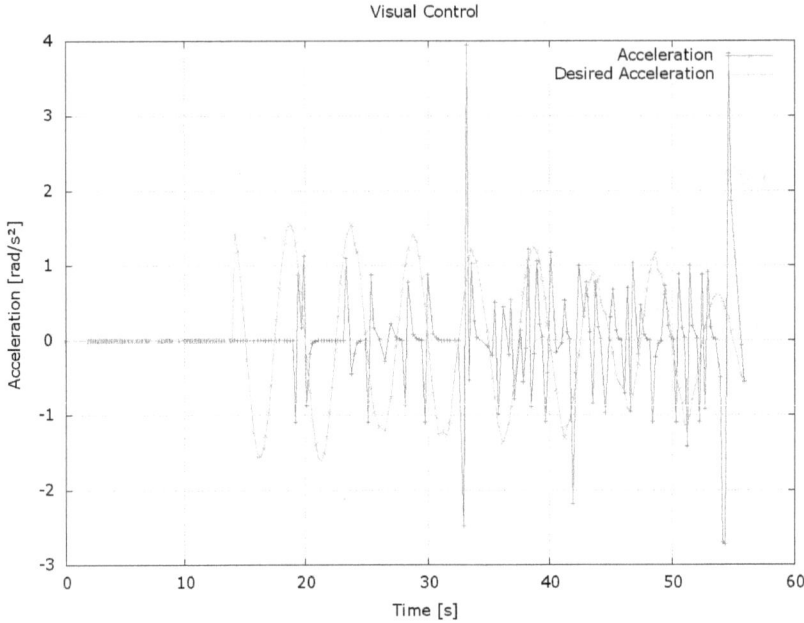

Figure 16.7:   Robot's two-paddle controlled oscillations tied to the network transmission delays.

16.3   Robot's dynamical analysis

In two previous sections the modelling and control of the actuators were described. Now, this section describes the robot's general movement with an approach to the Newton-Euler equations to model the tangential forces generated by the rotatory actuators. Further information on the underwater robot in discussion may be found ' . The forces generated by the paddle are decomposed into their components along the X and Y axes. Each force has a magnitude of f_{p_j} and a direction of λ_j. Therefore, the following equations are stated,

$$f_{x_j} = f_{p_j} \cos \lambda_j, \tag{16.34}$$

$$f_{y_j} = f_{p_j} \sin \lambda_j. \tag{16.35}$$

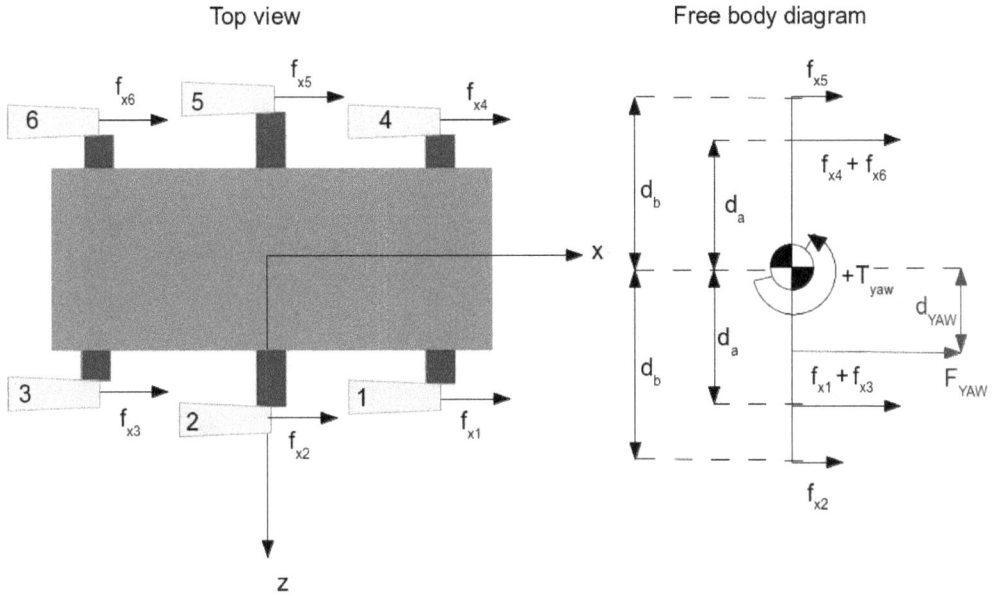

Figure 16.8: Robot's free body diagram of forces and torques with effects on the Y-axis (Yaw).

By substituting (16.31) in the last two equations, we obtain

$$f_{x_j} = \left(0.1963 \frac{(w_1 + 2w_2)l^2}{3} \rho \frac{A_j}{P_j} - 0.1554 \right) \cos \theta_j, \tag{16.36}$$

$$f_{y_j} = \left(0.1963 \frac{(w_1 + 2w_2)l^2}{3} \rho \frac{A_j}{P_j} - 0.1554 \right) \text{sen} \theta_j, \tag{16.37}$$

where A_j and P_j are the parameters of oscillation of the paddle j. Then, by using the free body diagram of the robot on each of its planes, we can obtain the torques around X, Y and Z axes.

In order to obtain the moment in *Yaw*, we sum all the moments generated by each paddle force

$$\tau_{yaw} = d_a(f_{x_1} + f_{x_3} - f_{x_4} - f_{x_6}) + d_b(f_{x_2} - f_{x_5}), \tag{16.38}$$

Likewise, the angular moments for *Roll* and *Pitch* are obtained

$$\tau_{roll} = d_a(f_{y_4} + f_{y_6} - f_{y_1} - f_{y_3}) + d_b(f_{y_5} - f_{y_2}), \tag{16.39}$$

$$\tau_{pitch} = d_c(f_{y_1} + f_{y_4} - f_{y_3} - f_{y_6}). \tag{16.40}$$

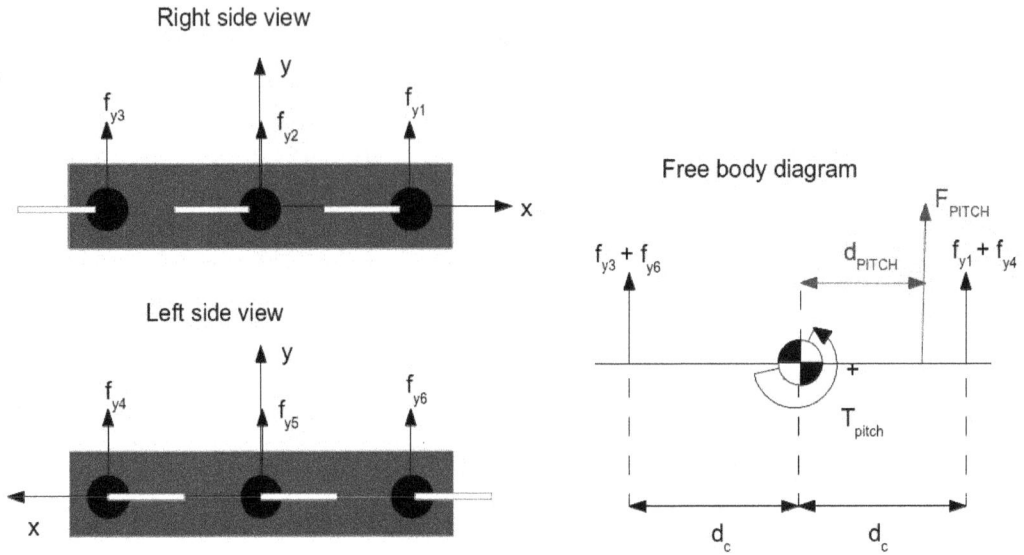

Figure 16.9: Robot's free body diagram of forces and torques with effects on the Z axis (Pitch).

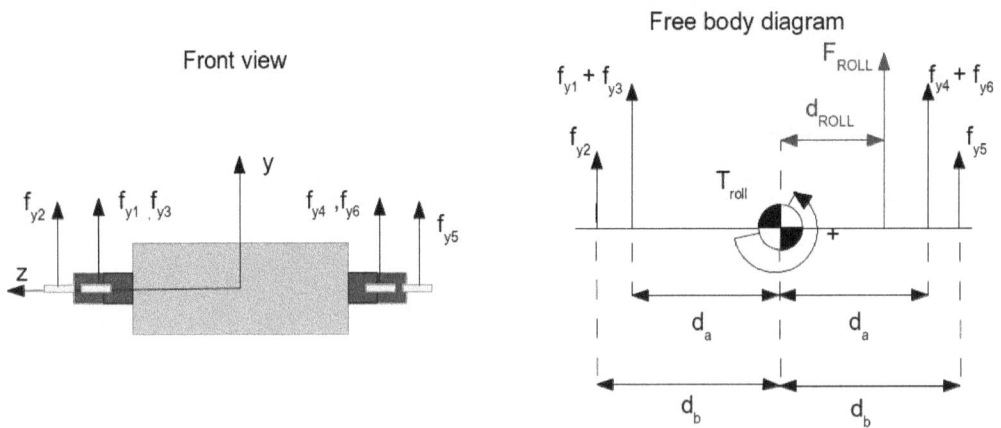

Figure 16.10: Robot's free body diagram of forces and torques with effects on the X axis (Roll).

Equations (16.38), (16.39) and (16.40) are expressed in the matrix form by

$$
\begin{pmatrix} \tau_{roll} \\ \tau_{yaw} \\ \tau_{pitch} \end{pmatrix} = \begin{pmatrix} d_a(f_{y_4} + f_{y_6} - f_{y_1} - f_{y_3}) + d_b(f_{y_5} - f_{y_2}) \\ d_a(f_{x_1} + f_{x_3} - f_{x_4} - f_{x_6}) + d_b(f_{x_2} - f_{x_5}) \\ d_c(f_{y_1} + f_{y_4} - f_{y_3} - f_{y_6}) \end{pmatrix},
\tag{16.41}
$$

and factorising the forces,

$$
\begin{pmatrix} \tau_{roll} \\ \tau_{yaw} \\ \tau_{pitch} \end{pmatrix} = \begin{pmatrix} 0 & 0 & 0 & 0 & 0 & 0 & -d_a & -d_b & -d_a & d_a & d_b & d_a \\ d_a & d_b & d_a & -d_a & -d_b & -d_a & 0 & 0 & 0 & 0 & 0 & 0 \\ 0 & 0 & 0 & 0 & 0 & 0 & d_c & 0 & -d_c & d_c & 0 & -d_c \end{pmatrix} \cdot \begin{pmatrix} f_{x_1} \\ f_{x_2} \\ f_{x_3} \\ f_{x_4} \\ f_{x_5} \\ f_{x_6} \\ f_{y_1} \\ f_{y_2} \\ f_{y_3} \\ f_{y_4} \\ f_{y_5} \\ f_{y_6} \end{pmatrix}
\tag{16.42}
$$

By simplifying the expression (16.42) iin terms of the resultant forces $(F_{yaw}, F_{pitch}, F_{roll})$, and the resultant distances $(d_{yaw}, d_{pitch}, d_{roll})$ as shown in the free body diagrams. The simplified expressions are

$$
\tau_{roll} = F_{roll} \cdot d_{roll},
\tag{16.43}
$$

$$
\tau_{pitch} = F_{pitch} \cdot d_{pitch},
\tag{16.44}
$$

$$
\tau_{yaw} = F_{yaw} \cdot d_{yaw}.
\tag{16.45}
$$

$$
\tag{16.46}
$$

With the torques around each robot's axis, the orientation is obtained. The angle α^R and torque around X axis are related by

$$
I_{cx}\ddot{\alpha}^R = \tau_{roll}
\tag{16.47}
$$

where I_{cx} is the moment of inertia w.r.t. the X axis, and it is known that

$$\ddot{\alpha}^R = \frac{d\dot{\alpha}^R}{dt},$$ (16.48)

then, by rearranging the terms, and integrating, we obtain

$$\ddot{\alpha}^R dt = d\dot{\alpha}^R$$

by completing the differentials with their respective integrals in both sides of the equation,

$$\int_{t_{i-1}}^{t_i} \ddot{\alpha}^R dt = \int_{\dot{\alpha}_{i-1}^R}^{\dot{\alpha}_i^R} d\dot{\alpha}^R$$

solving the defined integrals

$$\ddot{\alpha}^R t \big|_{t_{i-1}}^{t_i} = \dot{\alpha}^R \big|_{\dot{\alpha}_{i-1}^R}^{\dot{\alpha}_i^R}$$

a recursive equation is obtained,

$$\ddot{\alpha}^R(t_i - t_{i-1}) = \dot{\alpha}_i^R - \dot{\alpha}_{i-1}^R$$

or

$$\ddot{\alpha}^R \Delta t = \dot{\alpha}_i^R - \dot{\alpha}_{i-1}^R$$ (16.49)

Since equation (16.49) was provided in terms of $\dot{\alpha}$, then it now is deduced in terms of α. The angular velocity is the angle rate increment w.r.t. time as the next equation,

$$\dot{\alpha}^R = \frac{d\alpha^R}{dt}$$

by separating the differentials in both sides of the equation,

$$\dot{\alpha}^R dt = d\alpha^R$$

then, completing the differentials with their respective integrals,

$$\int_{t_{i-1}}^{t_i} \dot{\alpha}^R dt = \int_{\alpha_{i-1}^R}^{\alpha_i^R} d\alpha^R$$

by solving for the defined integrals,

$$\dot{\alpha}^R t \big|_{t_{i-1}}^{t_i} = \alpha^R \big|_{\alpha_{i-1}^R}^{\alpha_i^R}$$

a recursive model is obtained,

$$\dot{\alpha}^R (t_i - t_{i-1}) = \alpha_i^R - \alpha_{i-1}^R$$

or

$$\dot{\alpha}^R \Delta t = \alpha_i^R - \alpha_{i-1}^R$$

the next equation provides the actual angular velocity as the recursive angular positions,

$$\dot{\alpha}^R = \frac{\alpha_i^R - \alpha_{i-1}^R}{\Delta t} \tag{16.50}$$

Therefore, by substituting equation (16.50) in equation (16.49), a more complete recursive functional form of the angular acceleration is obtained,

$$\ddot{\alpha}^R \Delta t = \frac{\alpha_i^R - \alpha_{i-1}^R}{\Delta t} - \dot{\alpha}_{i-1}^R$$

algebraically rearranging Δt,

$$\ddot{\alpha}^R (\Delta t)^2 = \alpha_i^R - \alpha_{i-1}^R - \dot{\alpha}_{i-1}^R \Delta t$$

the recursive angle equation is obtained,

$$\alpha_i^R = \ddot{\alpha}^R (\Delta t)^2 + \dot{\alpha}_{i-1}^R \Delta t + \alpha_{i-1}^R \tag{16.51}$$

Finally, by substituting the equation (16.47) in equation (16.51), the following is stated,

Proposition 16.3.1 (amphibious angular movement). *The general recursive model α_{i-1} and $\dot{\alpha}_{i-1}$ is given as a function of angular and torsional moment I, and τ respectively:*

$$\alpha_i^R = \frac{\tau_{roll}}{I_{cx}} (\Delta t)^2 + \dot{\alpha}^R_{i-1} \Delta t + \alpha_{i-1}^R, \tag{16.52}$$

Therefore, by using the model of previous proposition for the three Euler axes, and by substituting the functional form of the torques in terms of tangential forces as provided by equation (16.39), the roll, pitch and yaw are stated consecutively,

Corollary 16.3.2 (recursive roll motion model). *The recursive model for the roll motion α_i in terms of α_{i-1}, $\dot{\alpha}_{i-1}$, and as a function of the angular moment I_{cx}, and involved forces is*

$$\alpha_i^R = \frac{d_a(f_{y_4} + f_{y_6} - f_{y_1} - f_{y_3}) + d_b(f_{y_5} - f_{y_2})}{I_{cx}}(\Delta t)^2 + \dot{\alpha}^R_{i-1}\Delta t + \alpha_{i-1}^R \qquad (16.53)$$

Similarly, he angles β^R (yaw) and γ^R (pitch) are obtained respectively.

Corollary 16.3.3 (recursive yaw motion model). *The recursive model for the yaw motion β_i in terms of β_{i-1}, $\dot{\beta}_{i-1}$, and as a function of the angular moment I_{cy}, and involved forces is*

$$\beta_i^R = \frac{d_a(f_{x_1} + f_{x_3} - f_{x_4} - f_{x_6}) + d_b(f_{x_2} - f_{x_5})}{I_{cy}}(\Delta t)^2 + \dot{\beta}^R_{i-1}\Delta t + \beta_{i-1}^R \qquad (16.54)$$

and

Corollary 16.3.4 (recursive pitch motion model). *The recursive model for the pitch motion γ_i in terms of γ_{i-1}, $\dot{\gamma}_{i-1}$, and as a function of the angular moment I_{cz}, and involved forces is*

$$\gamma_i^R = \frac{d_c(f_{y_1} + f_{y_4} - f_{y_3} - f_{y_6})}{I_{cz}}(\Delta t)^2 + \dot{\gamma}^R_{i-1}\Delta t + \gamma_{i-1}^R \qquad (16.55)$$

Hereafter, the models (16.53)-(16.55) stated in previous corollaries provide the relationships between the forces produced by the paddles, and the robot's orientations α^R, β^R, and γ^R.

16.4 Euler-Lagrange analysis

The underwater robot is considered as a rigid body (non deformable), with six degrees of freedom. Thus, the robot's dynamic model is algebraically developed by using the Euler-

Lagrange formulation. Some foundations about the application of the Euler-Lagrange equations on robotics has been reported . It is necessary to define the generalized coordinates by the vector

$$\mathbf{q} = (x, y, z, \alpha, \beta, \gamma)^\top \tag{16.56}$$

where x, y y z represent the position of the center of mass of the robot. α, β y γ represent the orientation of the robot with respect to the fixed frame in the roll-pitch-yaw parametrization. The angular speed $(\dot\alpha^R, \dot\beta^R$ and $\dot\gamma^R)$ of the robot are expressed in the robot's frame. To convert those speeds into the fixed frame (world frame) we use the rotation matrix that relates the robot frame with the fixed one.

$$\mathbf{R}_R^W = \mathbf{R} = \begin{pmatrix} \cos\gamma & -\sin\gamma & 0 \\ \sin\gamma & \cos\gamma & 0 \\ 0 & 0 & 1 \end{pmatrix} \begin{pmatrix} \cos\beta & 0 & \sin\beta \\ 0 & 1 & 0 \\ -\sin\beta & 0 & \cos\beta \end{pmatrix} \begin{pmatrix} 1 & 0 & 0 \\ 0 & \cos\alpha & -\sin\alpha \\ 0 & \sin\alpha & \cos\alpha \end{pmatrix},$$

$$= \begin{pmatrix} \cos\gamma\cos\beta & \cos\gamma\sin\beta\sin\alpha - \sin\gamma\cos\alpha & \cos\gamma\sin\beta\cos\alpha + \sin\gamma\sin\alpha \\ \sin\gamma\cos\beta & \sin\gamma\sin\beta\sin\alpha + \cos\gamma\cos\alpha & \sin\gamma\sin\beta\cos\alpha - \cos\gamma\sin\alpha \\ -\sin\beta & \cos\beta\sin\alpha & \cos\beta\cos\alpha \end{pmatrix} \tag{16.57}$$

Then, the relation between the translational velocities in the robot frame with the ones in the fixed frame is:

$$\begin{pmatrix} v_x \\ v_y \\ v_z \end{pmatrix} = \mathbf{R}^\top \cdot \begin{pmatrix} \dot{x} \\ \dot{y} \\ \dot{z} \end{pmatrix} \tag{16.58}$$

Thus, to convert the angular speed in the robot frame into the fixed one, we use the following property of the rotation matrices,

$$\dot{\mathbf{R}} \cdot \mathbf{R}^\top = \begin{pmatrix} 0 & -\omega_z & \omega_y \\ \omega_z & 0 & -\omega_x \\ -\omega_y & \omega_x & 0 \end{pmatrix} \tag{16.59}$$

By performing the product in the left side of the expression, we obtain:

$$
\begin{pmatrix}
0 & \sin\beta\dot\alpha - \dot\gamma & \sin\gamma\cos\beta\dot\alpha + \cos\gamma\dot\beta \\
-\sin\beta\dot\alpha + \dot\gamma & 0 & -\cos\gamma\cos\beta\dot\alpha + \sin\gamma\dot\beta \\
-\sin\gamma\cos\beta\dot\alpha - \cos\gamma\dot\beta\cos\gamma\cos\beta\dot\alpha - \sin\gamma\dot\beta & 0
\end{pmatrix}
=
\begin{pmatrix}
0 & -\omega_z^W & \omega_y^W \\
\omega_z^W & 0 & -\omega_x^W \\
-\omega_y^W & \omega_x^W & 0
\end{pmatrix}
$$

therefore, the angular speed is

$$
\begin{pmatrix}\omega_x^W\\ \omega_y^W\\ \omega_z^W\end{pmatrix}
=
\begin{pmatrix}\cos\gamma\cos\beta\dot\alpha - \sin\gamma\dot\beta\\ \sin\gamma\cos\beta\dot\alpha + \cos\gamma\dot\beta\\ -\sin\beta\dot\alpha + \dot\gamma\end{pmatrix}
=
\begin{pmatrix}\cos\gamma\cos\beta & -\sin\gamma & 0\\ \sin\gamma\cos\beta & \cos\gamma & 0\\ -\sin\beta & 0 & 1\end{pmatrix}
\cdot
\begin{pmatrix}\dot\alpha\\ \dot\beta\\ \dot\gamma\end{pmatrix}
\tag{16.60}
$$

By defining the matrix **G** as

$$
\mathbf{G} = \begin{pmatrix}\cos\gamma\cos\beta & -\sin\gamma & 0\\ \sin\gamma\cos\beta & \cos\gamma & 0\\ -\sin\beta & 0 & 1\end{pmatrix}.
\tag{16.61}
$$

Rotation matrix **R** converts a vector from the robot frame to the fixed one, then we have the following equation:

$$
\begin{pmatrix}\omega_x^W\\ \omega_y^W\\ \omega_z^W\end{pmatrix} = \mathbf{R}\begin{pmatrix}\dot\alpha^R\\ \dot\beta^R\\ \dot\gamma^R\end{pmatrix},
$$

then substituting (16.60)

$$
\mathbf{G}\cdot\begin{pmatrix}\dot\alpha\\ \dot\beta\\ \dot\gamma\end{pmatrix} = \mathbf{R}\cdot\begin{pmatrix}\dot\alpha^R\\ \dot\beta^R\\ \dot\gamma^R\end{pmatrix},
$$

and algebraically arranging

$$
\mathbf{R}^\top\cdot\mathbf{G}\cdot\begin{pmatrix}\dot\alpha\\ \dot\beta\\ \dot\gamma\end{pmatrix} = \begin{pmatrix}\dot\alpha^R\\ \dot\beta^R\\ \dot\gamma^R\end{pmatrix}
\tag{16.62}
$$

The transition matrix $\mathbf{R}^T\mathbf{G}$ relates the angular speeds in fixed frame with the robot's frame.

$$\mathbf{R}^\top \cdot \mathbf{G} = \begin{pmatrix} 1 & 0 & -\sin\beta \\ 0 & \cos\alpha & \sin\alpha\cos\beta \\ 0 & -\sin\alpha & \cos\alpha\cos\beta \end{pmatrix}. \tag{16.63}$$

Now, with all this information at hand, we can calculate the kinetic energy required by the robotic platform for its roto-translational movements,

$$K = \frac{1}{2}m\mathbf{v}^T\mathbf{v} + \frac{1}{2}\omega^T\mathbf{I_c}\omega, \tag{16.64}$$

and by substituting $\mathbf{v}$ and ω by (16.58) and (16.62),

$$K = \frac{1}{2}m\left(\dot{x}\dot{y}\dot{z}\right)\cdot\mathbf{R}\cdot\mathbf{R}^\top\cdot\begin{pmatrix}\dot{x}\\\dot{y}\\\dot{z}\end{pmatrix} + \frac{1}{2}\left(\dot{\alpha}\dot{\beta}\dot{\gamma}\right)\mathbf{G}^\top\cdot\mathbf{R}\cdot\mathbf{I_c}\cdot\mathbf{R}^\top\cdot\mathbf{G}\cdot\begin{pmatrix}\dot{\alpha}\\\dot{\beta}\\\dot{\gamma}\end{pmatrix}.$$

Considering that the robot has three planes of symmetry, then $\mathbf{I_c}$ can be simplified as a diagonal matrix $diag(I_{xx}, I_{yy}, I_{zz})$. Expand the last equation the following expression is produced

$$K = \frac{1}{2}m(\dot{x}^2 + \dot{y}^2 + \dot{z}^2) + \frac{1}{2}[I_{xx}(\dot{\alpha} - \dot{\gamma}\sin\beta)^2 + I_{yy}(\dot{\beta}\cos\alpha + \dot{\gamma}\sin\alpha\cos\beta)^2$$
$$+ I_{zz}(-\dot{\beta}\sin\alpha + \dot{\gamma}\cos\alpha\cos\beta)^2] \tag{16.65}$$

The potential energy of the robot is associated with its height, that is

$$P = -mgy \tag{16.66}$$

Once having the kinetic and potential energy models, the Lagrangian $L = K - P$ is stated,

$$L = \frac{1}{2}m(\dot{x}^2 + \dot{y}^2 + \dot{z}^2) + \frac{1}{2}[I_{xx}(\dot{\alpha} - \dot{\gamma}\sin\beta)^2 + I_{yy}(\dot{\beta}\cos\alpha + \dot{\gamma}\sin\alpha\cos\beta)^2$$
$$+ I_{zz}(-\dot{\beta}\sin\alpha + \dot{\gamma}\cos\alpha\cos\beta)^2] + mgy \tag{16.67}$$

Then, by using the Euler-Lagrange equation on each generalized coordinate:

$$\frac{d}{dt}\left(\frac{\partial L}{\partial \dot{q}_i}\right) - \frac{\partial L}{\partial q_i} = \tau_i \tag{16.68}$$

and by applying (16.68) for $q_i = x$,

$$\frac{d}{dt}\left(\frac{\partial L}{\partial \dot{x}}\right) - \frac{\partial L}{\partial x} = \tau_1$$
$$m\ddot{x} = \tau_1. \tag{16.69}$$

and for $q_i = y$,

$$\frac{d}{dt}\left(\frac{\partial L}{\partial \dot{y}}\right) - \frac{\partial L}{\partial y} = \tau_2$$
$$m\ddot{y} - mg = \tau_2. \tag{16.70}$$

finally, for $q_i = z$

$$\frac{d}{dt}\left(\frac{\partial L}{\partial \dot{z}}\right) - \frac{\partial L}{\partial z} = \tau_3$$
$$m\ddot{z} = \tau_3. \tag{16.71}$$

Thus, for $q_i = \alpha$,

$$\frac{\partial L}{\partial \dot{\alpha}} = I_{xx}(\dot{\alpha} - \dot{\gamma}\sin\beta)$$
$$\frac{d}{dt}\left(\frac{\partial L}{\partial \dot{\alpha}}\right) = I_{xx}(\ddot{\alpha} - \dot{\gamma}\dot{\beta}\cos\beta - \ddot{\alpha}\sin\beta)$$
$$\frac{\partial L}{\partial \alpha} = I_{yy}(\dot{\beta}\cos\alpha + \dot{\gamma}\sin\alpha\cos\beta)(-\dot{\beta}\sin\alpha + \dot{\gamma}\cos\alpha\cos\beta) +$$
$$I_{zz}(-\dot{\beta}\sin\alpha + \dot{\gamma}\cos\alpha\cos\beta)(-\dot{\beta}\cos\alpha - \dot{\gamma}\sin\alpha\cos\beta)$$

Therefore,

$$\frac{d}{dt}\left(\frac{\partial L}{\partial \dot{\alpha}}\right) - \frac{\partial L}{\partial \alpha} = \tau_4$$

$$I_{xx}(\ddot{\alpha} - \dot{\gamma}\dot{\beta}\cos\beta - \ddot{\alpha}\sin\beta) - (I_{yy} - I_{zz})(\dot{\beta}\cos\alpha + \dot{\gamma}\sin\alpha\cos\beta)$$

$$(-\dot{\beta}\sin\alpha + \dot{\gamma}\cos\alpha\cos\beta) = \tau_4 \tag{16.72}$$

Now, solving for $q_i = \beta$

$$\frac{\partial L}{\partial \dot{\beta}} = I_{yy}(\dot{\beta}\cos\alpha + \dot{\gamma}\sin\alpha\cos\beta)(\cos\alpha) + I_{zz}(-\dot{\beta}\sin\alpha + \dot{\gamma}\cos\alpha\cos\beta)(-\sin\alpha) =$$

$$I_{yy}\dot{\beta}\cos^2\alpha + I_{zz}\dot{\beta}\sin^2\alpha + (I_{yy} - I_{zz})\dot{\gamma}\sin\alpha\cos\alpha\cos\beta$$

$$\frac{d}{dt}\left(\frac{\partial L}{\partial \dot{\beta}}\right) = I_{yy}(\ddot{\beta}\cos^2\alpha - 2\dot{\alpha}\dot{\beta}\sin\alpha\cos\alpha) + I_{zz}(\ddot{\beta}\sin^2\alpha + 2\dot{\alpha}\dot{\beta}\sin\alpha\cos\alpha)$$

$$+(I_{yy} - I_{zz})(\ddot{\alpha}\sin\alpha\cos\alpha\cos\beta + \dot{\alpha}\dot{\gamma}\cos^2\alpha\cos\beta - \dot{\alpha}\dot{\gamma}\sin^2\alpha\cos\beta - \dot{\beta}\dot{\gamma}\sin\beta\sin\alpha\cos\alpha) \tag{16.73}$$

$$\frac{\partial L}{\partial \beta} = I_{xx}(\dot{\alpha} - \dot{\gamma}\sin\beta)(-\dot{\gamma}\cos\beta) + I_{yy}(\dot{\beta}\cos\alpha + \dot{\gamma}\sin\alpha\cos\beta)(-\dot{\gamma}\sin\alpha\sin\beta)+$$

$$I_{zz}(-\dot{\beta}\sin\alpha + \dot{\gamma}\cos\alpha\cos\beta)(-\dot{\gamma}\cos\alpha\sin\beta)$$

And, by arranging terms,

$$\frac{d}{dt}\left(\frac{\partial L}{\partial \dot{\beta}}\right) - \frac{\partial L}{\partial \beta} = \tau_5$$

hence,

$$I_{yy}(\ddot{\beta}\cos^2\alpha - 2\dot{\alpha}\dot{\beta}\sin\alpha\cos\alpha) + I_{zz}(\ddot{\beta}\sin^2\alpha + 2\dot{\alpha}\dot{\beta}\sin\alpha\cos\alpha)$$

$$+(I_{yy} - I_{zz})(\ddot{\alpha}\sin\alpha\cos\alpha\cos\beta + \dot{\alpha}\dot{\gamma}\cos^2\alpha\cos\beta - \dot{\alpha}\dot{\gamma}\sin^2\alpha\cos\beta - \dot{\beta}\dot{\gamma}\sin\beta\sin\alpha\cos\alpha) \tag{16.74}$$

$$-I_{xx}(-\dot{\alpha}\dot{\gamma}\cos\beta + \dot{\gamma}^2\sin\beta\cos\beta) - I_{yy}(-\dot{\beta}\dot{\gamma}\cos\alpha\sin\alpha\sin\beta - \dot{\gamma}^2\sin^2\alpha\cos\beta\sin\beta)-$$

$$I_{zz}(\dot{\beta}\dot{\gamma}\cos\alpha\sin\alpha\sin\beta - \dot{\gamma}^2\cos^2\alpha\sin\beta\cos\beta) = \tau_5$$

Now, for $q_i = \gamma$, and we obtain

$$\frac{\partial L}{\partial \dot\gamma} = I_{xx}(\dot\gamma \sin^2 \beta - \dot\alpha \sin\beta) + I_{yy}(\dot\beta \sin\alpha \cos\alpha \cos\beta + \dot\gamma \sin^2 \alpha \cos^2 \beta)$$

$$+ I_{zz}(-\dot\beta \sin\alpha \cos\alpha \cos\beta + \dot\gamma \cos^2 \alpha \cos^2 \beta)$$

$$\frac{d}{dt}\left(\frac{\partial L}{\partial \dot\gamma}\right) = I_{xx}(\ddot\alpha \sin^2 \beta + 2\dot\gamma\dot\beta \sin\beta \cos\beta - \ddot\alpha \sin\beta - \dot\alpha\dot\beta \cos\beta) +$$

$$I_{yy}(\ddot\gamma \sin^2 \alpha \cos^2 \beta + 2\dot\alpha\dot\gamma \sin\alpha \cos\alpha \cos^2 \beta - 2\dot\beta\dot\gamma \cos\beta \sin\beta \sin^2 \alpha)$$

$$+ I_{zz}(\ddot\gamma \cos^2 \alpha \cos^2 \beta - 2\dot\alpha\dot\gamma \cos\alpha \sin\alpha \cos^2 \beta - 2\dot\beta\dot\gamma \cos\beta \sin\beta \cos^2 \alpha)$$

$$+ (I_{yy} - I_{zz})(\ddot\beta \sin\alpha \cos\alpha \cos\beta + \dot\alpha\dot\beta \cos^2 \alpha \cos\beta - \dot\alpha\dot\beta \sin^2 \alpha \cos\beta - \dot\beta^2 \sin\beta \sin\alpha \cos\alpha)$$

$$\frac{\partial L}{\partial \gamma} = 0$$

and

$$\frac{d}{dt}\left(\frac{\partial L}{\partial \dot\gamma}\right) - \frac{\partial L}{\partial \gamma} = \tau_6,$$

$$I_{xx}(\ddot\alpha \sin^2 \beta + 2\dot\gamma\dot\beta \sin\beta \cos\beta - \ddot\alpha \sin\beta - \dot\alpha\dot\beta \cos\beta) + I_{yy}(\ddot\gamma \sin^2 \alpha \cos^2 \beta + 2\dot\alpha\dot\gamma \sin\alpha \cos\alpha \cos^2 \beta$$

$$- 2\dot\beta\dot\gamma \cos\beta \sin\beta \sin^2 \alpha) + I_{zz}(\ddot\gamma \cos^2 \alpha \cos^2 \beta - 2\dot\alpha\dot\gamma \cos\alpha \sin\alpha \cos^2 \beta - 2\dot\beta\dot\gamma \cos\beta \sin\beta \cos^2 \alpha)$$

$$+ (I_{yy} - I_{zz})(\ddot\beta \sin\alpha \cos\alpha \cos\beta + \dot\alpha\dot\beta \cos^2 \alpha \cos\beta - \dot\alpha\dot\beta \sin^2 \alpha \cos\beta - \dot\beta^2 \sin\beta \sin\alpha \cos\alpha) = \tau_6$$

(16.75)

Equations (16.69)-(16.75) represent the dynamics of the underwater robot. By factorising the second order derivatives of the generalized coordinates system in the dynamic equations, the inertia matrix is obtained by

$$\mathbf{H(q)} = \begin{pmatrix} \mathbf{H_1} & \mathbf{0}_{3\times 3} \\ \mathbf{0}_{3\times 3} & \mathbf{H_2} \end{pmatrix}$$

(16.76)

with

$$\mathbf{H_1} = \begin{pmatrix} m & 0 & 0 \\ 0 & m & 0 \\ 0 & 0 & m \end{pmatrix}$$

(16.77)

likewise

$$\mathbf{H}_2 = \begin{pmatrix} I_{xx} & 0 & -I_{xx}\sin\beta \\ 0 & I_{yy}\cos^2\alpha + I_{zz}\sin^2\alpha & (I_{yy} - I_{zz})\sin\alpha\cos\alpha\cos\beta \\ -I_{xx}\sin\beta(I_{yy} - I_{zz})\sin\alpha\cos\alpha\cos\beta & I_{xx}\sin^2\beta + I_{yy}\sin^2\alpha\cos^2\beta + I_{zz}\cos^2\alpha\cos^2\beta \end{pmatrix}$$

$$(16.78)$$

Thus, factorising the first derivatives of the generalized coordinates to obtain the Coriolis or centripetal matrix,

$$\mathbf{C}(\dot{\mathbf{q}}) = \begin{pmatrix} \mathbf{0}_{3\times3}\mathbf{0}_{3\times3} \\ \mathbf{0}_{3\times3} \ \mathbf{C}_1 \end{pmatrix}, \tag{16.79}$$

with

$$\mathbf{C}_1 = \begin{pmatrix} 0 \ c_{12} \ c_{13} \\ c_{21} \ 0 \ c_{c23} \\ c_{31} c_{32} \ 0 \end{pmatrix}, \tag{16.80}$$

and

$$c_{12} = \dot{\gamma}[-I_{xx}\cos\beta + (I_{yy} + I_{zz})(\sin^2\alpha - \cos^2\alpha)\cos\beta] + \\ \dot{\beta}(I_{yy} + I_{zz})\cos\alpha\sin\alpha$$

$$c_{13} = -\dot{\gamma}(I_{yy} + I_{zz})\sin\alpha\cos\alpha\cos^2\alpha$$

$$c_{21} = \dot{\beta}[(-2\sin\alpha\cos\alpha + \cos^2\alpha\cos\beta)(I_{yy} - I_{zz})] + \dot{\gamma}(I_{xx}\cos\beta - \sin^2\alpha\cos\beta)$$

$$c_{23} = \dot{\gamma}\sin\beta\cos\beta(I_{xx} + I_{yy}\sin^2\alpha + I_{zz}\cos^2\alpha)$$

$$c_{31} = \dot{\beta}[-I_{xx}\cos\beta + (I_{yy} - I_{zz})\cos\beta(\cos^2\alpha - \sin^2\alpha)] + 2\dot{\gamma}(I_{yy} - I_{zz})\sin\alpha\cos\alpha\cos^2\beta$$

$$c_{32} = \dot{\gamma}[2\sin\beta\cos\beta(I_{xx} - I_{yy}\sin^2\alpha - I_{zz}\cos^2\alpha)] - \dot{\beta}(I_{yy} - I_{zz})\sin\alpha\cos\alpha\sin\beta.$$

Finally, the gravity vector has the following form

$$\mathbf{g}(\mathbf{q}) = (0, -mg, 0, 0, 0, 0)^\top \tag{16.81}$$

Obtained matrices $\mathbf{H}(\mathbf{q})$, $\mathbf{C}(\dot{\mathbf{q}}, \mathbf{q})$, vector $\mathbf{g}(\mathbf{q})$, and vector $\mathbf{f}_d$, which represent diverse dissipative forces (out of this work's scope), then the robot's general dynamics equation is be expressed by

$$\mathbf{H}(\mathbf{q})\mathbf{q} + \mathbf{C}(\dot{\mathbf{q}}, \mathbf{q})\dot{\mathbf{q}} + \mathbf{g}(\mathbf{q}) - \mathbf{f}_d = \boldsymbol{\tau} \tag{16.82}$$

Bibliography

[1] Armstrong, B., *Friction: experimental determination, modeling and compensation*, IEEE Intl. Conf. on Robotics and Automation, Vol.3, pp.1422–1427, 1988.

[2] Hemati, N., Leu, Ming C., *A complete model characterization of brushless DC motors*, IEEE Transactions on Industry Applications, Vol.28(1), pp.172–180, 1992.

[3] Saha, S.K., *Introduction to Robotics*, Tata McGraw-Hill Education, 2008.

[4] Sagatun S. I., Fossen, T.I., *Lagrangian formulation of underwater vehicles' dynamics*, IEEE Intl. Conf. on Systems, Man, and Cybernetics, vol.2, pp. 1029–1034, 1991.

[5] Plamondon, N., *Modeling and Control of a Biomimetic Underwater Vehicle*, PhD thesis, Department of Mechanical Engineering, McGill University, Montreal, Canada, 2011.

[6] Georgiades, C., German A, Torres L.A., Dudek, G., *et al*, *AQUA: an aquatic walking robot*, Proc. on IEEE Intelligent Robots and Systems, vol.4, pp. 3525–353, 2004.

[7] Sattar, J., Dudek, G., *A Vision-Based Control and Interaction Framework for a Legged Underwater Robot*, Canadian Conference on Computer and Robot Vision, pp. 329–336, 2009.

Chapter 17

HOVER CRAFT DYNAMIC MODELLING

Marco Elizalde Ceballos and Edgar A. Martínez García

Laboratorio de Robótica, Institute of Engineering and Technology
Universidad Autónoma de Ciudad Juárez, Mexico.

This chapter discusses the analysis of a mathematical framework describing the dynamical motion of a hover craft. This work is a preliminary study of a first approach for future formulation of trajectory state estimation. The present chapter presents the formulation to model the vehicle's mobility with foundations on the interacting forces that provide the propulsive displacements. Two main aspects are analysed and modelled: the Euler speeds that are determined in terms of the propulsive devices on the aircraft, which provide the geometry of linear motion; and the angular motions around the roll, pitch and yaw axes usually associated to the aircraft instability. The direct and inverse dynamic equations of motion are disclosed based on the hovercraft physical design.

17.1 Translation velocities vector model

Based on the aerodynamic design of the aircraft (figure 17.1), there are some forces that influence the hover Cartesian motion. One force is produced by the air stream of the main fan propulsive system, which is located at the very back of the hovercraft. This propulsion produces the thrusting force f_{th} that is able to move the aircraft forward (depicted in figure 17.1). The angle of direction of the force f_{th} is called θ_{th}, and is controlled by the main rudder mechanism $^-$.

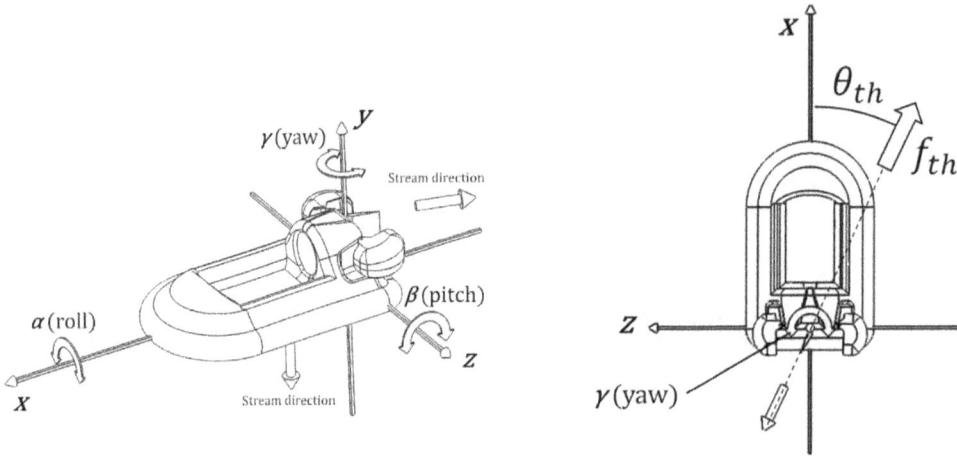

Figure 17.1: Hover craft diagram of local Euler axes, and main propulsive air streams direction.

In addition, the lifting force f_s applies underneath the hovercraft yielded by the air stream flowing from the inner craft's structure. This mass of air produces the air cushion to lift the vehicle from the ground surface. The sustenance or lifting force of the air cushion is described by f_s and modelled by the following expression,

$$f_s = p_s a \tag{17.1}$$

where p_a is the pressure within the *air cushion* chamber and a the effective cushion area.

This type of hovercraft integrates two auxiliary ailerons, left-back and right-back. These ailerons assist for steering, and for yielding braking effects (17.2). The forces f_{ar} and f_{al} represent the forces generated by the right and left ailerons' air stream, respectively $^-$. The thrust force is generated towards the opposite direction of the main air stream. The propulsion force generated by the air stream depends entirely on the fluid dynamics in the ducted fan. Thus, it is possible to describe the function of the hovercraft mobility by taking into account the four forces acting on the vehicle. By applying the Newton's second law of motion, we describe the acceleration $\ddot{x}$ along the X-axis by the next expression

$$\ddot{x} = \frac{1}{m}\left(f_{th}\cos(\theta_{th}) + f_{ar}\cos(\theta_{ar}) + f_{al}\cos(\theta_{al})\right) \tag{17.2}$$

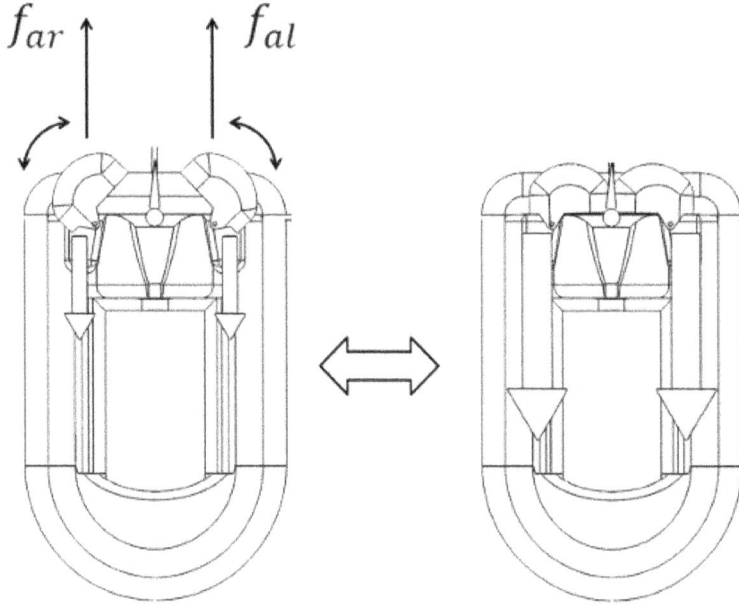

Figure 17.2: Top view of air streams and ailerons force direction.

Where θ_{th}, θ_{ar} y θ_{al} are the respective angles of each force on the plane XZ. The acceleration can be represented as a differential of the linear velocity over time in accordance with

$$\frac{d\dot{x}}{dt} = \frac{1}{m}\left(f_{th}\cos(\theta_{th}) + f_{ar}\cos(\theta_{ar}) + f_{al}\cos(\theta_{al})\right) \tag{17.3}$$

The differential can be represented by its discrete form. By assuming a change in time Δt, we can define

$$\frac{\dot{x}_t - \dot{x}_{t-1}}{\Delta t} = \frac{1}{m}\left(f_{th}\cos\theta_{th} + f_{ar}\cos\theta_{ar} + f_{al}\cos\theta_{al}\right) \tag{17.4}$$

By solving for $\dot{x}$ we have,

$$\dot{x}_t - \dot{x}_{t-1} = \frac{1}{m}\left(f_{th}\cos\theta_{th} + f_{ar}\cos\theta_{ar} + f_{al}\cos\theta_{al}\right)\Delta t \tag{17.5}$$

Definition 17.1.1 (instantaneous absolute velocity). The recursive model of the instantaneous absolute velocity as a function of the propulsive forces is defined by

$$\dot{x}_t = \frac{1}{m} \left(f_{th} \cos \theta_{th} + f_{ar} \cos \theta_{ar} + f_{al} \cos \theta_{al} \right) \Delta t + \dot{x}_{t-1} \tag{17.6}$$

From previous definition, the equation (17.6) states a model for the linear instantaneous velocity $\dot{x}_t$ in function of the main forces interacting over the hovercraft. In order to define the linear velocity along Y-axis, we need to take in account the lifting force f_s in accordance with

$$f_y = f_s - W \tag{17.7}$$

were W is the net weight of the vehicle. By applying Newton's second law of motion and solving for $\ddot{y}_t$ we have that

$$m\ddot{y}_{(t)} = f_s(t) - W \tag{17.8}$$

as well as,

$$\ddot{y}_{(t)} = \frac{1}{m} \left(f_{s(t)} - W \right) \tag{17.9}$$

The linear acceleration can be represented in its differential form. Then, by solving for $\dot{y}_t$ we have that

$$\frac{d\dot{y}_t}{dt} = \frac{1}{m} \left(f_{st} - W \right) \tag{17.10}$$

and

$$\frac{\dot{y}_{(t)} - \dot{y}_{(t-1)}}{\Delta t} = \frac{1}{m} \left(f_s - W \right) \tag{17.11}$$

$$\dot{y}_t - \dot{y}_{t-1} = \frac{1}{m} \left(f_s - W \right) \Delta t \tag{17.12}$$

$$\dot{y}_t = \frac{1}{m} \left(f_{st} - W \right) \Delta t + \dot{y}_{t-1} \tag{17.13}$$

Equation (17.13) describes the linear velocity of the vehicle along the Y-axis in terms of the lifting force f_s and the weight of the vehicle. Finally, a similar analysis can be performed to define the forces that interact on the vehicle along the Z-axis.

Thus, in terms of the second order derivative,

$$\ddot{z}_t = \frac{1}{m} (f_{th} \sin \theta_{th} + f_{ar} \sin \theta_{ar} + f_{al} \sin \theta_{al}) \tag{17.14}$$

or in terms of the first order derivative,

$$\frac{d\dot{z}_t}{dt} = \frac{1}{m} (f_{th} \sin(\theta_{th}) + f_{ar} \sin(\theta_{ar}) + f_{al} \sin(\theta_{al})) \tag{17.15}$$

thus, by stating the recursive form, which originally arose from the defined integrals,

$$\frac{\dot{z}_t - \dot{z}_{t-1}}{\Delta t} = \frac{1}{m} (f_{th} \sin(\theta_{th}) + f_{ar} \sin(\theta_{ar}) + f_{al} \sin(\theta_{al})) \tag{17.16}$$

and by arranging the time algebraically,

$$\dot{z}_t - \dot{z}_{t-1} = \frac{1}{m} (f_{th} \sin(\theta_{th}) + f_{ar} \sin(\theta_{ar}) + f_{al} \sin(\theta_{al})) \Delta t \tag{17.17}$$

hence, the recursive form of the instantaneous $\dot{z}_t$,

$$\dot{z}_t = \frac{1}{m} (f_{th} \sin(\theta_{th}) + f_{ar} \sin(\theta_{ar}) + f_{al} \sin(\theta_{al})) \Delta t + \dot{z}_{t-1} \tag{17.18}$$

By arranging equations (17.6), (17.13) and (17.18), the direct solution for the linear velocities of the hovercraft is described by the following postulate,

Postulate 17.1.2 (velocity vector). *The recursive velocity vector model is postulated as a function of the interacting propulsive forces.*

$$\begin{pmatrix} \dot{x}_t \\ \dot{y}_t \\ \dot{z}_t \end{pmatrix} = \frac{\Delta t}{m} \begin{pmatrix} \cos \theta_{th} & \cos(\theta_{ar}) & \cos(\theta_{al}) & 0 \\ 0 & 0 & 0 & 1 \\ \sin(\theta_{th}) & \sin(\theta_{ar}) & \sin(\theta_{al}) & 0 \end{pmatrix} \begin{pmatrix} f_{th} \\ f_{ar} \\ f_{al} \\ f_{s-W} \end{pmatrix} + \begin{pmatrix} \dot{x}_{t-1} \\ \dot{y}_{t-1} \\ \dot{z}_{t-1} \end{pmatrix} \tag{17.19}$$

17.1.1   Inverse solution

Let $\dot{\boldsymbol{\xi}}_t = (\dot{x}, \dot{y}, \dot{z})^\top$ be the velocity vector and, let $\mathbf{u}_t = (f_{th}, f_{ar}, f_{al}, f_{s-W})^\top$ be the input vector comprised of the propulsive forces. Thus, the equation (17.19) is now expressed in the matrix form by

$$\dot{\boldsymbol{\xi}}_t = \left(\frac{\Delta t}{m}\right) \mathbf{Q} \cdot \mathbf{u} + \dot{\boldsymbol{\xi}}_{t-1} \tag{17.20}$$

Likewise, let us represent the transition matrix $\mathbf{Q}$ by

$$\mathbf{Q} = \begin{pmatrix} \cos(\theta_{th})\cos(\theta_{ar})\cos(\theta_{al})0 \\ 0 \quad\quad 0 \quad\quad 0 \quad 1 \\ \sin(\theta_{th}) \; \sin(\theta_{ar}) \; \sin(\theta_{al})0 \end{pmatrix} \tag{17.21}$$

The equation (17.20) computes the linear velocities of the vehicle by controlling the forces interacting with it. For instance, by controlling the engine's acceleration throttle that speeds up/down the ducted fan, then it will generate propulsive forces surface-tangent over the vehicle. Those forces can be controlled by the rudder and the ailerons of the hovercraft to change the vehicle trajectory, Furthermore, to define the inverse solution to obtain the forces vector $\mathbf{u}_t$ as a function of the Cartesian velocities, by solving the equation (17.20),

Corollary 17.1.3 (vector of propulsive forces). *The instantaneous vector of propulsive forces is deduced as a function of the recursive velocity vectors.*

$$\mathbf{u}_t = \frac{m}{\Delta t} \mathbf{Q}^{-1} (\dot{\boldsymbol{\xi}}_t - \dot{\boldsymbol{\xi}}_{t-1}) \tag{17.22}$$

The proposed solution is fully completed by the equation (17.22), where it is necessary to firstly solve for the inverse closed form of $\mathbf{Q}$. It is achieved by using the Moore-Penrose pseudo inverse method (see section 1.2.5). Then, $\mathbf{Q}^{-1}$ can be defined as

$$\mathbf{Q}^{-1} = (\mathbf{Q}^\top \mathbf{Q})^{-1} \mathbf{Q}^\top \tag{17.23}$$

then solving the equation,

$$Q^\top Q = \begin{pmatrix} 1 & c(\theta_{th} - \theta_r)c(\theta_{th} - \theta_l) & 0 \\ c(\theta_{th} - \theta_r) & 1 & c(\theta_r - \theta_l) & 0 \\ c(\theta_{th} - \theta_l) & c(\theta_r - \theta_l) & 1 & 0 \\ 0 & 0 & 0 & 1 \end{pmatrix} \tag{17.24}$$

Hereafter, for readability purposes time indexes were omitted and cos and sin will be represented by c and s respectively. Then, we can define the determinant $k = \det(Q^\top Q)$ as

$$k = 2c(\theta_{th} - \theta_r)c(\theta_{th} - \theta_l)c(\theta_r - \theta_l) - c^2(\theta_{th} - \theta_r) - c^2(\theta_{th} - \theta_l) - c^2(\theta_r - \theta_l) + 1 \tag{17.25}$$

We can solve a matrix $P = (Q^\top Q)^{-1}$ using the determinant method as

$$P = (Q^\top Q)^{-1} = \begin{pmatrix} \frac{c^2(\theta_r - \theta_l)}{-k} & \frac{c(\theta_{th} - \theta_r) - c(\theta_{th} - \theta_l)c(\theta_r - \theta_l)}{-k} & \frac{c(\theta_{th} - \theta_l) - c(\theta_{th} - \theta_r)c(\theta_r - \theta_l)}{-k} & 0 \\ \frac{c(\theta_{th} - \theta_r) - c(\theta_{th} - \theta_l)c(\theta_r - \theta_l)}{-k} & \frac{c^2(\theta_{th} - \theta_l)}{-k} & \frac{c(\theta_r - \theta_l) - c(\theta_{th} - \theta_r)c(\theta_r - \theta_l)}{-k} & 0 \\ \frac{c(\theta_{th} - \theta_l) - c(\theta_{th} - \theta_r)c(\theta_r - \theta_l)}{-k} & \frac{c(\theta_r - \theta_l) - c(\theta_{th} - \theta_r)c(\theta_r - \theta_l)}{-k} & \frac{c^2(\theta_{th} - \theta_r)}{-k} & 0 \\ 0 & 0 & 0 & 1 \end{pmatrix} \tag{17.26}$$

Finally, the pseudo-inverse Q^{-1} can be defined as

$$Q^{-1} = PQ^\top = \begin{pmatrix} d & g & h & 0 \\ g & e & i & 0 \\ h & i & f & 0 \\ 0 & 0 & 0 & 1 \end{pmatrix} \cdot \begin{pmatrix} c(\theta_{th}) & 0 & s(\theta_{th}) \\ c(\theta_r) & 0 & s(\theta_r) \\ c(\theta_l) & 0 & s(\theta_l) \\ 0 & 1 & 0 \end{pmatrix} \tag{17.27}$$

and by developing the matrix operation the following matrix is obtained

$$Q^{-1} = \begin{pmatrix} c(\theta_{th})d + c(\theta_r)g + c(\theta_l)h & 0 & s(\theta_{th})d + s(\theta_r)g + s(\theta_l)h \\ c(\theta_{th})g + c(\theta_r)e + c(\theta_l)i & 0 & s(\theta_{th})g + s(\theta_r)e + s(\theta_l)i \\ c(\theta_{th})h + c(\theta_r)i + c(\theta_l)f & 0 & s(\theta_{th})h + s(\theta_r)i + s(\theta_l)f \\ 0 & 1 & 0 \end{pmatrix} \tag{17.28}$$

where the matrix elements are defined separately,

$$d = \frac{c^2(\theta_r - \theta_l)}{-k} \qquad e = \frac{c^2(\theta_{th} - \theta_l)}{-k} \qquad f = \frac{c^2(\theta_{th} - \theta_r)}{-k}$$

$$g = \frac{c(\theta_{th} - \theta_r) - c(\theta_{th} - \theta_l)c(\theta_r - \theta_l)}{-k}$$

$$h = \frac{c(\theta_{th} - \theta_l) - c(\theta_{th} - \theta_r)c(\theta_r - \theta_l)}{-k}$$

$$i = \frac{c(\theta_r - \theta_l) - c(\theta_{th} - \theta_r)c(\theta_r - \theta_l)}{-k}$$

17.2 Angular motion model

It is assumed that the angular velocities and perturbation (internals and externals) forces are closely related to each other. One example is the strong air stream hitting the lateral sides of the hovercraft. This air stream causes a set of oscillating turns over time to the vehicle's trajectory. Another example is the side inclination caused be the gravitational forces when the vehicle's swift effects occur. These types of motion affect the sensor readings that cause preventing the precise estimation of the vehicle displacements. In order to model the angular velocities of the vehicle we need to consider multiple factors that may cause a turn around al three axis of the hovercraft. Let assume that the sum of the perturbation forces will equal one perturbation force on each plane of a hovercraft. For example, we can define a sum force f_d on the XZ plane as shown in figure 17.3. This force will turn the vehicle a long a center of rotation generating an angular momentum or torque. Analysing the XZ plane we can define a torque τ as

$$\tau = I_{xz}\ddot{\gamma} \qquad\qquad (17.29)$$

Were I is the moment of inertia of the mass of the vehicle and $\ddot{\gamma}$ is the angular acceleration. We can express the inertial momentum as $I = \int r^2 dm$ and rewrite equation (17.29) as

$$\tau = \int r^2 dm \ddot{\gamma} \qquad\qquad (17.30)$$

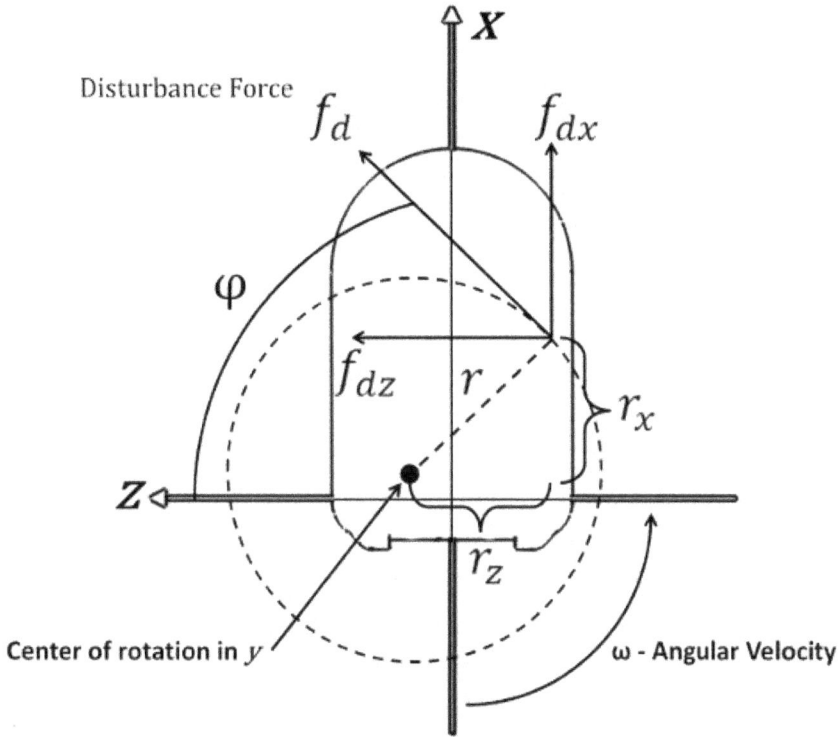

Figure 17.3: Free body diagram of perturbing forces f_d.

where r is the turn radius. In accordance with 17.3, we can also represent the torque as $\tau = rf_p$ and define an equality as

$$rf_p = \int r^2 dm \ddot{\gamma} \tag{17.31}$$

Solving for $\ddot{\gamma}$ we have

$$\ddot{\gamma} = \frac{rf_d}{r^2 \int dm} \tag{17.32}$$

and

$$\ddot{\gamma} = \frac{f_d}{rm} \tag{17.33}$$

Both, the perturbation force f_d and turn radius r vary over time. Hence, we represent equation (17.33) as

$$\ddot{\gamma}_t = \frac{1}{m}\left(\frac{f_{d_t}}{r_t}\right) \tag{17.34}$$

Expressing $\ddot{\gamma}_t$ in terms of the angular velocity, the following is defined,

$$\frac{d\dot{\gamma}_t}{dt} = \frac{1}{m}\left(\frac{f_{d_t}}{r_t}\right) \tag{17.35}$$

By establishing a constant time increment Δt, the differential in the equation (17.35) is in discrete form as

$$\frac{\dot{\gamma}_t - \dot{\gamma}_{t-1}}{\Delta t} = \frac{1}{m}\left(\frac{f_{d_t}}{r_t}\right) \tag{17.36}$$

Solving for $\dot{\gamma}_t$ we have

$$\dot{\gamma}_t - \dot{\gamma}_{t-1} = \frac{\Delta t}{m}\left(\frac{f_{d_t}}{r_t}\right) \tag{17.37}$$

or

$$\dot{\gamma}_t = \frac{\Delta t}{m}\left(\frac{f_{d_t}}{r_t}\right) + \dot{\gamma}_{t-1} \tag{17.38}$$

Equation (17.38) defines the angular velocity $\dot{\gamma}$ in terms of a perturbation force f_d. And extending the analysis to XY and YZ planes, in order to develop a model for the angular velocities of the vehicle:

$$\begin{pmatrix} \dot{\gamma}_t \\ \dot{\beta}_t \\ \dot{\alpha}_t \end{pmatrix} = \frac{\Delta t}{m} \begin{pmatrix} \eta_t/p_t \\ \kappa_t/q_t \\ \delta_t/r_r \end{pmatrix} + \begin{pmatrix} \dot{\gamma}_{t-1} \\ \dot{\beta}_{t-1} \\ \dot{\alpha}_{t-1} \end{pmatrix} \tag{17.39}$$

were η_t, κ_t and δ_t are the result perturbation forces in XZ, XY y YZ planes respectively; and p_t, q_t y r_t the turn radius generated by those forces. The inverse solution of (17.39) can be defined as

$$\begin{pmatrix} \eta_t \\ \kappa_t \\ \delta_t \end{pmatrix} = \frac{m}{\Delta t} \begin{pmatrix} p_t(\dot{\gamma}_t - \dot{\gamma}_{t-1}) \\ q_t(\dot{\beta}_t - \dot{\beta}_{t-1}) \\ r_t(\dot{\alpha}_t - \dot{\alpha}_{t-1}) \end{pmatrix} \tag{17.40}$$

Equation (17.40) is used to compute de perturbation forces acting on the hovercraft with the angular velocities as inputs. The angular velocities can be measure by an inertial sensor in order to describe the perturbation forces acting on the vehicle.

Bibliography

[1] Yun, L., Bliault, A., *Theory and Design of Air Cushion Craft*, J. Wiley & Sons. New York, USA, 2000.

[2] Niku, Saeed B., *Introduction to Robotics: Analysis, Control, Applications*, 2nd Ed. J. Wiley & Sons. USA, 2011.

[3] Ryota K., Osuka O., *Trajectory Control of an Air Cushion Vehicle*, Proc. of the IEEE/RSJ/GI Intl. Conf. on Intelligent Robots and Systems, China, pp. 1906–1913, 1994.

[4] Wang C., Liu Z., Fu M., Bian X., *Amphibious Hovercraft Course Control Based on Adaptive Multiple Model Approach*, Proc. of the 2010 IEEE Intl. Conf. on Mechatronics and Automation, China, pp. 601–604, 2010.

[5] Kim K., Lee Y., Oh S., Moroniti D., Mavris D., Vachtsevanos G., Papamarkos N., Georgoulas, G., *Guidance, Navigation and Control of an Unmanned Hovercraft*, 21st Mediterranean Conf. on Control & Automation, Grece, pp. 380–387, 2013.

[6] Fantoni I., Lozano R., Mazenc F., Pettersen K. Y., *Stabilization of a nonlinear underactuated hovercraft*, International Journal on Robust and Nonlinear Control, pp. 645–654, 2000.

[7] Wang C., Fu M., Bian X., Shi X., *Research on Heading Control of Air Cushion Vehicle*, Proc. of the IEEE Intl. Conf. on Automation and Logistics, China, pp. 1138–1142, 2008.

[8] Fu M., Ding F., Shi X., *Research on the Course and Turn-rate Coordinated Control for an Air Cushion Vehicle*, Proc. of the IEEE Intl. Conf. on Mechatronics and Automation, China, pp. 2424–2428, China.

[9] Wang C., Fu, M., Zhang, L., *Course Keeping of an Air Cushion Vehicle Based on Switching Control*, Proc. of the IEEE Intl. Conf. on Automation and Logistics, China, pp. 207–211, 2013.

[10] Lu J., Huang G., Li S., *Four-degree-of-freedom course stability of an air cushion vehicle*, Journal of Shanghai Jiaotong University (Science), Vol.15, pp. 163–167, 2010.

[11] Sira-Ramírez H., Aguilar-Ibáñez C., *On the Control of the Hovercraft System*, Dynamics and Control Journal, Kluwer Academic Publishers, Vol.10, pp. 151–163, 2000.

[12] Soe Myat H., Hwee Choo L., *Design and Development of a Compact Hovercraft Vehicle*, IEEE/ASME Intl. Conf. on Advanced Intelligent Mechatronics (AIM), Australia, pp. 2424–2428, 2013.

Index

www.ingramcontent.com/pod-product-compliance
Lightning Source LLC
Chambersburg PA
CBHW080130220326
41598CB00032B/5022